Communications and Control Engineering

Communications and Control Engineering is a high-level academic monograph series publishing research in control and systems theory, control engineering and communications. It has worldwide distribution to engineers, researchers, educators (several of the titles in this series find use as advanced textbooks although that is not their primary purpose), and libraries.

The series reflects the major technological and mathematical advances that have a great impact in the fields of communication and control. The range of areas to which control and systems theory is applied is broadening rapidly with particular growth being noticeable in the fields of finance and biologically inspired control. Books in this series generally pull together many related research threads in more mature areas of the subject than the highly specialised volumes of *Lecture Notes in Control and Information Sciences*. This series's mathematical and control-theoretic emphasis is complemented by *Advances in Industrial Control* which provides a much more applied, engineering-oriented outlook.

Indexed by SCOPUS and Engineering Index.

Publishing Ethics: Researchers should conduct their research from research proposal to publication in line with best practices and codes of conduct of relevant professional bodies and/or national and international regulatory bodies. For more details on individual ethics matters please see:
https://www.springer.com/gp/authors-editors/journal-author/journal-author-helpdesk/publishing-ethics/14214

More information about this series at https://link.springer.com/bookseries/61

Gianluigi Pillonetto · Tianshi Chen ·
Alessandro Chiuso · Giuseppe De Nicolao ·
Lennart Ljung

Regularized System Identification

Learning Dynamic Models from Data

 Springer

Gianluigi Pillonetto
Department of Information Engineering
University of Padova
Padova, Italy

Tianshi Chen
School of Data Science
The Chinese University of Hong Kong
Shenzhen, China

Alessandro Chiuso
Department of Information Engineering
University of Padova
Padova, Italy

Giuseppe De Nicolao
Electrical, Computer and Biomedical
Engineering
University of Pavia
Pavia, Italy

Lennart Ljung
Department of Electrical Engineering
Linköping University
Linköping, Sweden

ISSN 0178-5354   ISSN 2197-7119   (electronic)
Communications and Control Engineering
ISBN 978-3-030-95862-6   ISBN 978-3-030-95860-2   (eBook)
https://doi.org/10.1007/978-3-030-95860-2

MATLAB is a registered trademark of The MathWorks, Inc. See https://www.mathworks.com/trademarks for a list of additional trademarks

Mathematics Subject Classification: 93B30

This Springer imprint is published by the registered company Springer Nature Switzerland AG
The registered company address is: Gewerbestrasse 11, 6330 Cham, Switzerland

To my grandmother Rina

— Gianluigi Pillonetto

To my family, my parents Naimei and Wei, my wife Yu, and my little girl Yuening

—Tianshi Chen

To my wife Mascia, my daughter Angela and my mentor Giorgio

—Alessandro Chiuso

To my wife Elena and my sons Pietro and Laura

—Giuseppe De Nicolao

To my wife Ann-Kristin and my sons Johan and Arvid

—Lennart Ljung

Preface

System identification is concerned with estimating models of dynamical systems based on observed input and output signals. The term was coined in 1953 by Lotfi Zadeh, but various approaches had of course been suggested before that. One can distinguish two major routes in the development of system identification: (1) A *statistical route* relying on parameter estimation techniques such as Maximum Likelihood and (2) a *realization route*, based on techniques to realize (linear) dynamical systems from input/output descriptions, such as impulse responses. The literature on this in the past 70 years is extensive and impressive.

Mathematically, system identification is an *inverse problem* and may suffer from numerical instability. The Russian researcher Tikhonov suggested in the 1940s a general way to curb the number of solutions for inverse problems which he called *regularization*. A simple regularization method applied to linear regression became known as *ridge regression*. *Regularized system identification* was for a long time used as a term for ridge regression.

Around 2000 other ideas were put forward for achieving regularization. They had links to general function estimation with mathematical foundations in Reproducing Kernel Hilbert Spaces (RKHS) and kernel techniques. This resulted in intense research and extensive publications in the past 25 years. Regularized system identification has become also known as the *kernel approach* to identification.

It is the purpose of this book to give a comprehensive overview of this development. A flow diagram of the book's chapters is given in Fig. 1. It starts with the core of the regularization idea: To accept some bias in the estimates to achieve a smaller variance error and a better overall Mean Square Error (MSE) of the model. This is illustrated with the *Stein effect* discussed in Chap. 1.

Traditional System identification (the statistical route) is surveyed in Chap. 2. An archetypical model structure is the *linear regression* and Chap. 3 explains how regularization is handled in such models, while the Bayesian interpretation of this is given in Chap. 4. The linear regression perspective is lifted to general linear models of dynamical systems in Chap. 5.

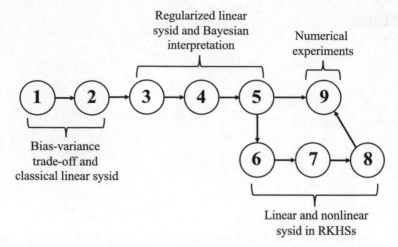

Fig. 1 *Chapter Dependencies* The first two chapters are introductory. They review the bias-variance trade-off, discussing the James–Stein estimator, and the classical approach to linear system identification. Regularized kernel-based approaches to linear system identification in finite-dimensional spaces are developed in Chaps. 3–5. The reader can directly skip to Chap. 9 where such techniques are illustrated via numerical experiments and real-world cases. A different flow to reach the final chapter moves along Chaps. 6–8 where regularization in reproducing kernel Hilbert spaces is described. These parts of the book address estimation of infinite-dimensional (discrete- or continuous-time) linear models and nonlinear system identification

With this, the basic techniques of practical regularization for linear models have been outlined and the readers may continue directly to Chap. 9 for numerical experiments and practical applications.

Chapters 6 and 7 lift the mathematical foundation of regularization with a treatment of how the techniques fit into the framework of RKHS, while Chap. 8 deals with applications to nonlinear models.

Sections marked with the symbol ⋆ contain quite technical material which can be skipped without interrupting the reading. Proofs of some of the theorems contained in the book are gathered in the Appendix present at the end of each chapter.

Padova, Italy   Gianluigi Pillonetto
Shenzhen, China   Tianshi Chen
Padova, Italy   Alessandro Chiuso
Pavia, Italy   Giuseppe De Nicolao
Linköping, Sweden   Lennart Ljung
July 2021

Acknowledgements

Many researchers have worked with authors of this book, as is clear from the list of references. Their support and ideas have been instrumental for our results and the contents of this book. We thank them all. We also acknowledge the financial support given to us by the Thousand Youth Talents Plan of China, the Natural Science Foundation of China (NSFC) under contract No. 61773329, the Shenzhen Science and Technology Innovation Council under contract No. Ji-20170189 and the Chinese University of Hong Kong, Shenzhen, under contract No. PF. 01.000249 and No. 2014.0003.23, the Department of Information Engineering, University of Padova (Italy), University of Pavia (Italy), University of Linköping (Sweden), the long time support by the Swedish research council (VR) and an advanced grant from the European Research Council (ERC).

Contents

1 Bias 1
  1.1   The Stein Effect 1
       1.1.1   The James–Stein Estimator 3
       1.1.2   Extensions of the James–Stein Estimator ★ 5
  1.2   Ridge Regression 7
  1.3   Further Topics and Advanced Reading 11
  1.4   Appendix: Proof of Theorem 1.1 12
  References 13

2 Classical System Identification 17
  2.1   The State-of-the-Art Identification Setup 17
  2.2 $\mathcal{M}$: Model Structures 18
       2.2.1   Linear Time-Invariant Models 19
            2.2.1.1   The McMillan Degree 20
            2.2.1.2   Black-Box Models 21
            2.2.1.3   Grey-Box Models 23
            2.2.1.4   Continuous-Time Models 24
       2.2.2   Nonlinear Models 24
  2.3 $\mathcal{I}$: Identification Methods—Criteria 24
       2.3.1   A Maximum Likelihood (ML) View 25
  2.4   Asymptotic Properties of the Estimated Models 26
       2.4.1   Bias and Variance 26
       2.4.2   Properties of the PEM Estimate as $N \to \infty$ 26
       2.4.3   Trade-Off Between Bias and Variance 28
  2.5 X: Experiment Design 29
  2.6 $\mathcal{V}$: Model Validation 29
       2.6.1   Falsifying Models: Residual Analysis 29
       2.6.2   Comparing Different Models 30
       2.6.3   Cross-Validation 30
  References 31

3 Regularization of Linear Regression Models 33
3.1 Linear Regression 33
3.2 The Least Squares Method 35
 3.2.1 Fundamentals of the Least Squares Method 35
  3.2.1.1 Normal Equations and LS Estimate 35
  3.2.1.2 Matrix Formulation 36
 3.2.2 Mean Squared Error and Model Order Selection 37
  3.2.2.1 Bias, Variance, and Mean Squared Error
       of the LS Estimate 37
  3.2.2.2 Model Order Selection 37
3.3 Ill-Conditioning 42
 3.3.1 Ill-Conditioned Least Squares Problems 42
  3.3.1.1 Singular Value Decomposition 42
  3.3.1.2 Condition Number 43
  3.3.1.3 Ill-Conditioned Matrix and LS Problem 43
  3.3.1.4 LS Estimate Exploiting the SVD of Φ 45
 3.3.2 Ill-Conditioning in System Identification 47
3.4 Regularized Least Squares with Quadratic Penalties 50
 3.4.1 Making an Ill-Conditioned LS Problem Well
     Conditioned 51
  3.4.1.1 Mean Squared Error 52
 3.4.2 Equivalent Degrees of Freedom 53
  3.4.2.1 Regularization Design: The Optimal
       Regularizer 56
3.5 Regularization Tuning for Quadratic Penalties 58
 3.5.1 Mean Squared Error and Expected Validation Error 58
  3.5.1.1 Minimizing the MSE 58
  3.5.1.2 Minimizing the EVE 59
 3.5.2 Efficient Sample Reuse 60
  3.5.2.1 Hold Out Cross-Validation 61
  3.5.2.2 k-Fold Cross-Validation 61
  3.5.2.3 Predicted Residual Error Sum of Squares
       and Variants 62
 3.5.3 Expected In-Sample Validation Error 63
  3.5.3.1 Expectation of the Sum of Squared
       Residuals, Optimism and Degrees
       of Freedom 64
  3.5.3.2 An Unbiased Estimator of the Expected
       In-Sample Validation Error 66
  3.5.3.3 Excess Degrees of Freedom* 67
3.6 Regularized Least Squares with Other Types
   of Regularizers ⋆ 69
 3.6.1 ℓ_1-Norm Regularization 69
  3.6.1.1 Computation of Sparse Solutions 70

3.6.1.2 LASSO Using an Orthogonal Regression
Matrix 70
3.6.1.3 LASSO Using a Generic Regression
Matrix: Geometric Interpretation 71
3.6.1.4 Sparsity Inducing Regularizers Beyond
the ℓ_1-Norm 73
3.6.1.5 Presence of Outliers and Robust
Regression 75
3.6.1.6 An Equivalence Between ℓ_1-Norm
Regularization and Huber Estimation 77
3.6.2 Nuclear Norm Regularization 78
3.6.2.1 Nuclear Norm Regularization for Matrix
Rank Minimization 78
3.6.2.2 Application in Covariance Matrix
Estimation with Low-Rank Structure 80
3.6.2.3 Vector Case: ℓ_1-Norm Regularization 81
3.7 Further Topics and Advanced Reading 82
3.8 Appendix 82
3.8.1 Fundamentals of Linear Algebra 82
3.8.1.1 QR Factorization and Singular Value
Decomposition 83
3.8.1.2 Vector and Matrix Norms 83
3.8.1.3 Matrix Inversion Lemma, Based on [49] 85
3.8.2 Proof of Lemma 3.1 85
3.8.3 Derivation of Predicted Residual Error Sum
of Squares (PRESS) 86
3.8.4 Proof of Theorem 3.7 88
3.8.5 A Variant of the Expected In-Sample Validation
Error and Its Unbiased Estimator 89
References 91

4 Bayesian Interpretation of Regularization 95
4.1 Preliminaries 95
4.2 Incorporating Prior Knowledge via Bayesian Estimation 97
4.2.1 Multivariate Gaussian Variables 99
4.2.2 The Gaussian Case 100
4.2.3 The Linear Gaussian Model 101
4.2.4 Hierarchical Bayes: Hyperparameters 104
4.3 Bayesian Interpretation of the James–Stein Estimator 105
4.4 Full and Empirical Bayes Approaches 107
4.5 Improper Priors and the Bias Space 109
4.6 Maximum Entropy Priors 110
4.7 Model Approximation via Optimal Projection ★ 114
4.8 Equivalent Degrees of Freedom 116
4.9 Bayesian Function Reconstruction 120

4.10   Markov Chain Monte Carlo Estimation 125
4.11   Model Selection Using Bayes Factors 127
4.12   Further Topics and Advanced Reading 129
4.13   Appendix 130
        4.13.1   Proof of Theorem 4.1 130
        4.13.2   Proof of Theorem 4.2 130
        4.13.3   Proof of Lemma 4.1 130
        4.13.4   Proof of Theorem 4.3 131
        4.13.5   Proof of Theorem 4.6 131
        4.13.6   Proof of Proposition 4.3 132
        4.13.7   Proof of Theorem 4.8 132
References 133

5 **Regularization for Linear System Identification** 135
5.1   Preliminaries 135
5.2   MSE and Regularization 137
5.3   Optimal Regularization for FIR Models 141
5.4   Bayesian Formulation and BIBO Stability 143
5.5   Smoothness and Contractivity: Time-
        and Frequency-Domain Interpretations 145
        5.5.1   Maximum Entropy Priors for Smoothness
                 and Stability: From Splines to Dynamical Systems 148
5.6   Regularization and Basis Expansion ⋆ 155
5.7   Hankel Nuclear Norm Regularization 159
5.8   Historical Overview 163
        5.8.1   The Distributed Lag Estimator: Prior Means
                 and Smoothing 163
        5.8.2   Frequency-Domain Smoothing and Stability 165
        5.8.3   Exponential Stability and Stochastic Embedding 166
5.9   Further Topics and Advanced Reading 168
5.10   Appendix 169
        5.10.1   Optimal Kernel 169
        5.10.2   Proof of Lemma 5.1 171
        5.10.3   Proof of Theorem 5.5 171
        5.10.4   Proof of Corollary 5.1 174
        5.10.5   Proof of Lemma 5.2 175
        5.10.6   Proof of Theorem 5.6 175
        5.10.7   Proof of Lemma 5.5 175
        5.10.8   Forward Representations of Stable-Splines
                 Kernels ⋆ 176
References 177

6 Regularization in Reproducing Kernel Hilbert Spaces 181
   6.1   Preliminaries 181
   6.2   Reproducing Kernel Hilbert Spaces 182
       6.2.1   Reproducing Kernel Hilbert Spaces Induced
               by Operations on Kernels ★ 189
   6.3   Spectral Representations of Reproducing Kernel Hilbert
       Spaces 191
       6.3.1   More General Spectral Representation ★ 195
   6.4   Kernel-Based Regularized Estimation 196
       6.4.1   Regularization in Reproducing Kernel Hilbert
               Spaces and the Representer Theorem 196
       6.4.2   Representer Theorem Using Linear and Bounded
               Functionals 199
   6.5   Regularization Networks and Support Vector Machines 200
       6.5.1   Regularization Networks 200
       6.5.2   Robust Regression via Huber Loss ★ 202
       6.5.3   Support Vector Regression ★ 202
       6.5.4   Support Vector Classification ★ 204
   6.6   Kernels Examples 205
       6.6.1   Linear Kernels, Regularized Linear Regression
               and System Identification 205
               6.6.1.1   Infinite-Dimensional Extensions ★ 206
       6.6.2   Kernels Given by a Finite Number of Basis
               Functions 208
       6.6.3   Feature Map and Feature Space ★ 208
       6.6.4   Polynomial Kernels 210
       6.6.5   Translation Invariant and Radial Basis Kernels 210
       6.6.6   Spline Kernels 211
       6.6.7   The Bias Space and the Spline Estimator 212
   6.7   Asymptotic Properties ★ 215
       6.7.1   The Regression Function/Optimal Predictor 215
       6.7.2   Regularization Networks: Statistical Consistency 216
       6.7.3   Connection with Statistical Learning Theory 218
   6.8   Further Topics and Advanced Reading 222
   6.9   Appendix 223
       6.9.1   Fundamentals of Functional Analysis 223
       6.9.2   Proof of Theorem 6.1 229
       6.9.3   Proof of Theorem 6.10 232
       6.9.4   Proof of Theorem 6.13 233
       6.9.5   Proofs of Theorems 6.15 and 6.16 235
       6.9.6   Proof of Theorem 6.21 236
   References 242

7   **Regularization in Reproducing Kernel Hilbert Spaces**
    for Linear System Identification 247
    7.1   Regularized Linear System Identification in Reproducing
           Kernel Hilbert Spaces 247
           7.1.1   Discrete-Time Case 247
                    7.1.1.1   FIR Case 248
                    7.1.1.2   IIR Case 250
           7.1.2   Continuous-Time Case 253
           7.1.3   More General Use of the Representer Theorem
                    for Linear System Identification ⋆ 258
           7.1.4   Connection with Bayesian Estimation of Gaussian
                    Processes 260
           7.1.5   A Numerical Example 263
    7.2   Kernel Tuning 266
           7.2.1   Marginal Likelihood Maximization 266
                    7.2.1.1   Numerical Example 267
           7.2.2   Stein's Unbiased Risk Estimator 271
           7.2.3   Generalized Cross-Validation 272
    7.3   Theory of Stable Reproducing Kernel Hilbert Spaces 274
           7.3.1   Kernel Stability: Necessary and Sufficient
                    Conditions 274
           7.3.2   Inclusions of Reproducing Kernel Hilbert Spaces
                    in More General Lebesque Spaces ⋆ 278
    7.4   Further Insights into Stable Reproducing Kernel Hilbert
           Spaces ⋆ 278
           7.4.1   Inclusions Between Notable Kernel Classes 279
           7.4.2   Spectral Decomposition of Stable Kernels 280
           7.4.3   Mercer Representations of Stable Reproducing
                    Kernel Hilbert Spaces and of Regularized
                    Estimators 282
           7.4.4   Necessary and Sufficient Stability Condition Using
                    Kernel Eigenvectors and Eigenvalues 284
    7.5   Minimax Properties of the Stable Spline Estimator ⋆ 286
           7.5.1   Data Generator and Minimax Optimality 287
           7.5.2   Stable Spline Estimator 288
           7.5.3   Bounds on the Estimation Error and Minimax
                    Properties 290
    7.6   Further Topics and Advanced Reading 292
    7.7   Appendix 296
           7.7.1   Derivation of the First-Order Stable Spline Norm 296
           7.7.2   Proof of Proposition 7.1 298
           7.7.3   Proof of Theorem 7.5 299
           7.7.4   Proof of Theorem 7.7 302
           7.7.5   Proof of Theorem 7.9 305
    References 307

8 Regularization for Nonlinear System Identification 313
   8.1   Nonlinear System Identification 313
   8.2   Kernel-Based Nonlinear System Identification 314
      8.2.1   Connection with Bayesian Estimation of Gaussian
             Random Fields 315
      8.2.2   Kernel Tuning 316
   8.3   Kernels for Nonlinear System Identification 319
      8.3.1   A Numerical Example 319
      8.3.2   Limitations of the Gaussian and Polynomial Kernel 323
      8.3.3   Nonlinear Stable Spline Kernel 326
      8.3.4   Numerical Example Revisited: Use of the Nonlinear
             Stable Spline Kernel 330
   8.4   Explicit Regularization of Volterra Models 331
   8.5   Other Examples of Regularization in Nonlinear System
      Identification 334
      8.5.1   Neural Networks and Deep Learning Models 334
      8.5.2   Static Nonlinearities and Gaussian Process (GP) 335
      8.5.3   Block-Oriented Models 335
      8.5.4   Hybrid Models 336
      8.5.5   Sparsity and Variable Selection 338
   References 340

9 Numerical Experiments and Real World Cases 343
   9.1   Identification of Discrete-Time Output Error Models 343
      9.1.1   Monte Carlo Studies with a Fixed Output Error
             Model 344
      9.1.2   Monte Carlo Studies with Different Output Error
             Models 347
             9.1.2.1   Results 348
      9.1.3   Real Data: A Robot Arm 350
      9.1.4   Real Data: A Hairdryer 353
   9.2   Identification of ARMAX Models 353
      9.2.1   Monte Carlo Experiment 356
      9.2.2   Real Data: Temperature Prediction 358
   9.3   Multi-task Learning and Population Approaches ★ 360
      9.3.1   Kernel-Based Multi-task Learning 362
      9.3.2   Numerical Example: Real Pharmacokinetic Data 364
   References 368

Index 371

Abbreviations and Notation

Notation

log	Natural logarithm
$\mathbb{N}$	The set of natural numbers
$\mathbb{R}^n$	The n-dimensional Euclidean space with $n \in \mathbb{N}$
$\mathbb{R}^+$	The set of nonnegative real numbers
$\min_\theta l(\theta)$	The problem of minimizing the objective $l(\theta)$ with respect to θ
$\arg\min_\theta l(\theta)$	The value of θ that minimizes $l(\theta)$
$\frac{\partial l(\theta)}{\partial \theta}$	The gradient of the function $l(\theta)$ with respect to θ
$\frac{\partial^2 l(\theta)}{\partial \theta \partial \theta}$	The Hessian matrix of the function $l(\theta)$ with respect to θ
X	The experimental conditions under which the data are generated
$\mathscr{D}$	The data
$\mathscr{D}_T$	Training/identification data set
N	Number of training/identification data
$\mathscr{D}_{test}$	Test data set
N_{test}	Number of test data
$\mathscr{M}$	The Model Structure and its parameters θ
$\mathscr{M}(\theta)$	A particular model corresponding to a particular parameter θ
$\mathscr{I}$	The identification method
$\mathscr{V}$	The validation process
$u(t)$	The input at time t
$y(t)$	The measurement output at time t
$\hat{y}(t)$	One-step-ahead output predictor at instant t
Z^{t-1}	Past input–output data up to instant $t-1$

$Y \in \mathbb{R}^N$	The vector collecting all the measurements in the single-output case
y_i	The i-th element of Y, $i = 1, \cdots, N$
y_v	Output for validation
u_{test}	Test input
y_{test}	Test output
$\Phi \in \mathcal{R}^{N \times n}$	The regression matrix
$\phi_i \in \mathbb{R}^n$	The i-th n-dimensional regressor, $i = 1, \cdots, N$
ϕ_v	Regressor for validation
$e(t)$	The noise at time t
$e_i \in \mathbb{R}$	The i-th measurement noise, $i = 1, \cdots, N$
σ^2	The variance of the measurement noise e_i
$E \in \mathbb{R}^N$	The measurement noise vector
$g(t)$	The impulse response at time t
$\theta \in \mathbb{R}^n$	A deterministic or stochastic parameter vector
θ_0	The true value of the deterministic parameter θ
$\hat{\theta}_N$	PEM estimate as a function of the data set size N
$\hat{\theta}^{LS}$	The least squares estimate
$\hat{\theta}^{JS}$	The James–Stein estimate
$\mathrm{MSE}(\hat{\theta}, \theta_0)$ or $\mathrm{MSE}_{\hat{\theta}}$	the mean square matrix of an estimator $\hat{\theta}$ of θ_0
q	The time shift operator, e.g., $qu(t) = u(t+1)$
$G(q, \theta)$	A transfer function model
$G_0(q)$	The true transfer function
$G(e^{i\omega}, \theta)$	The frequency response of the model
$H(q, \theta)$	A transfer function for the noise model
H_0	The true noise function
H_k	The Hankel matrix of the impulse response
ℓ	The loss function norm
$\kappa(\ell)$	The variance coefficient for a loss function ℓ in the fitting criterion
$\|\theta\|_0$	ℓ_0-norm of θ
$\|\theta\|_1$	ℓ_1-norm of θ
$\|\theta\|_2$	ℓ_2-norm of θ or the Euclidean norm
$A(i, :)$	The i-th row of a matrix A with $i \in \mathbb{N}$
$A(:, j)$	The j-th column of a matrix A with $j \in \mathbb{N}$
A^T	The transpose of a matrix A
$A \succeq B$	$A - B$ is a positive semidefinite matrix
A^{-1}	The inverse of a full-rank square matrix A
$I_n \in \mathbb{R}^{n \times n}$	The n-dimensional identity matrix
$\lambda_i(A)$	The i-th eigenvalue of a matrix A
$\sigma_1 \geq \sigma_2 \geq \cdots \sigma_n$ of a matrix A	the singular values of a matrix A
σ_{max} and σ_{min} of a matrix A	the largest and smallest singular value of a matrix A

$\text{cond}(A)$	The condition number of a matrix A with respect to the Euclidean norm and defined as $\text{cond}(A) = \sigma_{\max}/\sigma_{\min}$
$\text{diag}(a)$	The $n \times n$ diagonal matrix with diagonal elements equal to the vector $a \in \mathbb{R}^n$
A^+	The Moore–Penrose pseudoinverse of a matrix A
$\text{trace}(A)$	The trace of a matrix A
$\text{rank}(A)$	The rank of a matrix A
$\|A\|$	A general norm of a matrix
$\|A\|_F$	The Frobenius norm of a matrix A
$\|A\|_*$	The nuclear norm of a matrix A
$\hat{\theta}^{\text{R}}$	The regularized least squares estimate
$\hat{Y} = \Phi\hat{\theta}^{\text{R}}$	The predicted output of $\hat{\theta}^{\text{R}}$
H	The hat matrix of $\hat{\theta}^{\text{R}}$ linking Y to $\hat{Y}$, i.e., $\hat{Y} = HY$
$\text{dof}(\hat{\theta}^{\text{R}})$	The equivalent degrees of freedom of $\hat{\theta}^{\text{R}}$
γ	The regularization parameter
γ^{ML}	Maximum marginal likelihood estimate of γ
η	The hyperparameter vector
P	Inverse of a full-rank regularization matrix or of a full-rank kernel matrix if a kernel has been defined
$g^0(t)$	The true impulse response evaluated at instant t
$\hat{g}(t)$	Impulse response estimate evaluated at instant t
$\text{p}(\cdot)$	Probability density function
$h(\text{p})$	Differential entropy of the probability density function p
$\text{Pr}(A)$	Probability of the event A
$\mathscr{E}(X)$	The mathematical expectation of a random variable X
$\text{Var}(\cdot)$	Variance matrix
$\text{std}(\cdot)$	Standard deviation
$\text{Cov}(X, Y)$	The covariance matrix of two random vectors X and Y
μ_θ	The mean of θ
Σ_θ	The covariance of θ
$\theta|Y$	The random vector θ conditional on Y
$\Sigma_{\theta|Y}$	The posterior covariance of θ conditional on Y
θ^{B}	The Bayes estimate of θ, i.e., the mean of $\theta|Y$
θ^{MAP}	The maximum a posteriori estimate of θ, i.e., the value that maximizes the posterior of $\theta|Y$
$X \sim \mathcal{N}(m, \Sigma)$	$X \in \mathbb{R}^n$ is an n-dimensional Gaussian (normal) random vector with mean m and covariance matrix Σ
χ_n^2	Chi-square variable with n degrees of freedom

$\Gamma(g_1, g_2)$	Gamma probability density functions of parameters g_1, g_2
B_{12}	Bayes factor related to the models $\mathcal{M}^1$ and $\mathcal{M}^2$
$\langle f, g \rangle$	Inner product between functions f and g
$\|f\|$	Norm of the function f
$\mathcal{X}$	The domain of a function whose elements x are called input locations
K	Positive semidefinite kernel defined over $\mathcal{X} \times \mathcal{X}$
K_x	Kernel section centred at x defined over $\mathcal{X}$
$\mathcal{H}$	An RKHS
$\|f\|_{\mathcal{H}}$	The RKHS norm of a function $f \in \mathcal{H}$
$\langle f, g \rangle_{\mathcal{H}}$	The inner product between two functions $f, g \in \mathcal{H}$
ℓ_1	The space of absolutely summable sequences
ℓ_2	The space of squared summable sequences
ℓ_∞	The space of bounded sequences
$\mathcal{L}_1^\mu$	The space of absolutely summable functions on $\mathcal{X}$ equipped with the measure μ (often set to the classical Lebesque measure)
$\mathcal{L}_2^\mu$	The space of squared summable functions on $\mathcal{X}$ equipped with the measure μ (often set to the classical Lebesque measure)
$\mathcal{L}_\infty$	Space of essentially bounded functions
$\mathcal{C}$	Space of continuous functions
$\|f\|_\infty$	Sup-norm of the function f
δ_{ij}	Kronecker delta
$\dot{f}$	First-order derivative of f
$f^{(p)}$	Derivative of order p of f
$\chi_x(\cdot)$	Indicator function of the set $[0, x]$
$f_{\mathcal{Y}}$	The function f sampled on $\mathcal{Y}$
ρ_i	The i-th basis function or eigenfunction of a kernel K
ζ_i	The i-th eigenvalue of a kernel K
L_K	Linear operator defined by the kernel K
$\mathcal{V}_i(y_i, f(x_i))$	Loss function measuring the distance between y_i and $f(x_i)$
L_i	A linear and bounded functional from $\mathcal{H}$ to $\mathbb{R}$
η_i	Function in $\mathcal{H}$ that represents L_i
O	Matrix with i, j entry given by $\langle \eta_i, \eta_j \rangle_{\mathcal{H}}$
$\mathbf{K}$	Kernel matrix induced by the kernel K with i, j entry given by $K(x_i, x_j)$
$\| \cdot \|_\varepsilon$	Function associated to the Vapnik's ε-insensitive loss
G_p	Green's function of order p
$\mathrm{span}\{\phi_1, \ldots, \phi_m\}$	Subspace generated by the functions $\phi_1, \ldots, \phi_m$

S	A set or a subspace
$\overline{S}$	Completion of S
$S^\perp$	The orthogonal of S
$\mathrm{Err}(f)$	Least squares error associated to f
f_ρ	Regression function (minimizer of Err)
$\hat{g}_N$	Regularized estimator of the function g depending on the data set size N
$I(f)$	Expected risk associated to f
$\mathscr{S}_s$	Class of stable RKHSs
$\mathscr{S}_1$	Class of RKHSs induced by absolutely summable kernels
$\mathscr{S}_{ft}$	Class of RKHSs induced by finite-trace kernels
$\mathscr{S}_2$	Class of RKHSs induced by squared summable kernels
$\binom{k}{m}$	Binomial coefficient
$C(k, m)$	$\binom{k+m-1}{m-1}$, i.e., number of ways one can form the nonnegative integer k as the sum of m nonnegative integers

Abbreviations

ADMM	Alternating direction method of multipliers
AIC	Akaike's Information Criterion
AICc	The corrected AIC
AR	Autoregressive
ARMAX	Autoregressive moving average
ARX	Autoregressive with exogenous inputs
BIBO	Bounded-input bounded-output
BIC	The Bayesian information criterion
BJ	Box–Jenkins (model)
CV	Cross-validation
DC	Diagonal correlated
DI	Diagonal
EB	Empirical Bayes
ERM	Empirical risk minimization
EVE	Expected validation error
EVE_{in}	In-sample expected validation error
FIR	Finite impulse response
GCV	Generalized cross-validation
HOCV	Hold out cross-validation
IIR	Infinite impulse response
k-fold CV	k-fold cross-validation
LASSO	The least absolute shrinkage and selection operator

LOOCV	Leave-one-out cross-validation
LTI	Linear time invariant
MAP	Maximum a posteriori
MaxEnt	Maximum entropy
MCMC	Markov chain Monte Carlo
MDL	Rissanen's minimum description length
ME	Maximum entropy
MIMO	Multiple-input multiple-output
ML	Maximum likelihood
MLE	Maximum likelihood estimate
MLM	Marginal likelihood maximization
MSE	Mean squared error
OE	Output error
PEM	Prediction error method
PRESS	Predicted residual error sum of squares
ReLS	Regularized least squares
RKHS	Reproducing kernel Hilbert space
RMP	Rank minimization problems
SDP	Semidefinite program
SISO	Single-input single-output
SS	Stable spline
SURE	Stein's unbiased risk estimator
SVD	Singular value decomposition
TC	Tuned correlated
VC dimension	Vapnik–Chervonenkis dimension
WPSS	Weighted squared sum of estimated parameters
WRSS	Weighted squared sum of residuals

Chapter 1
Bias

Abstract Adopting a quadratic loss, the performance of an estimator can be measured in terms of its mean squared error which decomposes into a variance and a bias component. This introductory chapter contains two linear regression examples which describe the importance of designing estimators able to well balance these two components. The first example will deal with estimation of the means of independent Gaussians. We will review the classical least squares approach which, at first sight, could appear the most appropriate solution to the problem. Remarkably, we will instead see that this unbiased approach can be dominated by a particular biased estimator, the so-called James–Stein estimator. Within this book, this represents the first example of regularized least squares, an estimator which will play a key role in subsequent chapters. The second example will deal with a classical system identification problem: impulse response estimation. A simple numerical experiment will show how the variance of least squares can be too large, hence leading to unacceptable system reconstructions. The use of an approach, known as ridge regression, will give first simple intuitions on the usefulness of regularization in the system identification scenario.

1.1 The Stein Effect

Consider the following "basic" statistical problem. Starting from the realizations of N independent Gaussian random variables $y_i \sim \mathcal{N}(\theta_i, \sigma^2)$, our aim is to reconstruct the means θ_i, contained in the vector θ seen as a deterministic but unknown parameter vector.[1] The estimation performance will be measured in terms of mean squared error (MSE). In particular, let $\mathcal{E}$ and $\| \cdot \|$ denote expectation and Euclidean norm, respectively. Then, given an estimator $\hat{\theta}$ of an N-dimensional vector θ with ith

---

[1] In future chapters, θ_0 will be used to denote the true value of the deterministic vector that has generated the data, distinguishing it from the vector which parametrizes the model. In this introductory chapter, θ is instead used in both the cases to maintain the notation as simple as possible.

© The Author(s) 2022
G. Pillonetto et al., *Regularized System Identification*, Communications and Control Engineering, https://doi.org/10.1007/978-3-030-95860-2_1

component θ_i, one has

$$MSE_{\hat{\theta}} = \mathscr{E}\|\hat{\theta} - \theta\|^2$$

$$= \underbrace{\sum_{i=1}^{N} \mathscr{E}(\hat{\theta}_i - \mathscr{E}\hat{\theta}_i)^2}_{Variance} + \underbrace{\sum_{i=1}^{N} (\theta_i - \mathscr{E}\hat{\theta}_i)^2}_{Bias^2}, \qquad (1.1)$$

where in the last passage we have decomposed the error into two components. The first one is the *variance* of the estimator while the difference between the mean and the true parameter values measures the *bias*. If the mean coincides with θ, the estimator is said to be *unbiased*. The total error thus has two contributions: the variance and the (squared) bias.

Note that the mean estimation problem introduced above is a simple instance of linear Gaussian regression. In fact, letting I_N be the $N \times N$ identity matrix, the measurements model is

$$Y = \theta + E, \quad E \sim \mathscr{N}(0, \sigma^2 I_N), \qquad (1.2)$$

where Y is the N-dimensional (column) vector with ith component y_i. The most popular strategy to recover θ from data is least squares which also corresponds to maximum likelihood in this Gaussian scenario. The solution minimizes

$$\|Y - \theta\|^2$$

and is then given by

$$\hat{\theta}^{LS} = Y.$$

Apparently, the obtained estimator is the most reasonable one. A first intuitive argument supporting it is the fact that the random variables $\{y_j\}_{j \neq i}$ seem unable to carry any information on θ_i, since all the noises e_i are independent. Hence, the natural estimate of θ_i appears indeed its noisy observation y_i. This estimator is also unbiased: for any θ we have

$$\mathscr{E}\left(\hat{\theta}^{LS}\right) = \mathscr{E}(Y) = \theta.$$

Hence, from (1.1) we see that the MSE coincides with its variance, which is constant over θ and given by

$$MSE_{LS} = \mathscr{E}\|\hat{\theta}^{LS} - \theta\|^2 = N\sigma^2.$$

According to Markov's theorem $\hat{\theta}^{LS}$ is also efficient. This means that its variance is equal to the Cramér–Rao limit: no unbiased estimate can be better than the least squares estimate, e.g., see [9, 17].

1.1.1 The James–Stein Estimator

By introducing some bias in the inference process, it is easy to obtain estimators which dominate strictly least squares (in the MSE sense) over certain parameter regions. The most trivial example is the constant estimator $\hat{\theta} = a$. Its variance is null, so that its MSE reduces to the bias component $\|\theta - a\|^2$. Hence, even if the behaviour of $\hat{\theta}$ is unacceptable in most of the parameter space, this estimator outperforms least squares in the region

$$\{\theta \text{ s.t. } \|\theta - a\|^2 < N\sigma^2\}.$$

Note a feature common to least squares and the constant estimator. Both of them do not attempt to trade bias and variance, they just set to zero one of the two MSE components in (1.1). An alternative route is the design of estimators which try to balance bias and variance. Rather surprisingly, we will now see that this strategy can dominate $\hat{\theta}^{LS}$ over the entire parameter space.

The first criticisms about least squares were introduced by Stein in the '50s [23] and can be so summarized. A good mean estimator $\hat{\theta}$ should also lead to a good estimate of the Euclidean norm of θ. Thus, one should have

$$\|\hat{\theta}\| \approx \|\theta\|.$$

But, if we consider the "natural" estimator $\hat{\theta}^{LS} = Y$, in view of the independence of the errors e_i, one obtains

$$\mathscr{E}\|Y\|^2 = N\sigma^2 + \|\theta\|^2.$$

This shows that the least squares estimator tends to overestimate $\|\theta\|$. It thus seems desirable to correct $\hat{\theta}^{LS}$ by shrinking the estimate towards the origin, e.g., adopting estimators of the form $\hat{\theta}^{LS}(1 - r)$, where r is a positive scalar. The most famous example is the James–Stein estimator [15] where r is determined from data as follows:

$$r = \frac{(N - 2)\sigma^2}{\|Y\|^2},$$

hence leading to

$$\hat{\theta}^{JS} = Y - \frac{(N - 2)\sigma^2}{\|Y\|^2}Y.$$

Note that, even if all the components of Y are mutually independent, $\hat{\theta}^{JS}$ exploits all of them to estimate each θ_i. The surprising outcome is that $\hat{\theta}^{JS}$ outperforms $\hat{\theta}^{LS}$ over all the parameter space, as illustrated in the next theorem.

Theorem 1.1 (James–Stein's MSE, based on [15]) *Consider N Gaussian and independent random variables $y_i \sim \mathscr{N}(\theta_i, \sigma^2)$. Let also $\hat{\theta}^{JS}$ denote the James–Stein estimator of the means, i.e.,*

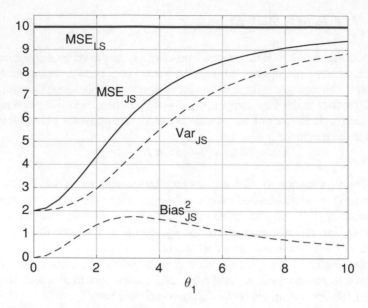

Fig. 1.1 Estimation of the mean $\theta \in \mathbb{R}^{10}$ of a Gaussian with covariance equal to the identity matrix. The plot displays the mean squared error of least squares (MSE_{LS}) and of the James–Stein estimator (MSE_{JS}), including its bias-variance decomposition, as a function of θ_1 with $\theta = [\theta_1\ 0\ldots 0]$

$$\hat{\theta}^{JS} = Y - \frac{(N-2)\sigma^2}{\|Y\|^2} Y.$$

Then, if $N \geq 3$, the MSE of $\hat{\theta}^{JS}$ satisfies

$$MSE_{JS} < N\sigma^2 \quad \forall \theta.$$

We say that an estimator dominates another estimator if for all the θ its MSE is not larger and for some θ it is smaller. In statistics an estimator is then said to be *admissible* if no other estimator exists that dominates it in terms of MSE. The above theorem then shows that the least squares estimator of the mean of a multivariate Gaussian is not admissible if the dimension exceeds two. The reason is that, even when the Gaussians are independent, the global MSE can be reduced uniformly by adding some bias to the estimate. This is also graphically illustrated in Fig. 1.1 where MSE_{JS}, along with its decomposition, is plotted as a function of the component θ_1 of the ten-dimensional vector $\theta = [\theta_1\ 0\ldots 0]$ (noise variance is equal to one). One can see that $MSE_{JS} < MSE_{LS}$ since the bias introduced by $\hat{\theta}^{JS}$ is compensated by a greater reduction in the variance of the estimate. Note however that James–Stein improves the overall MSE and not the individual errors affecting the θ_i. This aspect can be important in certain applications where it is not desirable to trade a higher individual MSE for a smaller overall MSE.

It is easy to check that the James–Stein estimator admits the following interesting reformulation:

$$\hat{\theta}^{JS} = \arg\min_{\theta} \|Y - \theta\|^2 + \gamma \|\theta\|^2$$

$$= Y \frac{1}{1 + \gamma},$$
(1.3)

where the positive scalar γ is determined from data as follows:

$$\gamma = \frac{(N-2)\sigma^2}{\|Y\|^2 - (N-2)\sigma^2}.$$
(1.4)

Equation (1.3) thus reveals that $\hat{\theta}^{JS}$ is a particular version of *regularized least squares*, an estimator which will play a central role in this book. In particular, the objective in (1.3) contains two contrasting terms. The first one, $\|Y - \theta\|^2$, is a quadratic loss which measures the adherence to experimental data. The second one, $\|\theta\|^2$, is a regularizer which shrinks the estimate towards the origin by penalizing the energy of the solution. The role of the regularization parameter γ is then to balance these two components via a simple scalar adjustment. Equation (1.4) shows that James–Stein's strategy is to set its value to the inverse of an estimate of the signal-to-noise ratio.

1.1.2 Extensions of the James–Stein Estimator ★

We have seen that the James–Stein estimator corrects each component of $\hat{\theta}^{LS}$ shifting it towards the origin. This implies that the MSE improvement will be better when the components of θ are close to zero. Actually, there is nothing special in the origin. If the true θ is expected to be close to $a \in \mathbb{R}^N$, one can modify the original $\hat{\theta}^{JS}$ as follows:

$$\hat{\theta}^{JS} = Y - \frac{(N-2)\sigma^2}{\|Y - a\|^2} (Y - a).$$

The result is an estimator which still dominates least squares, with the origin's role now played by a. The estimator thus concentrates the MSE improvement around a.

Now, let us consider a non-orthonormal scenario where Gaussian linear regression now amounts to estimating θ from the N measurements

$$y_i = d_i\theta_i + e_i \quad e_i \sim \mathcal{N}(0, 1),$$

with all the noises e_i mutually independent. The least squares (maximum likelihood) estimator is now

$$\hat{\theta}_i^{LS} = \frac{y_i}{d_i}, \quad i = 1, \ldots, N,$$

and its MSE is the sum of the variances of $\hat{\theta}_i^{LS}$, i.e.,

$$MSE_{LS} = \sum_{i=1}^{N} \frac{1}{d_i^2}.$$

Note that the MSE can be large when just one of the d_i is small. In this case, the problem is said to be *ill-conditioned*: even a moderate measurement error can lead to a large reconstruction error.

Also in this non-orthonormal scenario, it is possible to design estimators whose MSE is uniformly smaller than MSE_{LS}. The number of possible choices is huge, depending on which region of the parameter space one wants to concentrate the improvement. There is however an important limitation shared by all of Stein-type estimators: in general they are not much effective against ill-conditioning. This is illustrated in the following example. It illustrates an estimator whose negative features are well representative of some drawbacks of Stein's estimation in non-orthogonal settings.

Example 1.2 (*A generalization of James–Stein*) Consider the estimator $\hat{\theta}$ whose ith component is given by

$$\hat{\theta}_i = \left[1 - \frac{N-2}{S}d_i^2\right]\frac{y_i}{d_i}, \quad i = 1, \ldots, N, \tag{1.5}$$

where

$$S = \sum_{i=1}^{N} d_i^2 y_i^2.$$

It is now shown that $\hat{\theta}$ is a generalization of James–Stein able to outperform least squares over the entire parameter space. In fact, defining

$$h_i(Y) = -d_i^2 \frac{N-2}{S}y_i,$$

after simple computations we obtain

$$MSE_{\hat{\theta}} = \sum_{i=1}^{N} \frac{1}{d_i^2} + \mathscr{E}\left[2\sum_{i=1}^{N} \frac{(y_i - d_i\theta_i)h_i(Y)}{d_i^2} + \sum_{i=1}^{N} \frac{h_i^2(Y)}{d_i^2}\right]$$

$$= \sum_{i=1}^{N} \frac{1}{d_i^2} + \mathscr{E}\left[2\sum_{i=1}^{N} \frac{1}{d_i^2}\frac{\partial h_i(Y)}{\partial y_i} + \sum_{i=1}^{N} \frac{h_i^2(Y)}{d_i^2}\right],$$

where the last equality comes from Lemma 1.1 reported in Sect. 1.4. Since

$$\frac{\partial h_i(Y)}{\partial y_i} = -d_i^2 \frac{N-2}{S} + d_i^4 \frac{N-2}{S^2}2y_i^2,$$

one has

$$
\mathscr{E}\left[2\sum_{i=1}^{N}\frac{1}{d_i^2}\frac{\partial h_i(Y)}{\partial y_i}+\sum_{i=1}^{N}\frac{h_i^2(Y)}{d_i^2}\right]
$$

$$
=\mathscr{E}\left[-\frac{2(N-2)N}{S}+\frac{2(N-2)}{S^2}\sum_{i=1}^{N}2d_i^2\,y_i^2+\frac{(N-2)^2}{S^2}\sum_{i=1}^{N}d_i^2\,y_i^2\right]
$$

$$
=-\mathscr{E}\frac{(N-2)^2}{S}<0
$$

which implies

$$
MSE_{\hat{\theta}}<MSE_{\hat{\theta}LS}\ \ \forall\theta.
$$

However, assume that the problem is ill-conditioned. Then, if one d_i is small and the values of d_i are quite spread, we could well have $d_i^2/S\approx 0$. Hence, (1.5) essentially reduces to

$$
\hat{\theta}_i=\left[1-\frac{N-2}{S}d_i^2\right]\frac{y_i}{d_i}\approx\frac{y_i}{d_i},
$$

which is the least squares estimate of θ_i. This means that the signal components mostly influenced by the noise, i.e., associated with small d_i, are not regularized. Thus, in presence of ill-conditioning, $\hat{\theta}$ will likely return an estimate affected by large errors. $\qquad\square$

1.2 Ridge Regression

Consider now one of the fundamental problems in system identification. The task is to estimate the impulse response g^0 of a discrete-time, linear and causal dynamic system, starting from noisy output data. The measurements model is

$$
y(t)=\sum_{k=1}^{\infty}g_k^0 u(t-k)+e(t),\ \ t=1,\dots,N, \tag{1.6}
$$

where t denotes time, the sampling interval is one time unit for simplicity, the g_k^0 indicate the impulse response coefficients, $u(t)$ is the known system input while $e(t)$ is the noise.

To determine the impulse response from input–output measurements, one of the main questions is how to parametrize the unknown g^0. The classical approach, which will be also reviewed in the next chapter, introduces a collection of impulse response models $g(\theta)$, each parametrized by a different vector θ. In particular, here we will adopt an FIR model of order m, i.e., $g_k(\theta)=\theta_k$ for $k=1,\dots,m$ and zero elsewhere. This permits to reformulate (1.6) as a linear regression: we stack all the elements

$y(t)$ and $e(t)$ to form the vectors Y and E and obtain the model

$$Y = \Phi\theta + E$$

with the regression matrix $\Phi \in \mathbb{R}^{N \times m}$ given by

$$\Phi = \begin{pmatrix} u(0) & u(-1) & u(-2) & \dots & u(-m+1) \\ u(1) & u(0) & u(-1) & \dots & u(-m) \\ \dots & & & & \\ u(N-1) & u(N-2) & u(N-3) & \dots & u(N-m) \end{pmatrix}.$$

We can now use least squares to estimate θ. Assuming $\Phi^T \Phi$ of full rank, we obtain

$$\hat{\theta}^{LS} = \arg\min_\theta \|Y - \Phi\theta\|^2 \tag{1.7a}$$

$$= (\Phi^T \Phi)^{-1} \Phi^T Y. \tag{1.7b}$$

Note that the impulse response estimate is function of the FIR order which corresponds to the dimension m of θ. The choice of m is a trade-off between bias (a large m is needed to describe slowly decaying impulse responses without too much error) and variance (large m requires estimation of many parameters leading to large variance). This can be illustrated with a numerical experiment. The unknown impulse response g^0 is defined by the following rational transfer function:

$$\frac{(z+1)^2}{z(z-0.8)(z-0.7)},$$

which, in practice, is equal to zero after less than 50 samples (g^0 is the red line in Fig. 1.3). We estimate the system from 1000 outputs corrupted by white and Gaussian noises $e(t)$ of variance equal to the variance of the noiseless output divided by 50, see Fig. 1.2 (bottom panel). Data come from the system initially at rest and then fed at $t = 0$ with white noise low-pass filtered by $z/(z-0.99)$, see Fig. 1.2 (top panel). The reconstruction error is very large if we try to estimate g^0 with $m = 50$: linear models are easy to estimate but the drawback is that high-order FIR may suffer from high variance. Hence, it is important to select a model order which well balances bias and variance. To do that one needs to try different values of m then using some validation procedures to determine the "optimal" one. In this case, since the true g^0 is known, we can obtain the best value by selecting that $m \in [1, \dots, 50]$ which minimizes the MSE. This is an example of *oracle-based* procedure not implementable in practice: the optimal order is selected exploiting the knowledge of the true system. We obtain $m = 18$ which corresponds to $MSE_{LS} = 70.7$ and leads to the impulse response estimate displayed in Fig. 1.3. Even if the data set size is large and the signal-to-noise ratio is good, the estimate is far from satisfactory. The reason is that the low-pass input has poor excitation and leads to an ill-conditioned problem. This means that the condition number of the regression matrix Φ is large so that also a small output error can produce a large reconstruction error.

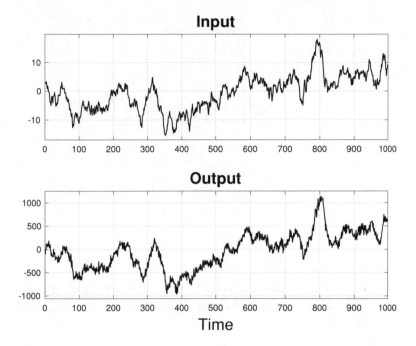

Fig. 1.2 Input–output data

Fig. 1.3 True impulse
response g^0 (thick red line)
and least squares estimate

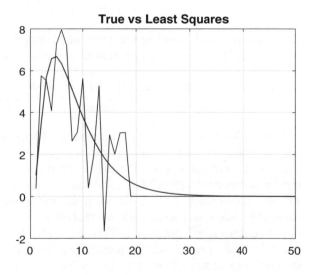

Fig. 1.4 MSE of ridge
regression and its
bias-variance decomposition
as a function of the
regularization parameter

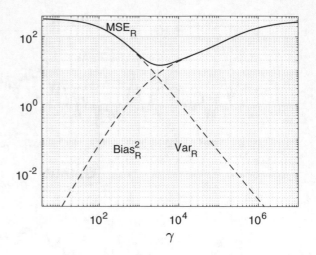

An alternative to the classical paradigm, where different model structures are introduced, is the following straightforward generalization of (1.3), known as *ridge regression* [13, 14]:

$$\hat{\theta}^R = \arg\min_\theta \|Y - \Phi\theta\|^2 + \gamma\|\theta\|^2 \tag{1.8a}$$

$$= (\Phi^T\Phi + \gamma I_m)^{-1}\Phi^T Y, \tag{1.8b}$$

where we set $m = 50$ to solve our problem. Letting $A = (\Phi^T\Phi + \gamma I_m)^{-1}\Phi^T$, it is easy to derive the MSE decomposition associated with $\hat{\theta}^R$:

$$MSE_R = \underbrace{\sigma^2\text{Trace}(AA^T)}_{Variance} + \underbrace{\|\theta - A\Phi\theta\|^2}_{Bias^2}. \tag{1.9}$$

Figure 1.4 displays MSE_R for the particular system identification problem at hand as a function of the regularization parameter. Note that γ plays the role of the model order in the classical scenario but can be tuned in a continuous manner to reach a good bias-variance trade-off. It is also interesting to see its influence on the variance and bias components. The variance is a decreasing function of the regularization parameter. Hence, its maximum is reached for $\gamma = 0$ where $\hat{\theta}^R$ reduces to the least squares estimator $\hat{\theta}^{LS}$ given by (1.7) with $m = 50$. Instead, the bias increases with γ. At the limit, for $\gamma \to \infty$, the penalty $\|\theta\|^2$ is so overweighted that $\hat{\theta}^R$ becomes the constant estimator centred on the origin (it returns all null impulse response coefficients).

In Fig. 1.5, we finally display the ridge regularized estimate with γ set to the value minimizing the error and leading to $MSE_R = 16.8$. It is evident that ridge regression provides a much better bias-variance trade-off than selecting the FIR order.

Fig. 1.5 True impulse response g^0 (thick red line) and ridge regularized estimate

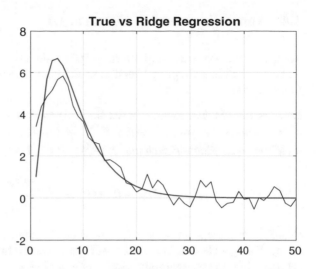

1.3 Further Topics and Advanced Reading

Stein's intuition on the development of an estimator able to dominate least squares in terms of global MSE can be found in [23], while the specific shape of $\hat{\theta}^{JS}$ has been obtained in [15]. From then, a large variety of different estimators outperforming least squares, also under different losses, have been designed. It has been proved that there exists estimators which dominate James–Stein, even if the MSE improvement is not large, as described in [12, 16, 25]. Extensions and applications can be found in [5, 6, 11, 22, 24, 26]. A James–Stein version of the Kalman filter is derived in [18]. For interesting discussions on the limitations of Stein-type estimators in facing ill-conditioning see [8] but also [19] for new outcomes with better numerical stability properties. Other developments are reported in [7] where generalizations of Stein's lemma are also described.

The paper [10] describes connections between James–Stein estimation and the so-called empirical Bayes approaches which will be treated later on in this book. The interplay between Stein-type estimators and the Bayes approach is also discussed in [2]. Here, one can also find an estimator which dominates least squares concentrating the MSE improvement in an ellipsoid that can be chosen by the user in the parameter space. This approach is deeply connected with robust Bayesian estimation concepts, e.g., see [1, 3].

The term ridge regression has been popularized by the works [13, 14]. This approach, introduced to guard against ill-conditioning and numerical instability, is an example of Tikhonov regularization for ill-posed problems. Among the first classical works on regularization and inverse problems, it is worth already citing [4, 20, 27–29]. A recent survey on the use of regularization for system identification can be instead found in [21]. The literature on this topic is huge and other relevant works will be cited in the next chapters.

1.4   Appendix: Proof of Theorem 1.1

To discuss the properties of the James–Stein estimator, first it is useful to introduce a result which is a simplified version of Lemma 3.2 reported in Chap. 3, known as Stein's lemma.

Lemma 1.1 (Stein's lemma, based on [24]) *Consider N Gaussian and independent random variables $y_i \sim \mathcal{N}(\theta_i, \sigma^2)$. For $i = 1, \ldots, n$, let also $h : \mathbb{R}^N \to \mathbb{R}$ be a differentiable function such that $\mathscr{E}\left|\frac{\partial h(Y)}{\partial y_i}\right| < \infty$. Then, it holds that*

$$\mathscr{E}(y_i - \theta_i)h(Y) = \sigma^2 \mathscr{E}\frac{\partial h(Y)}{\partial y_i}.$$

Proof During the proof, we use $\mathscr{E}_{j \neq i}$ to denote the expectation conditional on $\{y_j\}_{j \neq i}$. Also, abusing notation, $h(x)$ with $x \in \mathbb{R}$ indicates the function h with ith argument set to x while the other arguments are set to y_j $j \neq i$.

Note that, in view of the independence assumptions, each y_i conditional on $\{y_j\}_{j \neq i}$ is still Gaussian with mean θ_i and variance σ^2. Then, using integration by parts, one has

$$\begin{aligned}
\mathscr{E}_{j \neq i}\left(\frac{\partial h(Y)}{\partial y_i}\right) &= \int_{-\infty}^{+\infty} \frac{\partial h(x)}{\partial x} \frac{\exp(-(x - \theta_i)^2/(2\sigma^2))}{\sqrt{2\pi}\sigma} dx \\
&= \left[h(x)\frac{\exp(-(x - \theta_i)^2/(2\sigma^2))}{\sqrt{2\pi}\sigma}\right]_{-\infty}^{+\infty} \\
&\quad + \int_{-\infty}^{+\infty} \frac{(x - \theta_i)}{\sigma^2}h(x)\frac{\exp(-(x - \theta_i)^2/(2\sigma^2))}{\sqrt{2\pi}\sigma} dx \\
&= \int_{-\infty}^{+\infty} \frac{(x - \theta_i)}{\sigma^2}h(x)\frac{\exp(-(x - \theta_i)^2/(2\sigma^2))}{\sqrt{2\pi}\sigma} dx \\
&= \frac{\mathscr{E}_{j \neq i}\left((y_i - \theta_i)h(Y)\right)}{\sigma^2}.
\end{aligned}$$

Note that the penultimate equality exploits the fact that $h(x)\exp(-(x - \theta_i)^2/(2\sigma^2))$ must be infinitesimal as $x \to \infty$, otherwise the assumption $\mathscr{E}\left|\frac{\partial h(Y)}{\partial y_i}\right| < \infty$ would not hold. Using the above result, we obtain

$$\begin{aligned}
\mathscr{E}\left((y_i - \theta_i)h(Y)\right) &= \mathscr{E}\left[\mathscr{E}_{j \neq i}\left((y_i - \theta_i)h(Y)\right)\right] \\
&= \sigma^2 \mathscr{E}\left[\mathscr{E}_{j \neq i}\left(\frac{\partial h(Y)}{\partial y_i}\right)\right] \\
&= \sigma^2 \mathscr{E}\frac{\partial h(Y)}{\partial y_i}
\end{aligned}$$

and this completes the proof.   □

We now show that the MSE of the James–Stein estimator is uniformly smaller than the MSE of least squares. One has

$$MSE_{JS} = \mathcal{E}\left(\|\theta - \hat{\theta}^{JS}(Y)\|^2\right)$$

$$= \mathcal{E}\left(\left\|\theta - Y + \frac{(N-2)\sigma^2}{\|Y\|^2}Y\right\|^2\right)$$

$$= \mathcal{E}\left(\|\theta - Y\|^2 + \frac{(N-2)^2\sigma^4}{\|Y\|^4}\|Y\|^2 + 2(\theta - Y)^T Y \frac{(N-2)\sigma^2}{\|Y\|^2}\right)$$

$$= N\sigma^2 + \mathcal{E}\left(\frac{(N-2)^2\sigma^4}{\|Y\|^2} + 2(\theta - Y)^T Y \frac{(N-2)\sigma^2}{\|Y\|^2}\right).$$

As for the last term inside the expectation, exploiting Stein's lemma with

$$h_i(Y) = \frac{y_i}{\|Y\|^2}, \quad \frac{\partial h_i(Y)}{\partial y_i} = \frac{1}{\|Y\|^2} - 2\frac{y_i^2}{\|Y\|^4},$$

one has

$$\mathcal{E}\left(\frac{(\theta - Y)^T Y}{\|Y\|^2}\right) = \mathcal{E}\left(\sum_{i=1}^N (\theta_i - y_i)h_i(Y)\right)$$

$$= -\sigma^2 \mathcal{E}\left(\sum_{i=1}^N \left(\frac{1}{\|Y\|^2} - 2\frac{y_i^2}{\|Y\|^4}\right)\right)$$

$$= -\sigma^2 \mathcal{E}\left(\frac{N-2}{\|Y\|^2}\right).$$

Using this equality in the MSE expression, we finally obtain

$$MSE_{JS} = N\sigma^2 + \mathcal{E}\left(\frac{(N-2)^2\sigma^4}{\|Y\|^2} - 2\frac{(N-2)^2\sigma^4}{\|Y\|^2}\right)$$

$$= N\sigma^2 - (N-2)^2\sigma^4 \mathcal{E}\left(\frac{1}{\|Y\|^2}\right) < N\sigma^2.$$

References

1. Berger JO (1980) A robust generalized Bayes estimator and confidence region for a multivariate normal mean. Ann Stat 8:716–761
2. Berger JO (1982) Selecting a minimax estimator of a multivariate normal mean. Ann Stat 10:81–92

3. Berger JO (1994) An overview of robust Bayesian analysis. Test 3:5–124
4. Bertero M (1989) Linear inverse and ill-posed problems. Adv Electron Electron Phys 75:1–120
5. Bhattacharya PK (1966) Estimating the mean of a multivariate normal population with general quadratic loss function. Ann Math Stat 37:1819–1824
6. Bock ME (1975) Minimax estimators of the mean of a multivariate normal distribution. Ann Stat 3:209–218
7. Brandwein AC, Strawderman WE (2012) Stein estimation for spherically symmetric distributions: recent developments. Stat Sci 27:11–23
8. Casella G (1980) Minimax ridge regression estimation. Ann Stat 8:1036–1056
9. Casella G, Berger R (2001) Statistical inference. Cengage Learning
10. Efron B, Morris C (1973) Stein's estimation rule and its competitors - an empirical Bayes approach. J Am Stat Assoc 68(341):117–130
11. Greenberg E, Webster CE (1983) Advanced econometrics: a bridge to the literature. Wiley
12. Guo YY, Pal N (1992) A sequence of improvements over the James–Stein estimator. J Multivar Anal 42:302–317
13. Hoerl AE (1962) Application of ridge analysis to regression problems. Chem Eng Prog 58:54–59
14. Hoerl AE, Kennard RW (1970) Ridge regression: biased estimation for nonorthogonal problems. Technometrics 12:55–67
15. James W, Stein C (1961) Estimation with quadratic loss. In: Proceedings of the 4th Berkeley symposium on mathematical statistics and probability, vol. I. University of California Press, pp 361–379
16. Kubokawa T (1991) An approach to improving the James–Stein estimator. J Multivar Anal 36:121–126
17. Ljung L (1999) System identification - theory for the user, 2nd edn. Prentice-Hall, Upper Saddle River
18. Manton JH, Krishnamurthy V, Poor HV (1998) James–Stein state filtering algorithms. IEEE Trans Signal Process 46(9):2431–2447
19. Maruyama Y, Strawderman WE (2005) A new class of generalized Bayes minimax ridge regression estimators. Ann Stat 1753–1770
20. Phillips DL (1962) A technique for the numerical solution of certain integral equations of the first kind. J Assoc Comput Mach 9:84–97
21. Pillonetto G, Dinuzzo F, Chen T, De Nicolao G, Ljung L (2014) Kernel methods in system identification, machine learning and function estimation: a survey. Automatica 50
22. Shinozaki N (1974) A note on estimating the mean vector of a multivariate normal distribution with general quadratic loss function. Keio Eng Rep 27:105–112
23. Stein C (1956) Inadmissibility of the usual estimator for the mean of a multivariate distribution. In: Proceedings of the 3rd Berkeley symposium on mathematical statistics and probability, vol I. University of California Press, pp 197–206
24. Stein C (1981) Estimation of the mean of a multivariate normal distribution. Ann Stat 9:1135–1151
25. Strawderman WE (1971) Proper Bayes minimax estimators of the multivariate normal mean. Ann Math Stat 42:385–388
26. Strawderman WE (1978) Minimax adaptive generalized ridge regression estimators. J Am Stat Assoc 73:623–627
27. Tikhonov AN, Arsenin VY (1977) Solutions of ill-posed problems. Winston/Wiley, Washington, D.C
28. Tikhonov AN (1943) On the stability of inverse problems. Dokl Akad Nauk SSSR 39:195–198
29. Tikhonov AN (1963) On the solution of incorrectly formulated problems and the regularization method. Dokl Akad Nauk SSSR 151:501–504

Chapter 2
Classical System Identification

Abstract System identification as a field has been around since the 1950s with roots from statistical theory. A substantial body of concepts, theory, algorithms and experience has been developed since then. Indeed, there is a very extensive literature on the subject, with many text books, like [5, 8, 12]. Some main points of this "classical" field are summarized in this chapter, just pointing to the basic structure of the problem area. The problem centres around four main pillars: (1) the observed data from the system, (2) a parametrized set of candidate models, "the Model structure", (3) an estimation method that fits the model parameters to the observed data and (4) a validation process that helps taking decisions about the choice of model structure. The crucial choice is that of the model structure. The archetypical choice for linear models is the ARX model, a linear difference equation between the system's input and output signals. This is a universal approximator for linear systems—for sufficiently high orders of the equations, arbitrarily good descriptions of the system are obtained. For a "good" model, proper choices of structural parameters, like the equation orders, are required. An essential part of the classical theory deals with asymptotic quality measures, bias and variance, that aim at giving the best mean square error between the model and the true system. Some of this theory is reviewed in this chapter for estimation methods of the maximum likelihood character.

2.1 The State-of-the-Art Identification Setup

System Identification is characterized by five basic concepts:

- X: the experimental conditions under which the data is generated;
- $\mathscr{D}$: the data;
- $\mathscr{M}$: the model structure and its parameters θ;
- $\mathscr{I}$: the identification method by which a parameter value $\hat{\theta}$ in the model structure $\mathscr{M}(\theta)$ is determined based on the data $\mathscr{D}$;
- $\mathscr{V}$: the validation process that generates confidence in the identified model.

G. Pillonetto et al., *Regularized System Identification*, Communications and Control Engineering, https://doi.org/10.1007/978-3-030-95860-2_2

Fig. 2.1 The identification
work loop

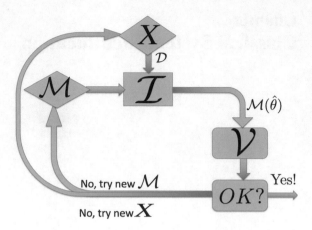

See Fig. 2.1. It is typically an iterative process to navigate to a model that passes through the validation test ("is not falsified"), involving revisions of the necessary choices. For several of the steps in this loop helpful support tools have been developed. It is however not quite possible or desirable to fully automate the choices, since subjective perspectives related to the intended use of the model are very important.

2.2 $\mathcal{M}$: Model Structures

A model structure $\mathcal{M}$ is set of a parametrized models that describe the relations between the inputs u and outputs y of the system. The parameters are denoted by θ so a particular model will be denoted by $\mathcal{M}(\theta)$. The set of models then is

$$\mathcal{M} = \{\mathcal{M}(\theta)|\theta \in D_{\mathcal{M}}\}. \tag{2.1}$$

The models may be expressed and formalized in many different ways. The most common model is linear and time-invariant linear (LTI), but possible models include both nonlinear and time-varying cases, so a list of actually used concrete model will be both very long and diverse.

It is useful to take the general view that a model gives a rule to predict (one-step-ahead) the output at time t, i.e., $y(t)$ (a p-dimensional column vector), based on observations of previous input–output data up to time $t-1$ (denoted by $Z^{t-1} = \{y(t-1), u(t-1), y(t-2), u(t-2), \ldots\}$). Here $u(t)$ is the input at time t and we assume here that the data are collected in discrete time and denote for simplicity the samples as enumerated by t.

The predicted output will then be

$$\hat{y}(t|\theta) = g(t, \theta, Z^{t-1}) \tag{2.2}$$

for a certain function g of past data. This covers a very wide variety of model descriptions, sometimes in a somewhat abstract way. The descriptions become much more explicit when we specialize to linear models.

A note on "inputs" All measurable disturbances that affect y should be included among the inputs u to the system, even if they cannot be manipulated as control inputs. In some cases, the system may entirely lack measurable inputs, so the model (2.2) then just describes how future outputs can be predicted from past ones. Such models are called *time series*, and correspond to systems that are driven by unobservable disturbances. Most of the techniques described in this book apply also to such models.

A note on disturbances A complete model involves both a description of the input–output relations and a description of how various disturbance or noise sources affect the measurements. The noise description is essential both to understand the quality of the model predictions and the model uncertainty. Proper control design also requires a picture of the disturbances in the system.

2.2.1 Linear Time-Invariant Models

For linear time-invariant (LTI) systems, a general model structure is given by the transfer function G from input u to output y and with an additive disturbance—or noise—$v(t)$:

$$y(t) = G(q, \theta)u(t) + v(t). \tag{2.3a}$$

This model is in discrete time and q denotes the shift operator $qy(t) = y(t + 1)$. The sampling interval is set to one time unit. The expansion of $G(q, \theta)$ in the inverse (backwards) shift operator gives the *impulse response* of the system:

$$G(q, \theta)u(t) = \sum_{k=1}^{\infty} g_k(\theta)q^{-k}u(t) = \sum_{k=1}^{\infty} g_k(\theta)u(t - k). \tag{2.3b}$$

The discrete-time Fourier transform (or the z-transform of the impulse response, evaluated in $z = e^{i\omega}$) gives the *frequency response* of the system:

$$G(e^{i\omega}, \theta) = \sum_{k=1}^{\infty} g_k(\theta)e^{-ik\omega}. \tag{2.3c}$$

The function G describes how an input sinusoid shifts phase and amplitude when it passes through the system.

The additive noise term v can be described as white noise $e(t)$, filtered through another transfer function H:

$$v(t) = H(q, \theta)e(t) \tag{2.3d}$$
$$\mathscr{E}e^2(t) = \sigma^2 \tag{2.3e}$$
$$\mathscr{E}e(t)e^T(k) = 0 \text{ if } k \neq t \tag{2.3f}$$

($\mathscr{E}$ denotes mathematical expectation).

This noise characterization is quite versatile and with a suitable choice of H it can describe a disturbance with a quite arbitrary spectrum. It is useful to normalize (2.3d) by making H *monic*:

$$H(q, \theta) = 1 + h_1(\theta)q^{-1} + \cdots . \tag{2.3g}$$

To think in terms of the general model description (2.2) with the predictor as a unifying model concept, assuming H to be inversely stable [5, Sect. 3.2] it is useful to rewrite (2.3) as

$$H^{-1}(q, \theta)y(t) = H^{-1}(q, \theta)G(q, \theta)u(t) + e(t)$$
$$y(t) = [1 - H^{-1}(q, \theta)]y(t) + H^{-1}(q, \theta)Gu(t) + e(t) =$$
$$y(t) = G(q, \theta)u(t) + [1 - H^{-1}(q, \theta)][y(t) - G(q, \theta)u(t)] + e(t).$$

Note that the expansion of H^{-1} starts with "1", so the first term starts with $\tilde{h}_1q^{-1}$ so there is a delay in y. That means that the right-hand side is known at time $t - 1$ except for the term $e(t)$, which is unpredictable at time $t - 1$ and must be estimated with its mean 0. All this means that the predictor for (2.3) (the conditional mean of $y(t)$ given past data) is

$$\hat{y}(t|\theta) = G(q, \theta)u(t) + [1 - H^{-1}(q, \theta)][y(t) - G(q, \theta)u(t)]. \tag{2.4}$$

It is easy to interpret the first term as a simulation using the input u, adjusted with a prediction of the additive disturbance $v(t)$ at time t, based on past values of v. The predictor is thus an easy reformulation of the basic transfer functions G and H. The question now is how to parametrize these.

2.2.1.1   The McMillan Degree

Given just the sequence of impulse responses g_k, with $k = 1, 2, \ldots$, one may consider different ways of representing the system in a more compact form, like rational transfer functions or state-space models, to be considered below. A quite useful concept is then *the McMillan degree*:

From a given impulse response sequence, g_k (that could be $p \times m$ matrices that describe a system with m inputs and p outputs) form the *Hankel matrix*

$$H_k = \begin{bmatrix} g_1 & g_2 & g_3 & \cdots & g_k \\ g_2 & g_3 & g_4 & \cdots & g_{k+1} \\ \cdots & \cdots & \cdots & \cdots & \cdots \\ g_k & g_{k+1} & g_{k+2} & \cdots & g_{2k-1} \end{bmatrix}. \tag{2.5}$$

Then as k increases, the McMillan degree n of the impulse response is the maximal rank of H_k:

$$n = \max_k \text{rank } H_k. \tag{2.6}$$

This means that the impulse response can be generated from an nth-order state-space model, but not from any lower-order model.

2.2.1.2 Black-Box Models

A *black-box* model uses no physical insight or interpretation, but is just a general and flexible parameterization. It is natural to let G and H be rational in the shift operator:

$$G(q, \theta) = \frac{B(q)}{F(q)}; \quad H(q, \theta) = \frac{C(q)}{D(q)} \tag{2.7a}$$

$$B(q) = b_1 q^{-1} + b_2 q^{-2} + \ldots b_{nb} q^{-nb} \tag{2.7b}$$

$$F(q) = 1 + f_1 q^{-1} + \ldots + f_{nf} q^{-nf}, \tag{2.7c}$$

with then C and D *monic* like F, i.e., start with a "1", and the vector collecting all the coefficients

$$\theta = [b_1, b_2, \ldots, f_{nf}]. \tag{2.7d}$$

Common black-box structures of this kind are FIR (finite impulse response model, $F = C = D = 1$), ARMAX (autoregressive moving average with exogenous input, $F = D$), and BJ (Box–Jenkins, all four polynomials different).

A Very Common Case: The ARX Model
A very common case is that $F = D = A$ and $C = 1$ which gives the *ARX model* (autoregressive with exogenous input):

$$y(t) = A^{-1}(q)B(q)u(t) + A^{-1}(q)e(t) \text{ or} \tag{2.8a}$$

$$A(q)y(t) = B(q)u(t) + e(t) \text{ or} \tag{2.8b}$$

$$y(t) + a_1 y(t-1) + \ldots + a_{n_a} y(t-n_a) \tag{2.8c}$$

$$= b_1 u(t-1) + \ldots + b_{n_b} u(t-n_b). \tag{2.8d}$$

This means that the expression for the predictor (2.4) becomes very simple:

$$\hat{y}(t|\theta) = \varphi^T(t)\theta \tag{2.9}$$

$$\varphi^T(t) = \begin{bmatrix} -y(t-1) & -y(t-2) & \ldots & -y(t-n_a) & u(t-1) & \ldots & u(t-n_b) \end{bmatrix} \tag{2.10}$$

$$\theta^T = \begin{bmatrix} a_1 & a_2 & \ldots & a_{n_a} & b_1 & b_2 & \ldots & b_{n_b} \end{bmatrix}. \tag{2.11}$$

In statistics, such a model is known as a *linear regression*.

We note that as n_a and n_b increase to infinity the predictor (2.9) may approximate any linear model predictor (2.4). This points to a *very important general approximation property of ARX models*:

Theorem 2.1 (based on [6]) *Suppose a true linear system is given by*

$$y(t) = G_0(q)u(t) + H_0(q)e(t), \tag{2.12}$$

where $G_0(q)$ and $H_0^{-1}(q)$ are stable filters,

$$G_0(q) = \sum_{k=1}^{\infty} g_k(q^{-k})$$

$$H_0^{-1}(q) = \sum_{k=1}^{\infty} \tilde{h}_k(q^{-k})$$

$$d(n) = \sum_{k=n}^{\infty} |g_k| + |\tilde{h}_K|$$

and e is a sequence of independent zero-mean random variables with bounded fourth-order moments.

Consider an ARX model (2.8) with orders $n_a, n_b = n$, estimated from N observations. Assume that the order n depends on the number of data as $n(N)$, and tends to infinity such that $n(N)^5/N \to 0$. Assume also that the system is such that $d(n(N))\sqrt{N} \to 0$ as $N \to \infty$. Then the ARX model estimates $\hat{A}_{n(N)}(q)$ and $\hat{B}_{n(N)}(q)$ of order $n(N)$ obey

$$\frac{\hat{B}_{n(N)}(q)}{\hat{A}_{n(N)}(q)} \to G_0(q), \quad \frac{1}{\hat{A}_{n(N)}(q)} \to H_0(q) \text{ as } N \to \infty. \tag{2.13}$$

Intuitively, the above result follows from the fact that the true predictor for the system

$$\hat{y}(t|\theta) = (1 - H_0^{-1})y(t) + H_0^{-1}G_0u(t) = \sum_{k=1}^{\infty} \tilde{h}_k y(t-k) + \tilde{g}_k u(t-k)$$

is stable. Hence, it can be truncated at any n with arbitrary accuracy, and the truncated sum is the predictor of an nth-order ARX model.

This is quite a useful result saying that ARX models can approximate any linear system, if the orders are sufficiently large. ARX models are easy to estimate. The estimates are calculated by linear least squares (LS) techniques, which are convex and numerically robust. Estimating a high-order ARX model, possibly followed by some model order reduction, could thus be an alternative to the numerically more demanding general PEM criterion minimization (2.22) introduced later on. This has been extensively used, e.g., by [14, 15]. The only drawback with high-order ARX models is that they may suffer from high variance.

2.2.1.3 Grey-Box Models

If some physical facts are known about the system, these could be incorporated in the model structure. Such a model that is based on physical insights and has a built-in behaviour that mimics known physics is known as a *Grey-Box Model*. For example, it could that for an airplane whose motion equations are known from Newton's laws, but certain parameters are unknown, like the aerodynamical derivatives. Then it is natural to build a continuous-time state-space model from known physical equations:

$$\begin{aligned} \dot{x}(t) &= A(\theta)x(t) + B(\theta)u(t) \\ y(t) &= C(\theta)x(t) + D(\theta)u(t) + v(t). \end{aligned} \tag{2.14}$$

Here θ are simply some entries of the matrices A, B, C, D, corresponding to unknown physical parameters, while the other matrix entries signify known physical behaviour. This model can be sampled with well-known sampling formulas (obeying the input inter-sample properties, zero-order hold or first-order hold) to give

$$\begin{aligned} x(t+1) &= \mathcal{F}(\theta)x(t) + \mathcal{G}(\theta)u(t) \\ y(t) &= C(\theta)x(t) + D(\theta)u(t) + w(t). \end{aligned} \tag{2.15}$$

The model (2.15) has the transfer function from u to y

$$G(q, \theta) = C(\theta)[qI - \mathcal{F}(\theta)]^{-1}\mathcal{G}(\theta) + D(\theta) \tag{2.16}$$

so we have achieved a particular parameterization of the general linear model (2.3).

2.2.1.4   Continuous-Time Models

The general model description (2.2) describes how the predictions evolve in discrete time. But in many cases we are interested in continuous-time (CT) models, like models for physical interpretation and simulation. But CT model estimation is contained in the described framework, as the linear state-space model (2.14) illustrates.

2.2.2   Nonlinear Models

A nonlinear model is a relation (2.2), where the function g is nonlinear in the input–output data Z. There is a rich variation in how to specify the function g more explicitly. A quite general way is the nonlinear state-space equation, which is a counterpart to (2.15):

$$x(t+1) = f(x(t), u(t), v(t), \theta)$$
$$y(t) = h(x(t), e(t), \theta),$$

(2.17)

where v and e are white noises.

2.3 $\mathscr{I}$: Identification Methods—Criteria

The goal of identification is to match the model to the data. Here the basic techniques for such matching will be discussed. Suppose we have collected a data record in the time domain

$$\mathscr{D}_T = \{u(1), y(1), \ldots, u(N), y(N)\}$$

(2.18)

which will be called in this book *identification set* or *training set*, with N being its size. A natural way to evaluate a model is to see how well it is able to predict the measured output since the model is in essence a predictor. It is thus quite natural to form the prediction errors for (2.2):

$$\varepsilon(t, \theta) = y(t) - \hat{y}(t|\theta).$$

(2.19)

The "size" of this error can be measured by some scalar norm:

$$\ell(\varepsilon(t, \theta))$$

(2.20)

and the performance of the predictor over the whole data record $\mathscr{D}_T$ is given by

$$V_N(\theta) = \sum_{t=1}^{N} \ell(\varepsilon(t, \theta)). \tag{2.21}$$

A natural parameter estimate is the value that minimizes this prediction fit:

$$\hat{\theta}_N = \arg\min_{\theta \in D_{\mathscr{M}}} V_N(\theta). \tag{2.22}$$

This is the *Prediction Error Method (PEM)* and it is applicable to general model structures. See, e.g., [5] or [7] for more details.

The PEM approach can be embedded in a statistical setting. The ML methodology below offers a systematic framework to do so.

2.3.1 A Maximum Likelihood (ML) View

If the system innovations e have a probability density function (pdf) $f(x)$, then the criterion function (2.21) with $\ell(x) = -\log f(x)$ will be the logarithm of the *Likelihood function*. See Lemma 5.1 in [5]. More specifically, let the system have p outputs, and let the innovations be Gaussian with zero mean and covariance matrix Λ, so that

$$y(t) = \hat{y}(t|\theta_0) + e(t), \quad e(t) \in N(0, \Lambda) \tag{2.23}$$

for the θ_0 that generated the data. Then it follows that the negative logarithm of the likelihood function for estimating θ from y is

$$L_N(\theta) = \frac{1}{2}[V_N(\theta) + N \log \det \Lambda + Np \log 2\pi], \tag{2.24}$$

where $V_N(\theta)$ is defined by (2.21), with

$$\ell(\varepsilon(t, \theta)) = \varepsilon^T(t, \theta)\Lambda^{-1}\varepsilon(t, \theta). \tag{2.25}$$

That means that the maximum likelihood model estimate (MLE) for known Λ is obtained by minimizing $V_N(\theta)$. If Λ is not known, it can be included among the parameters and estimated, ([5], p. 218), which results in a criterion

$$D_N(\theta) = \det \sum_{t=1}^{N} \varepsilon(t, \theta)\varepsilon^T(t, \theta) \tag{2.26}$$

to be minimized.

A Bayesian interpretation of (2.22) as well as a regularized version will be given in Chap. 4.

2.4   Asymptotic Properties of the Estimated Models

As we have seen in the first chapter, bias and variance play important roles in estimation problems. We will here give a short account of how these concepts are treated in classical system identification.

2.4.1   Bias and Variance

The observations, certainly of the output from the system, are affected by noise and disturbances. That means that the estimated model parameters (2.22) also will be affected by disturbances. These disturbances are typically described as stochastic processes, which makes the estimate $\hat{\theta}_N$ a *random variable*. This has a certain probability density function, which could be complicated to compute. Often the analysis is restricted to its mean and variance only. The difference between the mean and a true description of the system measures the *bias* of the model. If the mean coincides with the true system, the estimate is said to be *unbiased*. As already pointed out in (1.1), the total error in a model thus has two contributions: the bias and the variance.

2.4.2   Properties of the PEM Estimate as $N \rightarrow \infty$

Except in simple special cases it is quite difficult to compute the pdf of the estimate $\hat{\theta}_N$. However, its *asymptotic properties* as $N \rightarrow \infty$ are easier to establish. The basic results can be summarized as follows (see [5, Chaps. 8 and 9] for a more complete treatment):

- **Limit model:**

$$\hat{\theta}_N \rightarrow \theta^* = \arg\min \left[\lim_{N \to \infty} \frac{1}{N} V_N(\theta) \approx \mathscr{E}\ell(\varepsilon(t, \theta)) \right]. \qquad (2.27)$$

  Here $\mathscr{E}$ denotes mathematical expectation. So the estimate will converge to the best possible model, in the sense that it gives the smallest average prediction error.
- **Asymptotic covariance matrix for scalar output models:**
  In case the prediction errors $e(t) = \varepsilon(t, \theta^*)$ for the limit model are approximately white, the covariance matrix of the parameters is asymptotically given by

$$\mathrm{Cov}\hat{\theta}_N \sim \frac{\kappa(\ell)}{N} \left[\mathrm{Cov}\frac{d}{d\theta}\hat{y}(t|\theta) \right]^{-1}. \qquad (2.28)$$

  That means that the covariance matrix of the parameter estimate is given by the inverse covariance matrix of the gradient of the predictor w.r.t. the parameters.

Here (prime denoting derivatives)

$$\kappa(\ell) = \frac{\mathscr{E}[\ell'(e(t))]^2}{\mathscr{E}\ell''(e(t))^2}.$$ (2.29)

Note that

$$\kappa(\ell) = \sigma^2 = \mathscr{E}e^2(t) \quad \text{if} \quad \ell(e) = e^2/2.$$

If the model structure contains the true system, it can be shown that this covariance matrix is the smallest that can be achieved by any unbiased estimate, in case the norm ℓ is chosen as the logarithm of the pdf of e. That is, it fulfils the *the Cramér–Rao inequality*, [2]. These results are valid for quite general model structures.

• **Results for LTI models:**

Now, specialize to linear models (2.3) and assume that the true system is described by

$$y(t) = G_0(q)u(t) + H_0(q)e(t),$$ (2.30)

which could be general transfer functions, possibly much more complicated than the model. Then

–

$$0^* = \arg\min_\theta \int_{-\pi}^{\pi} |G(e^{i\omega}, \theta) - G_0(e^{i\omega})|^2 \frac{\Phi_u(\omega)}{|H(e^{i\omega}, \theta)|^2} d\omega.$$ (2.31)

That is, the frequency function of the limiting model will approximate the true frequency function as well as possible in a frequency norm given by the input spectrum Φ_u and the noise model.

– For a linear black-box model, the covariance of the estimated frequency function is

$$\text{Cov}G(e^{i\omega}, \hat{\theta}_N) \sim \frac{n}{N} \frac{\Phi_v(\omega)}{\Phi_u(\omega)} \quad \text{as } n, N \to \infty,$$ (2.32)

where n is the model order and Φ_v is the noise spectrum $\sigma^2|H_0(e^{i\omega})|^2$. The variance of the estimated frequency function at a given frequency is thus, for a high-order model, proportional to the noise-to-signal ratio at that frequency. That is a natural and intuitive result.

2.4.3   Trade-Off Between Bias and Variance

The quality of the model depends on the quality of the measured data and the flexibility of the chosen model structure (2.1). A more flexible model structure typically has smaller bias, since it is easier to come closer to the true system. At the same time, it will have a higher variance: with higher flexibility it is easier to be fooled by disturbances and this may lead to data overfitting. So the trade-off between bias and variance to reach a small total error is a choice of balanced flexibility of the model structure.

As the model gets more flexible, the fit to the estimation data in (2.22), given by $V_N(\hat{\theta}_N)$, will always improve. To account for the variance contribution, it is thus necessary to modify this fit to assess the total quality of the model. A much used technique for this is Akaike's criterion (AIC), [1],

$$\hat{\theta}_N = \underset{\mathcal{M},\theta \in D_{\mathcal{M}}}{\arg\min}\ 2L_N(\theta) + 2\dim\theta, \qquad (2.33)$$

where L_N is the negative log likelihood function. The minimization also takes place over a family of model structures with different number of parameters (dim θ).

For Gaussian innovations e with unknown and estimated variance, the criterion AIC takes the form

$$\hat{\theta}_N = \underset{\mathcal{M},\theta \in D_{\mathcal{M}}}{\arg\min}\ \left[\log \det \left[\frac{1}{N} \sum_{t=1}^{N} \varepsilon(t,\theta)\varepsilon^T(t,\theta) \right] + 2\frac{m}{N} \right] \quad \text{AIC} \qquad (2.34)$$

with $m = \dim\theta$ and after normalization and omission of model-independent quantities.

There is also a small-sample version, described in [4] and known in the literature as corrected Akaike's criterion (AICc), defined by

$$\hat{\theta}_N = \arg\min_{\theta}\ \left[\log \det \left[\frac{1}{N} \sum_{t=1}^{N} \varepsilon(t,\theta)\varepsilon^T(t,\theta) \right] + 2\frac{m}{(N-m-1)} \right], \quad \text{AICc.}$$
$$\qquad (2.35)$$

Another variant places a larger penalty on the model flexibility:

$$\hat{\theta}_N = \arg\min_{\theta}\ \left[\log \det \left[\frac{1}{N} \sum_{t=1}^{N} \varepsilon(t,\theta)\varepsilon^T(t,\theta) \right] + \log(N)\frac{m}{N} \right], \quad \text{BIC, MDL.}$$
$$\qquad (2.36)$$

This is known as Bayesian information criterion (BIC) or Rissanen's Minimum Description Length (MDL) criterion, see, e.g., [10, 11] and [5, pp. 505–507].

Section 2.6 contains further aspects on the choice of model structure.

2.5 X: Experiment Design

Experiment design involves all questions that concern the collection of estimation data, such as selecting which signals to measure, which sampling rate to use, and also the design of the input including possible feedback configurations.

The theory of experiment design primarily relies upon analysis of how the asymptotic parameter covariance matrix (2.28) depends on the design variables: so the essence of experiment design can be symbolized as

$$\min_{X} \text{trace}\{C[E\psi(t)\psi^{T}(t)]^{-1}\},$$

where ψ is the gradient of the prediction w.r.t. the parameters and the matrix C is used to weight variables reflecting the intended use of the model.

For linear systems, the input design is often expressed as selecting the spectrum (frequency contents) of u.

This leads to the following recipe: let the input's power be concentrated to frequency regions where a good model fit is essential, and where disturbances are dominating.

The measurement setup, like if band limited inputs are used to estimate continuous-time models and how the experiment equipment is instrumented with band-pass filters, e.g., see [8, Sects. 13.2–3], also belongs to the important experiment design questions.

2.6 $\mathscr{V}$: Model Validation

Model validation is about obtaining a model that, at least for the time being, can be accepted. It amounts to examining and scrutinizing the model to check if it can be used for its purpose. These methods are of course problem dependent and contain several subjective elements, Therefore, no conclusive procedure for validation can be given. A few useful techniques will be listed here. Basically it is a matter of trying to falsify a model under the conditions it will be used for and also to gain confidence in its ability to reproduce new data from the system.

2.6.1 Falsifying Models: Residual Analysis

An estimated model is never a correct description of a true system. In that sense, a model cannot be "validated", i.e., proved to be correct. Instead it is instructive to try and *falsify* it, i.e., confront it with facts that may contradict its correctness. A good principle is to look for the *simplest unfalsified model*, see, e.g., [9].

Residual analysis is the leading technique for falsifying models: the residuals or one-step-ahead prediction errors $\hat{\varepsilon}(t) = \varepsilon(t, \hat{\theta}_N) = y(t) - \hat{y}(t|\hat{\theta}_N)$ should ideally not contain any traces of past inputs or past residuals. If they did, it means that the predictions are not ideal. So, it is natural to test the correlation functions

$$\hat{r}_{\hat{\varepsilon},u}(k) = \frac{1}{N} \sum_{t=1}^{N} \hat{\varepsilon}(t+k)u(t) \tag{2.37}$$

$$\hat{r}_{\hat{\varepsilon}}(k) = \frac{1}{N} \sum_{t=1}^{N} \hat{\varepsilon}(t+k)\hat{\varepsilon}(t) \tag{2.38}$$

and check that they are not larger than certain thresholds. Here N is the length of the data record and k typically ranges over a fraction of the interval $[-N, N]$. See, e.g., [5, Sect. 16.6] for more details.

2.6.2 Comparing Different Models

When several models have been estimated it is a question to choose the "best one". Then, models that employ more parameters naturally show a better fit to the data, and it is necessary to outweigh that. The model selection criteria AIC (2.34) and BIC (2.36) are examples of how such decisions can be taken. They can be extended to regular hypothesis tests where more complex models are accepted or rejected at various test levels, see, e.g., [5, Sect. 16.4].

Making comparisons in the frequency domain is a very useful complement for domain experts used to think in terms of natural frequencies, natural damping, etc.

2.6.3 Cross-Validation

Cross-validation (CV) is an important statistical concept that loosely means that the model performance is tested on a data set (*validation data*) other than the estimation data. There is an extensive literature on cross-validation, e.g., [13] and many ways to split up available data into estimation and validation parts have been suggested. The goal is to obtain an estimate of the prediction capability of future data of the model in correspondence with different choices of θ. Parameter selection is thus performed by optimizing the estimated prediction score. *Hold out validation* is the simplest form of CV: the available data are split in two parts, where one of them (*estimation set*) is used to estimate the model, and the other one (*validation set*) is used to assess the prediction capability. By ensuring independence of the model fit from the validation data, the estimate of the prediction performance is approximately unbiased. For models that do not require estimation of initial states, like FIR and ARX models, CV can be applied efficiently in more sophisticated ways by splitting the data into more portions, as described in [3].

References

1. Akaike H (1974) A new look at the statistical model identification. IEEE Trans Autom Control AC-19:716–723
2. Cramér H (1946) Mathematical methods of statistics. Princeton University Press, Princeton, N.J
3. Hastie T, Tibshirani R, Friedman J (2001) The elements of statistical learning. Springer
4. Hurvich C, Tsai C (1989) Regression and time series model selection in small samples. Biometrika 76:297–307
5. Ljung L (1999) System identification - theory for the user, 2nd edn. Prentice-Hall, Upper Saddle River, NJ
6. Ljung L, Wahlberg B (1992) Asymptotic properties of the least-squares method for estimating transfer functions and disturbance spectra. Adv Appl Prob 24:412–440
7. Ljung L (2002) Prediction error estimation methods. Circuits Syst Signal Process 21(1):11–21
8. Pintelon R, Schoukens J (2012) System identification: a frequency domain approach, 2nd edn. Wiley
9. Popper KR (1934) The logic of scientific discovery. Basic Books, New York
10. Rissanen J (1978) Modelling by shortest data description. Automatica 14:465–471
11. Schwarz G (1978) Estimating the dimension of a model. Ann Stat 6:461–464
12. Söderström T, Stoica P (1989) System identification. Prentice-Hall Int., London
13. Stone M (1977) Asymptotics for and against cross-validation. Biometrika 64(1)
14. Zhu YC (1989) Asymptotic properties of prediction error methods. Int J Adapt Control Signal Process 3:357–373
15. Zhu Y, Backx T (1993) Identification of multivariable industrial processes for diagnosis and control. Springer, London

Chapter 3
Regularization of Linear Regression Models

Abstract Linear regression models are widely used in statistics, machine learning and system identification. They allow to face many important problems, are easy to fit and enjoy simple analytical properties. The simplest method to fit linear regression models is least squares whose systematic treatment is available in many textbooks, e.g., [35, Chap. 4], [12]. Linear regression models can be fitted also in different way and a class of methods that we will consider in this chapter is the so-called regularized least squares. It is an extension of least squares which minimizes the sum of the square loss function and a regularization term. This latter can take various forms, leading to several variants which have been applied extensively in theory as well as in practical applications. In this chapter, we will focus on these methods and introduce their fundamentals. In the first part of the appendix to this chapter, we also report some basic results of linear algebra useful for the reading.

3.1 Linear Regression

Regression theory is concerned with modelling relationships among variables. It is used for predicting one dependent variable based on the information provided by one or more independent variables. In linear regression, the relationship among variables is given by linear functions. To illustrate this, we start from the function estimation problem because it is intuitive and easy to understand.

The aim of function estimation is to reconstruct a function $g : \mathbb{R}^n \to \mathbb{R}$ with $n \in \mathbb{N}$ from a collection of N measured values of $g(x)$ and x which we denote, respectively, by y_i and x_i for $i = 1, \ldots, N$. For generic values of x, the estimate $\hat{g}$ should give a good prediction $\hat{g}(x)$ of $g(x)$. The variables x and $g(x)$ are often called the input and the output variable or simply the input and the output, respectively. The collection of measured values of x and $g(x)$, given by the couples $\{x_i, y_i\}$, is called the data set or also the training set. In practical applications, the measurement y_i is often not precise and subject to some disturbance, i.e., for a given input x_i there is often discrepancy between $g(x_i)$ and its measured value y_i. To describe this phenomenon, it is natural

© The Author(s) 2022
G. Pillonetto et al., *Regularized System Identification*, Communications and Control Engineering, https://doi.org/10.1007/978-3-030-95860-2_3

to introduce a disturbance variable $e \in \mathbb{R}$ and assume that, for any given $x \in \mathbb{R}^n$, the measured value of $g(x)$ is

$$y = g(x) + e. \tag{3.1}$$

Hence, y is the measured output and $g(x)$ is the noise-free or true output. Accordingly, the training data $\{x_i, y_i\}_{i=1}^N$ are collected as follows:

$$y_i = g(x_i) + e_i, \quad i = 1, \ldots, N. \tag{3.2}$$

We are interested in linear regression models for estimation of g. For illustration, an example is now introduced.

Example 3.1 (*Polynomial regression*) We consider $g : [0, 1] \to \mathbb{R}$ and assume that such function is smooth. Then, g can be well approximated by polynomials with a certain order. In this case, a linear regression model for the function estimation problem takes the following form:

$$y_i = \theta_1 + \sum_{k=2}^{n} \theta_k x_i^{k-1} + e_i, \quad i = 1, \ldots, N, \tag{3.3}$$

where $\theta_k \in \mathbb{R}$ for $k = 1, \ldots, n$. Defining

$$\phi(x_i) = [\, 1 \ x_i \ \ldots \ x_i^{n-1} \,]^T, \quad \theta = [\, \theta_1 \ \theta_2 \ \ldots \ \theta_n \,]^T, \tag{3.4}$$

where, for a real-valued matrix A, the notation A^T denotes its matrix transpose, we rewrite (3.3) as

$$y_i = \phi(x_i)^T \theta + e_i, \quad i = 1, \ldots, N \tag{3.5}$$

obtaining a more compact expression. □

Although (3.5) is derived from Example 3.1, it is the general linear regression model studied in the theory of regression. For convenience, we remove the dependence of $\phi(x_i)$ on x_i and simply write $\phi(x_i)$ as ϕ_i, when the context is clear. In addition, all the vectors are column vectors. Then, model (3.5) becomes

$$y_i = \phi_i^T \theta + e_i, \quad i = 1, \ldots, N, \quad y_i \in \mathbb{R}, \ \phi_i \in \mathbb{R}^n, \ \theta \in \mathbb{R}^n, \ e_i \in \mathbb{R}. \tag{3.6}$$

In what follows, we will focus on (3.6) and introduce the linear regression problem, the methods of least squares and regularized least squares. We will call $y_i \in \mathbb{R}$ the measured output, $\phi_i \in \mathbb{R}^n$ the regressor, $\theta \in \mathbb{R}^n$ the model parameter, n the model order, and e_i the measurement noise.

Before proceeding, it should be noted that the choice of the model order n is a critical problem in practical applications. The rule of thumb is to set n to a large enough value such that g can be represented by the proposed model structure. In system identification, this corresponds to introducing a model structure flexible enough to contain the true system. Consider, e.g., Example 3.1 again and assume that the function g is actually a polynomial of order 5. Clearly, if the dimension of θ does not satisfy $n \geq 6$, then x^5 cannot be represented and some model bias will affect the estimation process. However, the order n should not be chosen larger than necessary, because this can increase the variance of the model estimate. This problem is actually the same as model selection complexity in the classical system identification and is connected with the bias-variance trade-off illustrated in the first two chapters and also discussed in more detail shortly.

Also in light of the above discussion, we often assume that the model order n is either large enough for g to be adequately represented by the proposed model or even that a true model parameter that has generated the data exists, denoted by $\theta_0 \in \mathbb{R}^n$. Hence, we can formulate linear regression as the problem of obtaining an estimate $\hat{\theta}$ such that, given a new regressor $\phi \in \mathbb{R}^n$, the prediction $\phi^T \hat{\theta}$ is close to $\phi^T \theta_0$.

3.2 The Least Squares Method

There are many methods to estimate θ in the linear regression model (3.6). In this section, we consider the least squares (LS) method.

3.2.1 *Fundamentals of the Least Squares Method*

Given the data y_i, ϕ_i for $i = 1, \ldots, N$, one way to estimate θ is to minimize the least squares (LS) criterion:

$$\hat{\theta}^{\mathrm{LS}} = \arg\min_{\theta} l(\theta), \qquad l(\theta) = \sum_{i=1}^{N} (y_i - \phi_i^T \theta)^2, \qquad (3.7)$$

where $l(\theta)$ is the LS criterion and $\hat{\theta}^{\mathrm{LS}}$ is the LS estimate of θ. Then, the predicted output $\hat{y}$ for the value of $\phi^T \theta_0$ with $\phi \in \mathbb{R}^n$ is obtained as

$$\hat{y} = \phi^T \hat{\theta}^{\mathrm{LS}}. \qquad (3.8)$$

3.2.1.1   Normal Equations and LS Estimate

The LS estimate $\hat{\theta}^{\mathrm{LS}}$ given by (3.7) has a closed-form expression. To see this, note that the first- and second-order derivatives of $l(\theta)$ with respect to θ are

$$\frac{\partial l(\theta)}{\partial \theta} = 2 \sum_{i=1}^{N} \phi_i \phi_i^T \theta - 2 \sum_{i=1}^{N} \phi_i y_i, \quad \frac{\partial^2 l(\theta)}{\partial \theta \partial \theta} = 2 \sum_{i=1}^{N} \phi_i \phi_i^T \succcurlyeq 0, \quad (3.9)$$

where $A \succcurlyeq 0$ means that A is a positive semidefinite matrix. Then all $\hat{\theta}^{LS}$ that satisfy

$$\left[\sum_{i=1}^{N} \phi_i \phi_i^T \right] \hat{\theta}^{LS} = \sum_{i=1}^{N} \phi_i y_i \qquad (3.10)$$

are global minima of $l(\theta)$. The set of Eqs. (3.10) is known as the normal equations. For the time being, we assume that $\sum_{i=1}^{N} \phi_i \phi_i^T$ is full rank.[1] Then

$$\hat{\theta}^{LS} = \left[\sum_{i=1}^{N} \phi_i \phi_i^T \right]^{-1} \sum_{i=1}^{N} \phi_i y_i. \qquad (3.11)$$

3.2.1.2 Matrix Formulation

It is often convenient to rewrite the LS method in matrix form. To this goal, let

$$Y = \begin{bmatrix} y_1 \\ y_2 \\ \vdots \\ y_N \end{bmatrix}, \quad \Phi = \begin{bmatrix} \phi_1^T \\ \phi_2^T \\ \vdots \\ \phi_N^T \end{bmatrix}, \quad E = \begin{bmatrix} e_1 \\ e_2 \\ \vdots \\ e_N \end{bmatrix}. \qquad (3.12)$$

We can then rewrite (3.6) with the θ_0 that generated the data, the LS criterion (3.7), the normal Eqs. (3.10) and the LS estimate (3.11) in matrix form, respectively:

$$Y = \Phi \theta_0 + E \qquad (3.13)$$

$$\hat{\theta}^{LS} = \arg\min_{\theta} l(\theta), \quad l(\theta) = \|Y - \Phi\theta\|_2^2 \qquad (3.14)$$

$$\Phi^T \Phi \hat{\theta}^{LS} = \Phi^T Y \qquad (3.15)$$

$$\hat{\theta}^{LS} = (\Phi^T \Phi)^{-1} \Phi^T Y, \qquad (3.16)$$

where $\| \cdot \|_2$ is the Euclidean norm, i.e., the 2-norm, and Φ is called the regression matrix.

---

[1] Recall that the column rank (resp., the row rank) of a matrix is the dimension of the space spanned by the columns (resp., the rows) of the matrix. It is a fundamental result in linear algebra that the column rank and the row rank of a matrix are always equal and this number is called the rank of the matrix. A matrix is said to be full rank if its rank is equal to the lesser of the number of rows and columns and a matrix is said to be rank deficient otherwise.

3.2.2 Mean Squared Error and Model Order Selection

3.2.2.1 Bias, Variance, and Mean Squared Error of the LS Estimate

We study the linear regression problem in a probabilistic framework, assuming that data are generated according to (3.13) and that

the measurement noises e_i, for $i = 1, \ldots, N$, are i.i.d. with mean 0 and variance σ^2.

$$(3.17)$$

Due to this assumption, the LS estimator $\hat{\theta}^{LS}$, as well as any estimator of θ dependent on the data, becomes random variables. Then, it is interesting to study the statistical properties of $\hat{\theta}^{LS}$, such as the bias, variance and mean squared error (MSE).

All the expectations reported below are computed with respect to the noises e_i with the regressors ϕ_i assumed to be deterministic. Simple calculations lead to

$$\mathcal{E}(\hat{\theta}^{LS}) = \theta_0 \tag{3.18a}$$

$$\hat{\theta}^{LS}_{\text{bias}} = \mathcal{E}(\hat{\theta}^{LS}) - \theta_0 = 0 \tag{3.18b}$$

$$\text{Cov}(\hat{\theta}^{LS}, \hat{\theta}^{LS}) = \mathcal{E}[(\hat{\theta}^{LS} - \mathcal{E}(\hat{\theta}^{LS}))(\hat{\theta}^{LS} - \mathcal{E}(\hat{\theta}^{LS}))^T] = \sigma^2(\Phi^T\Phi)^{-1} \tag{3.18c}$$

$$\begin{aligned} \text{MSE}(\hat{\theta}^{LS}, \theta_0) &= \mathcal{E}[(\hat{\theta}^{LS} - \theta_0)(\hat{\theta}^{LS} - \theta_0)^T] \\ &= \text{Cov}(\hat{\theta}^{LS}, \hat{\theta}^{LS}) + \hat{\theta}^{LS}_{\text{bias}}(\hat{\theta}^{LS}_{\text{bias}})^T \\ &= \sigma^2(\Phi^T\Phi)^{-1}, \end{aligned} \tag{3.18d}$$

where $\text{Cov}(\hat{\theta}^{LS}, \hat{\theta}^{LS})$ is the covariance matrix of $\hat{\theta}^{LS}$ and $\text{MSE}(\hat{\theta}^{LS}, \theta_0)$ is the MSE matrix of $\hat{\theta}^{LS}$ function of the true model parameter θ_0.

3.2.2.2 Model Order Selection

The issue of model order selection is essentially the same as that of model complexity selection in the classical system identification scenario. Therefore, the techniques introduced in Sect. 2.4.3 can be used to choose the model order n, e.g., Akaike's information criterion (AIC) [1], the Bayesian Information criterion (BIC) or Minimum Description Length (MDL) approach [25, 39].

The quality of the LS estimate $\hat{\theta}^{LS}$ depends on the adopted model order n. In practical applications, model complexity is in general unknown and needs to be determined from data. As the model order n gets larger, the fit to the data $\|Y - \Phi\hat{\theta}^{LS}\|_2^2$ in (3.14) will become smaller, but the variances along the diagonal of the MSE matrix (3.18d) of $\hat{\theta}^{LS}$ will become larger at the same time. When assessing the quality of $\hat{\theta}^{LS}$, one way to account for the increasing variance is to introduce criteria that suitably modify the plain data fit. AIC and BIC are techniques following this idea and can be used for model order selection. More specifically, besides (3.17), further assuming that the errors are independent and Gaussian, i.e.,

$$e_i \sim \mathcal{N}(0, \sigma^2), \quad i = 1, \ldots, N \tag{3.19}$$

with known noise variance σ^2, we obtain

$$\text{AIC:} \quad \hat{\theta}^{\text{LS}} = \arg\min_{\theta \in \mathbb{R}^n} \frac{1}{N} \|Y - \Phi\theta\|_2^2 + 2\sigma^2 \frac{n}{N}, \tag{3.20}$$

$$\text{BIC or MDL:} \quad \hat{\theta}^{\text{LS}} = \arg\min_{\theta \in \mathbb{R}^n} \frac{1}{N} \|Y - \Phi\theta\|_2^2 + \log(N)\sigma^2 \frac{n}{N}, \tag{3.21}$$

where the minimization also takes place over a family of model structures with different dimension n of θ.

Another way is to estimate the prediction capability of the model on some unseen data which are not used for model estimation. As briefly seen in Sect. 2.6.3, cross-validation (CV) exploits this idea and is among the most widely used techniques for model selection. Recall that hold out CV is the simplest form of CV with data divided into two parts. One part is used to estimate the model with different model orders and the other part is used to assess the prediction capability of each model through the prediction score $\|Y_v - \Phi_v \hat{\theta}^{\text{LS}}\|_2^2$. Here, Y_v, Φ_v are the validation data which are different from those used to derive $\hat{\theta}^{\text{LS}}$. The model order giving the best prediction score will be chosen.

The noise variance σ^2 of the measurement noises e_i plays an important role in statistical modelling, e.g., in the assessment of the variance of $\hat{\theta}^{\text{LS}}$ and in the model order selection using, e.g., AIC (3.20) or BIC (3.21). In practical applications, the noise variance σ^2 is in general unknown and needs to be estimated from the data Y and Φ. It can be estimated in different ways based on the maximum likelihood estimation (MLE) method or the statistical property of $\hat{\theta}^{\text{LS}}$.

Under (3.17) and the Gaussian assumption (3.19), the ML estimate of σ^2, as given in [25, p. 506], is

$$\hat{\sigma}^{2,\text{ML}} = \frac{1}{N} \|Y - \Phi\hat{\theta}^{\text{LS}}\|_2^2. \tag{3.22}$$

Using only assumption (3.17), an unbiased estimator of σ^2, as given in [25, p. 554], turns out

$$\hat{\sigma}^2 = \frac{1}{N - n} \|Y - \Phi\hat{\theta}^{\text{LS}}\|_2^2. \tag{3.23}$$

AIC and BIC were reported, respectively, in (3.20) and (3.21) assuming known noise variance. When σ^2 is unknown, the use of the ML estimate (3.22) leads to the widely used AIC and BIC for Gaussian innovations, e.g., [25, pp. 506–507]:

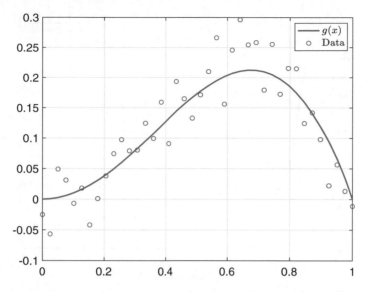

Fig. 3.1 Polynomial regression: the function $g(x)$ (blue curve) and the data $\{x_i, y_i\}_{i=1}^{40}$ (red circles)

$$\text{AIC:}\quad \hat{\theta}^{LS} = \underset{\theta \in \mathbb{R}^n}{\arg\min}\ \log\left(\frac{1}{N}\|Y - \Phi\theta\|_2^2\right) + 2\frac{n}{N}, \tag{3.24}$$

$$\text{BIC or MDL:}\quad \hat{\theta}^{LS} = \underset{\theta \in \mathbb{R}^n}{\arg\min}\ \log\left(\frac{1}{N}\|Y - \Phi\theta\|_2^2\right) + \log(N)\frac{n}{N}. \tag{3.25}$$

Example 3.2 (*Polynomial regression using LS and discrete model order selection*)
We apply the LS method and the model order selection techniques to polynomial
regression as sketched in Example 3.1. Let the function g be

$$g(x) = \sin^2(x)(1 - x^2), \quad x \in [0, 1]. \tag{3.26}$$

Then, we generate the data as follows:

$$y_i = \sin^2(x_i)(1 - x_i^2) + e_i, \quad i = 1, \ldots, 40, \tag{3.27}$$

where $x_1 = 0$, $x_{40} = 1$, the $x_2, \ldots, x_{39}$ are evenly spaced points between x_1 and
x_{40}, and the noises e_i are i.i.d. Gaussian distributed with zero mean and standard
deviation 0.034. The function g and the generated data are shown in Fig. 3.1.

The function g is smooth and can be well approximated by polynomials. However,
it is unclear which order should be chosen. Hence, we test the values $n = 1, \ldots, 15$
and, for each order n, we form the regressor (3.4), the linears regression model (3.13)
and derive the LS estimate $\hat{\theta}^{LS}$. As shown in Fig. 3.2, as the order n increases the
data fit $\|Y - \Phi\hat{\theta}^{LS}\|_2^2$ keeps decreasing.

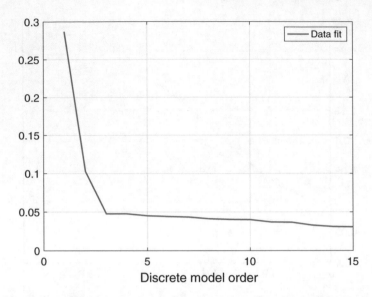

Fig. 3.2 Polynomial regression: profile of the LS data fit as a function of the discrete model order n

For model order selection, we use AIC (3.24), BIC (3.25) and hold out CV with $x_i, y_i, i = 1, 3, \ldots, 39$ for estimation and $x_i, y_i, i = 2, 4, \ldots, 40$ for validation. Figure 3.3 plots the values of AIC (3.24), BIC (3.25) and the prediction score of hold out CV. The order n selected by AIC and BIC are the same and equal to 3 while that selected by hold out CV is 7.

To evaluate the performance of models of different complexity, we compute the fit measure

$$\mathscr{F} = 100 \left(1 - \left[\frac{\sum_{k=1}^{40} |g(x_k) - \hat{g}(x_k)|^2}{\sum_{k=1}^{40} |g(x_k) - \bar{g}^0|^2} \right]^{1/2} \right), \quad \bar{g}^0 = \frac{1}{40} \sum_{k=1}^{40} g(x_k). \quad (3.28)$$

Note that $\mathscr{F} = 100$ means a perfect agreement between $g(x)$ and the corresponding estimate. The model fits for $n = 1, \ldots, 15$ are shown in Fig. 3.4: the order $n = 3$ gives the best prediction. Figure 3.5 plots the estimates of $g(x)$ for $n = 3, 7, 15$ over the $x_i, i = 1, \ldots, 40$. Overfitting occurs when $n = 15$, indicating that the corresponding model is too flexible and fooled by the noise. □

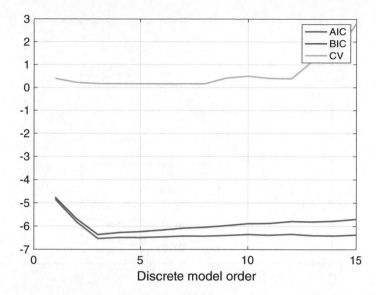

Fig. 3.3 Polynomial regression: model order selection with $n = 1, \dots, 15$ using LS. The blue curve, the red curve and the yellow curve show the values of AIC (3.24), BIC (3.25) and the prediction score of hold out CV, respectively

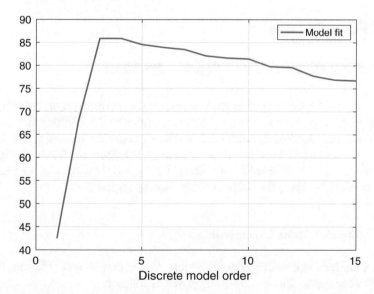

Fig. 3.4 Polynomial regression: profile of the model fit (3.28) as a function of the order n using LS. The most accurate estimate is obtained with model order equal to 3 which corresponds to a second-order polynomial

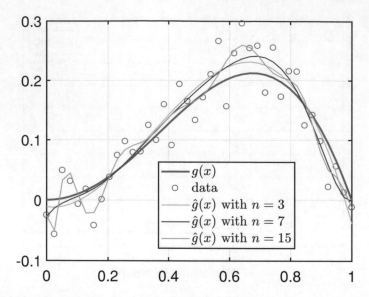

Fig. 3.5 Polynomial regression: true function (blue line) and LS estimates obtained using three different model orders given by $n = 3, 7$ and 15

3.3 Ill-Conditioning

3.3.1 Ill-Conditioned Least Squares Problems

When $\Phi \in \mathbb{R}^{N \times n}$ with $N \geq n$ is rank deficient, i.e., $\text{rank}(\Phi) < n$, or "close" to rank deficient, the corresponding LS problem is said to be ill-conditioned. Examples were already encountered in Sect. 1.1.2 to discuss some limitations of the James–Stein estimators and in Sect. 1.2 in the context of FIR models. There are different ways to handle ill-conditioned LS problems. Below, we show how to calculate $\hat{\theta}^{\text{LS}}$ more accurately by using the singular value decomposition (SVD).

3.3.1.1 Singular Value Decomposition

SVD is a fundamental matrix decomposition. Any matrix $\Phi \in \mathbb{R}^{N \times n}$, with $N \geq n$ to simplify the exposition, can be decomposed as follows:

$$\Phi = U \Lambda V^T, \tag{3.29}$$

where Λ is a rectangular diagonal matrix with nonnegative diagonal entries σ_i, $i = 1, \ldots, n$ and $U \in \mathbb{R}^{N \times N}$ and $V \in \mathbb{R}^{n \times n}$ are orthogonal matrices, i.e., such that $U^T U = U U^T = I_N$ and $V^T V = V V^T = I_n$. The factorization (3.29) is called the singular value decomposition of Φ and the σ_i are called the singular values of Φ. Without loss of generality, they can be assumed to be ordered according to their magnitude:

$$\sigma_1 \geq \sigma_2 \geq \cdots \geq \sigma_n \geq 0.$$

Since $\Phi^T \Phi = V \Lambda^T \Lambda V^T = V D^2 V^T$, where D is a square diagonal matrix whose diagonal entries are the σ_i, it follows that

$$\sigma_i = \sqrt{\lambda_i(\Phi^T \Phi)}, \quad i = 1, \ldots, n, \tag{3.30}$$

where $\lambda_i(A)$ denotes the ith eigenvalue of the matrix A.

3.3.1.2 Condition Number

The condition number of a matrix is a measure of how "close" is the matrix to rank deficient. When Φ is an invertible square matrix, it is denoted by $\text{cond}(\Phi)$ below and defined as

$$\text{cond}(\Phi) = \|\Phi^{-1}\| \|\Phi\|, \tag{3.31}$$

where $\| \cdot \|$ is a matrix norm, with the convention that $\text{cond}(\Phi) = \infty$ for singular Φ. For a generic $\Phi \in \mathbb{R}^{N \times n}$, with SVD in the form (3.29), its condition number with respect to the 2-norm $\| \cdot \|_2$ is defined as

$$\text{cond}(\Phi) = \frac{\sigma_{\max}}{\sigma_{\min}}, \tag{3.32}$$

where $\sigma_{\max} = \sigma_1$ and $\sigma_{\min} = \sigma_n$ are the largest and smallest singular values of Φ, respectively. If we use the 2-norm $\| \cdot \|_2$ in (3.31), then (3.31) coincides with (3.32). Hereafter, the condition number of a matrix will be defined by (3.32).

3.3.1.3 Ill-Conditioned Matrix and LS Problem

The condition number of a matrix is important since it can be used to measure the sensitivity of the LS estimate to perturbations in the data. To be specific, let $\Phi \in \mathbb{R}^{N \times n}$ be full rank and let δY denote a small componentwise perturbation in Y. The solution of the perturbed LS criterion becomes

$$\tilde{\theta}_2^{\text{LS}} = \arg\min_\theta \|(Y + \delta Y) - \Phi\theta\|_2^2. \tag{3.33}$$

Then, it can be shown, e.g., [17, Chap. 5], [10, Chap. 3], that

$$\frac{\|\tilde{\theta}_2^{LS} - \hat{\theta}_2^{LS}\|_2}{\|\hat{\theta}_2^{LS}\|_2} \leq \text{cond}(\Phi)\varepsilon + O\left(\varepsilon^2\right), \quad \varepsilon = \frac{\|\delta Y\|_2}{\|Y\|_2}. \tag{3.34}$$

So, the relative error bound depends on $\text{cond}(\Phi)$: the larger $\text{cond}(\Phi)$, the larger the relative error. One can thus say that the matrix Φ (and the LS problem) with a small condition number is well conditioned, while the matrix Φ (and the LS problem) with a large condition number is ill-conditioned. The condition number enters also more complex bounds on the relative error due to perturbations on the matrix Φ [10, 17].

Example 3.3 (*Effect of ill-conditioning on LS*) Consider the linear regression model (3.13). Let

$$\Phi = \frac{1}{2}\begin{bmatrix} 1 & 1 \\ 1 + 10^{-8} & 1 - 10^{-8} \end{bmatrix}, \quad Y = \begin{bmatrix} 1 \\ 1 \end{bmatrix}. \tag{3.35}$$

The two singular values of Φ are $\sigma_{\max} = 1$ and $\sigma_{\min} = 5 \times 10^{-9}$, implying that $\text{cond}(\Phi) = 2 \times 10^8$. Thus, Φ and the LS problem (3.14) are ill-conditioned.

Using the normal Eq. (3.15), we obtain the LS estimate $\hat{\theta}_1^{LS}$ in closed form:

$$\hat{\theta}_1^{LS} = (\Phi^T \Phi)^{-1}\Phi^T Y = \Phi^{-1}Y = \begin{bmatrix} 1 \\ 1 \end{bmatrix}. \tag{3.36}$$

Now, suppose that there is a small perturbation δY in Y

$$\delta Y = \begin{bmatrix} 0.01 \\ 0 \end{bmatrix}. \tag{3.37}$$

Solving the normal Eq. (3.15) with Y replaced by $Y + \delta Y$ now gives

$$\hat{\theta}_2^{LS} = \begin{bmatrix} 1.01 - 10^6 \\ 1.01 + 10^6 \end{bmatrix}. \tag{3.38}$$

So, when the LS problem (3.14) is ill-conditioned, a small perturbation in Y could cause a significant change in the LS estimate derived by solving the normal Eq. (3.15) directly.   □

Example 3.4 (*Polynomial regression: ill-conditioned LS Problem*) We revisit the polynomial regression Examples (3.26) and (3.27) stressing the dependence of the condition number on the polynomial complexity. In particular, Fig. 3.6 shows that the ill-conditioning of the regression matrix Φ constructed according to (3.4) and (3.12) augments as the dimension n increases. This further points out the importance of a careful selection of the discrete model order to control the estimator's variance when using LS.   □

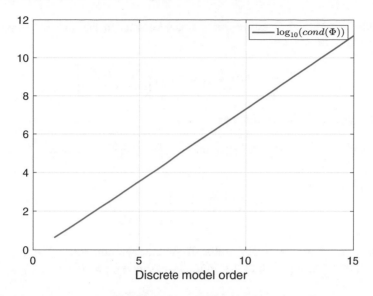

Fig. 3.6 Polynomial regression: profile of the base 10 logarithm of the condition number of Φ as a function of the order n

3.3.1.4 LS Estimate Exploiting the SVD of Φ

In order to obtain more accurate LS estimates for ill-conditioned problems, one can use the SVD of Φ. Given $\Phi \in \mathbb{R}^{N \times n}$ with $N \geq n$, we consider two cases:

- Φ is rank deficient, i.e., rank$(\Phi) < n$.
- Φ is full rank but has a very large condition number, i.e., rank$(\Phi) = n$ but cond(Φ) is very large.

For the rank-deficient case, we assume without loss of generality that rank$(\Phi) = m < n$. In this case, the LS problem does not have a unique solution. To get a special solution, we have to impose extra conditions on the solutions of the LS problem.

Let the singular value decomposition of Φ be

$$\Phi = U \Lambda V^T = \begin{bmatrix} U_1 & U_2 \end{bmatrix} \begin{bmatrix} \Lambda_1 & 0 \\ 0 & 0 \end{bmatrix} \begin{bmatrix} V_1 & V_2 \end{bmatrix}^T, \tag{3.39}$$

where $\Lambda_1 \in \mathbb{R}^{m \times m}$ is diagonal and positive definite while $U_1 \in \mathbb{R}^{N \times m}$ and $V_1 \in \mathbb{R}^{n \times m}$.

We now perform a change of coordinates in both the output and parameter space

$$\tilde{Y} = U^T Y = \begin{bmatrix} U_1^T Y \\ U_2^T Y \end{bmatrix} = \begin{bmatrix} \tilde{Y}_1 \\ \tilde{Y}_1 \end{bmatrix}, \qquad \tilde{\theta} = V^T \theta = \begin{bmatrix} V_1^T \theta \\ V_2^T \theta \end{bmatrix} = \begin{bmatrix} \tilde{\theta}_1 \\ \tilde{\theta}_1 \end{bmatrix}.$$

Note that both $\tilde{Y}_1$ and $\tilde{\theta}_1$ are m-dimensional vectors. In the new coordinates, the residual vector is

$$U^T (Y - \Phi\theta) = \tilde{Y} - \Lambda\tilde{\theta} = \begin{bmatrix} \tilde{Y}_1 - \Lambda_1\tilde{\theta}_1 \\ \tilde{Y}_2 \end{bmatrix}.$$

The LS criterion can be rewritten as

$$\|Y - \Phi\theta\|^2 = (Y - \Phi\theta)^T UU^T (Y - \Phi\theta) = \|\tilde{Y} - \Lambda\tilde{\theta}\|^2 = \|\tilde{Y}_1 - \Lambda_1\tilde{\theta}_1\|^2 + \|\tilde{Y}_2\|^2$$

and is minimized by

$$\tilde{\theta}^{\mathrm{LS}} = \begin{bmatrix} \tilde{\theta}_1^{\mathrm{LS}} \\ \tilde{\theta}_2^{\mathrm{LS}} \end{bmatrix} = \begin{bmatrix} \Lambda^{-1}\tilde{Y}_1 \\ \tilde{\theta}_2 \end{bmatrix}, \tag{3.40}$$

where $\tilde{\theta}_2 \in \mathbb{R}^{n-m}$ is an arbitrary vector. To get the minimum norm solution, one can set $\tilde{\theta}_2 = 0$ that, turning back to the original coordinates, yields

$$\hat{\theta}^{\mathrm{LS}} = V\tilde{\theta}^{\mathrm{LS}} = V_1 \Lambda_1^{-1} U_1^T Y. \tag{3.41}$$

Interestingly, for the rank-deficient case, the special solution (3.41) relates to the Moore–Penrose pseudoinverse of Φ, defined as

$$\Phi^+ = V\Sigma^+U^T = \begin{bmatrix} V_1 & V_2 \end{bmatrix} \begin{bmatrix} \Lambda_1^{-1} & 0 \\ 0 & 0 \end{bmatrix} \begin{bmatrix} U_1 & U_2 \end{bmatrix}^T = V_1\Sigma_1^{-1}U_1^T.$$

So, given a matrix Σ, its pseudoinverse Σ^+ is obtained by replacing all the nonzero diagonal entries by their reciprocal and transposing the resulting matrix. When $\mathrm{rank}(\Phi) = n$, the pseudoinverse returns the usual (unique) LS solution

$$\Phi^+ = \left(\Phi^T\Phi\right)^{-1}\Phi^T.$$

It follows that the minimum norm solution among the general solutions of the LS problem (3.14) can be always written as

$$\hat{\theta}^{\mathrm{LS}} = \Phi^+Y.$$

For the rank-deficient case, due to roundoff errors, Φ may have some very small computed singular values other than the m singular values contained in Λ_1 in (3.39). The situation is similar to the case where Φ is full rank but with a very large condition number. Note also that the rank of Φ needs to be known beforehand to compute the SVD of Φ. However, numerical determination of the rank of a matrix is nontrivial (and out of scope of this book). Here, we just mention a simple way to deal with these issues by using the so-called truncated SVD.

Consider the SVD (3.39) and, without loss of generality, assume

$$\Lambda = \mathrm{diag}(\sigma_1, \sigma_2, \ldots, \sigma_n) \quad \text{with} \quad \sigma_1 \geq \sigma_2 \geq \cdots \geq \sigma_n \geq 0.$$

Now set $\hat{\sigma}_i = \sigma_i$ if $\sigma_i > tol$ and $\hat{\sigma}_i = 0$ otherwise. Then

$$\hat{\Phi} = U\hat{\Lambda}V^T, \tag{3.42}$$

where $\hat{\Lambda} \in \mathbb{R}^{N \times n}$ is diagonal with entries $\hat{\sigma}_1, \hat{\sigma}_2, \ldots, \hat{\sigma}_n$, is called the truncated SVD of Φ. So, the truncated SVD (3.42) can be used to handle the case where Φ has full rank but large condition number: for a given tol, it suffices to replace Φ with $\hat{\Phi}$ and then to compute the LS estimate of θ by means of $\hat{\Phi}^+ Y$.

Example 3.5 (*Truncated SVD*) We revisit Example 3.3 by making use of the truncated SVD of Φ. We take the user-supplied measure of uncertainty tol to be 1e-7. Then the LS estimate $\hat{\theta}_3^{LS}$ computed by (3.41) with Y replaced by $Y + \delta Y$ becomes

$$\hat{\theta}_3^{LS} = \hat{\Phi}^+(Y + \delta Y) = \begin{bmatrix} 1.0050 \\ 1.0049 \end{bmatrix}. \tag{3.43}$$

One can thus see that the estimate is now very close to $[1\ 1]^T$ which was the one obtained in absence of the perturbation δY. □

3.3.2 Ill-Conditioning in System Identification

In Sect. 1.2 we have illustrated an ill-conditioned system identification problem. Below, we will see that the difficulty was due to the fact that low-pass filtered inputs may induce regression matrices with large cond(Φ).

Consider the FIR model of order n:

$$y(t) = \sum_{k=1}^{n} g_k u(t - k) + e(t), \quad t = 1, \ldots, N, \tag{3.44}$$

which can be written in the form (3.13) as follows:

$$Y = \Phi\theta_0 + E$$

$$Y = \begin{bmatrix} y(1) \\ y(2) \\ \vdots \\ y(N) \end{bmatrix}, \quad \Phi = \begin{bmatrix} u(0) & u(-1) & \cdots & u(1-n) \\ u(1) & u(2) & \cdots & u(2-n) \\ \vdots & \vdots & \cdots & \vdots \\ u(N-1) & u(N-2) & \cdots & u(N-n) \end{bmatrix},$$

(3.45)

$$\theta_0 = \begin{bmatrix} g_1 \\ g_2 \\ \vdots \\ g_n \end{bmatrix}, \quad E = \begin{bmatrix} e_1 \\ e_2 \\ \vdots \\ e_N \end{bmatrix}.$$

Then we have

$$\Phi^T\Phi = \begin{bmatrix} \sum_{t=0}^{N-1} u(t)^2 & \sum_{t=0}^{N-1} u(t)u(t-1) & \cdots & \sum_{t=0}^{N-1} u(t)u(t-n+1) \\ \sum_{t=0}^{N-1} u(t)u(t-1) & \sum_{t=-1}^{N-2} u(t)^2 & \cdots & \sum_{t=-1}^{N-2} u(t)u(t-n+2) \\ \vdots & \vdots & \cdots & \vdots \\ \sum_{t=0}^{N-1} u(t)u(t-n+1) & \sum_{t=-n+1}^{N-n} u(t)u(t+n-2) & \cdots & \sum_{t=-n+1}^{N-n} u(t)^2 \end{bmatrix}.$$

(3.46)

Since $\text{cond}(\Phi^T\Phi) = (\text{cond}(\Phi))^2$, we study $\text{cond}(\Phi^T\Phi)$ in what follows. In addition, while so far we have assumed deterministic regressors, now we work in a more structured probabilistic framework where the system input is a stochastic process. This implies that Φ is a random matrix. In particular, $u(t)$ is filtered white noise, with the filter assumed to be stable and given by

$$H(q) = \sum_{k=0}^{\infty} h(k)q^{-k}.$$

(3.47a)

Hence,

$$u(t) = \sum_{k=0}^{\infty} h(k)v(t-k) = H(q)v(t),$$

(3.47b)

where $v(t)$ is zero-mean white noise of variance σ^2 with bounded fourth moments. It comes that $u(t)$ is a zero-mean stationary stochastic process with covariance function $k_u(t,s) = \mathcal{E}[u(t)u(s)] = R_u(t-s)$ with $R_u(\tau)$ defined as follows:

$$\mathcal{E}[u(t)u(t-\tau)] = \sum_{k=0}^{\infty}\sum_{l=0}^{\infty} h(k)h(l)\mathcal{E}[v(t-k)v(t-\tau-l)]$$

$$= \sum_{k=0}^{\infty} h(k)h(k-\tau) \triangleq R_u(\tau).$$

From the ergodic theory, e.g., [25, Theorem 3.4], it also follows that

$$\frac{1}{N}\sum_{t=1}^{N} u(t)u(t-\tau) \rightarrow R_u(\tau), \quad N \rightarrow \infty, \quad \text{a.s.} \tag{3.48}$$

From (3.46) and (3.48), one obtains the following almost sure convergence:

$$\frac{1}{N}\Phi^T\Phi \rightarrow \begin{bmatrix} R_u(0) & R_u(1) & \cdots & R_u(n-1) \\ R_u(1) & R_u(0) & \cdots & R_u(n-2) \\ \vdots & \vdots & \cdots & \vdots \\ R_u(n-1) & R_u(n-2) & \cdots & R_u(0) \end{bmatrix}, \quad N \rightarrow \infty, \quad \text{a.s.} \tag{3.49}$$

So, $\lim_{N\rightarrow\infty} \frac{1}{N}\Phi^T\Phi$ is the covariance matrix of $\begin{bmatrix} u(1) \ldots u(n) \end{bmatrix}^T$ whose condition number thus provides insights on the ill-conditioning affecting the system identification problem.

Since the covariance matrix is real and symmetric, its condition number is the ratio between the largest and the smallest of its eigenvalues. An important result of O. Toeplitz, e.g., [44], [20, Chap. 5], says that *as $n \rightarrow \infty$, the eigenvalues of the covariance matrix of the infinite-dimensional vector $\begin{bmatrix} u(1) \ u(2) \ldots \end{bmatrix}^T$ coincide with the set of values assumed by the power spectrum of $u(t)$, which is given by*

$$\Psi_u(\omega) = \sum_{\tau=-\infty}^{+\infty} R_u(\tau)e^{-i\omega\tau}. \tag{3.50}$$

Hence, considering also that $\Psi_u(-\omega) = \Psi_u(\omega)$, one has

$$\text{cond}\left(\lim_{n\rightarrow\infty} \lim_{N\rightarrow\infty} \frac{1}{N}\Phi^T\Phi \right) = \frac{\max_{\omega\in[0,\pi]} \Psi_u(\omega)}{\min_{\omega\in[0,\pi]} \Psi_u(\omega)}. \tag{3.51}$$

In addition, since $u(t)$ is a filtered white noise (3.47) and $H(q)$ is stable, one also has [see, e.g., [25, p. 37] for details]:

$$\Psi_u(\omega) = \sigma^2 |H(e^{i\omega})|^2, \tag{3.52}$$

where $H(e^{i\omega})$ is the frequency function of the filter $H(q)$, i.e.,

$$H(e^{i\omega}) = \sum_{k=0}^{\infty} h(k)e^{-i\omega k}. \tag{3.53}$$

Finally, combining the results (3.49)–(3.53) yields

$$\mathrm{cond}\left(\lim_{n\to\infty}\lim_{N\to\infty}\frac{1}{N}\Phi^T\Phi\right) = \frac{\max_{\omega\in[0,\pi]}|H(e^{i\omega})|^2}{\min_{\omega\in[0,\pi]}|H(e^{i\omega})|^2}. \tag{3.54}$$

When the maximum of $|H(e^{i\omega})|$ is significantly larger than the minimum of $|H(e^{i\omega})|$, the matrix $\lim_{n\to\infty}\lim_{N\to\infty}\frac{1}{N}\Phi^T\Phi$ could be very ill-conditioned. For instance, if we consider the stable filter

$$H(q) = \frac{1}{(1-aq^{-1})^2}, \quad 0 \le a < 1, \tag{3.55}$$

then one has

$$\frac{\max_{\omega\in[0,\pi]}|H(e^{i\omega})|^2}{\min_{\omega\in[0,\pi]}|H(e^{i\omega})|^2} = \frac{(1+a)^4}{(1-a)^4}. \tag{3.56}$$

As a varies from 0.01 to 0.99, input power is more concentrated at low frequencies and the ill-conditioning affecting the system identification problem augments. In fact, the above quantity increases from about 1 to 1.6×10^9.

3.4   Regularized Least Squares with Quadratic Penalties

One way to handle ill-conditioning is to use regularized least squares (ReLS). Such method will play a special role in this book to control overfitting by encoding prior knowledge. First insights on these aspects are provided below.

ReLS adds a regularization term $J(\theta)$ into the LS criterion (3.14), yielding the following problem:

$$\hat\theta^{\mathrm{R}} = \arg\min_{\theta} \|Y - \Phi\theta\|_2^2 + \gamma J(\theta), \tag{3.57}$$

where $\gamma \ge 0$ is often called the regularization parameter. It has to balance the adherence to the data $\|Y - \Phi\theta\|_2^2$ and the penalty $J(\theta)$. There are many choices for the regularization term which can be connected with the prior knowledge on the true model parameter θ_0 that needs to be estimated.

In this section, we consider regularization terms $J(\theta)$ which are quadratic functions of θ. The resulting estimator will be denoted by ReLS-Q in this chapter. In particular, we let $J(\theta) = \theta^T P^{-1}\theta$ so that the ReLS criterion (3.57) becomes

$$\hat\theta^{\mathrm{R}} = \arg\min_{\theta} \|Y - \Phi\theta\|_2^2 + \gamma\theta^T P^{-1}\theta \tag{3.58a}$$

$$= (\Phi^T\Phi + \gamma P^{-1})^{-1}\Phi^T Y \tag{3.58b}$$

$$= P\Phi^T(\Phi P\Phi^T + \gamma I_N)^{-1}Y \tag{3.58c}$$

$$= (P\Phi^T\Phi + \gamma I_n)^{-1}P\Phi^T Y, \tag{3.58d}$$

where $P \in \mathbb{R}^{n \times n}$ is a positive semidefinite matrix, here assumed invertible, often called the regularization matrix, and I_n is the n-dimensional identity matrix.[2]

Remark 3.1 The regularization matrix P could be singular. In this case, (3.58a) is not well defined but, with a suitable arrangement, we can use the Moore–Penrose pseudoinverse P^+ instead of P^{-1}. In particular, let the SVD of P be

$$P = \begin{bmatrix} U_1 & U_2 \end{bmatrix} \begin{bmatrix} \Lambda_P & 0 \\ 0 & 0 \end{bmatrix} \begin{bmatrix} U_1 & U_2 \end{bmatrix}^T ,$$

where Λ_P is a diagonal matrix with the positive singular values of P as diagonal elements and $U = \begin{bmatrix} U_1 & U_2 \end{bmatrix}$ is an orthogonal matrix with U_1 having the same number of columns as that of Λ_P. Recall also that $P^+ = U_1 \Lambda_P^{-1} U_1$. In order to find how (3.58a) should be modified for singular P, let us consider

$$P_\varepsilon = \begin{bmatrix} U_1 & U_2 \end{bmatrix} \begin{bmatrix} \Lambda_P & 0 \\ 0 & \varepsilon I \end{bmatrix} \begin{bmatrix} U_1 & U_2 \end{bmatrix}^T , \quad \varepsilon > 0.$$

By replacing P with P_ε in (3.58a), we obtain

$$\hat{\theta}^{\mathrm{R}} = \arg \min_\theta \|Y - \Phi\theta\|_2^2 + \gamma \theta^T U_1 \Lambda_P^{-1} U_1^T \theta + \frac{\gamma}{\varepsilon} \theta^T U_2 U_2^T \theta. \qquad (3.59)$$

If we let $\varepsilon \to 0$, it follows that the parameter vector must satisfy $U_2^T \theta = 0$. Therefore, we may conveniently associate to a singular P the modified regularization problem

$$\hat{\theta}^{\mathrm{R}} = \arg \min_\theta \|Y - \Phi\theta\|_2^2 + \gamma \theta^T P^+ \theta \qquad (3.60a)$$

$$\mathrm{subj.\,to} \quad U_2^T \theta = 0. \qquad (3.60b)$$

If P^{-1} is replaced by P^+, it is easy to verify that (3.58c) or (3.58d) is still the optimal solution of (3.60). Instead, this does not hold for (3.58b). For convenience, we will use (3.58a) in the sequel and refer to (3.60) for its rigorous meaning.

3.4.1 Making an Ill-Conditioned LS Problem Well Conditioned

The ReLS-Q can make the ill-conditioned LS problem well conditioned. Consider ridge regression which, as discussed in Sect. 1.2, corresponds to setting $P = I_n$, hence obtaining

---

[2] The step from (3.58c) to (3.58d) follows from the matrix equality $A(I_j + BA)^{-1} = (I_k + AB)^{-1}A$ which holds for every $A \in \mathbb{R}^{k \times j}$ and $B \in \mathbb{R}^{j \times k}$.

$$\hat{\theta}^{\text{R}} = \arg\min_{\theta} \|Y - \Phi\theta\|_2^2 + \gamma\|\theta\|_2^2 \tag{3.61a}$$

$$= (\Phi^T\Phi + \gamma I_n)^{-1}\Phi^T Y. \tag{3.61b}$$

The parameter γ directly affects the condition number of $(\Phi^T\Phi + \gamma I_n)$ whose inverse defines the regularized estimate. In fact, the positive definite square matrix $(\Phi^T\Phi + \gamma I_n)$ has eigenvalues (coincident with its singular values) equal to $\sigma_i^2 + \gamma$. Therefore,

$$\text{cond}(\Phi^T\Phi + \gamma I_n) = \frac{\sigma_1^2 + \gamma}{\sigma_n^2 + \gamma}$$

which can be adjusted by tuning the regularization parameter γ. This means that regularization can make the LS problem well conditioned even when Φ is rank deficient: if the smallest singular value is null one has

$$\text{cond}(\Phi^T\Phi + \gamma I_n) = \frac{\sigma_1^2 + \gamma}{\gamma}.$$

3.4.1.1 Mean Squared Error

Simple calculations of expectations with respect to the errors e_i, with the regressors ϕ_i assumed to be deterministic, lead to

$$\mathscr{E}(\hat{\theta}^{\text{R}}) = (\Phi^T\Phi + \gamma P^{-1})^{-1}\Phi^T\Phi\theta_0 \tag{3.62a}$$

$$\hat{\theta}^{\text{R}}_{\text{bias}} = \mathscr{E}(\hat{\theta}^{\text{R}}) - \theta_0 = -(\Phi^T\Phi + \gamma P^{-1})^{-1}\gamma P^{-1}\theta_0 \tag{3.62b}$$

$$\text{Cov}(\hat{\theta}^{\text{R}}, \hat{\theta}^{\text{R}}) = \mathscr{E}[(\hat{\theta}^{\text{R}} - \mathscr{E}(\hat{\theta}^{\text{R}}))(\hat{\theta}^{\text{R}} - \mathscr{E}(\hat{\theta}^{\text{R}}))^T]$$

$$= (\Phi^T\Phi + \gamma P^{-1})^{-1}\sigma^2\Phi^T\Phi(\Phi^T\Phi + \gamma P^{-1})^{-1} \tag{3.62c}$$

$$\text{MSE}(\hat{\theta}^{\text{R}}, \theta_0) = \mathscr{E}(\hat{\theta}^{\text{R}} - \theta_0)(\hat{\theta}^{\text{R}} - \theta_0)^T$$

$$= \text{Cov}(\hat{\theta}^{\text{R}}, \hat{\theta}^{\text{R}}) + \hat{\theta}^{\text{R}}_{\text{bias}}(\hat{\theta}^{\text{R}}_{\text{bias}})^T$$

$$= (\Phi^T\Phi + \gamma P^{-1})^{-1}(\sigma^2\Phi^T\Phi + \gamma^2 P^{-1}\theta_0\theta_0^T P^{-1})(\Phi^T\Phi + \gamma P^{-1})^{-1}, \tag{3.62d}$$

where $\text{Cov}(\hat{\theta}^{\text{R}}, \hat{\theta}^{\text{R}})$ is the covariance matrix of $\hat{\theta}^{\text{R}}$ and $\text{MSE}(\hat{\theta}^{\text{R}}, \theta_0)$ is the MSE matrix of $\hat{\theta}^{\text{R}}$ function of the true model parameter θ_0. Expression (3.62) shows clearly regularization's influence on the statistical properties of $\hat{\theta}^{\text{R}}$:

- when $\gamma = 0$, i.e., there is no regularization, $\hat{\theta}^{\text{R}}$ reduces to $\hat{\theta}^{\text{LS}}$ and $\text{MSE}(\hat{\theta}^{\text{R}}, \theta_0)$ reduces to $\sigma^2(\Phi^T\Phi)^{-1}$;
- when $\gamma > 0$, the regularized estimator $\hat{\theta}^{\text{R}}$ is biased and the MSE matrix of $\hat{\theta}^{\text{R}}$ is decomposed into two components: the bias $\hat{\theta}^{\text{R}}_{\text{bias}}(\hat{\theta}^{\text{R}}_{\text{bias}})^T$ and the variance $\text{Cov}(\hat{\theta}^{\text{R}}, \hat{\theta}^{\text{R}})$. By a suitable choice of the regularization matrix P and the regularization parameter γ, the variance of $\hat{\theta}^{\text{R}}$ can be made "smaller" and, if the resulting

increase in the bias is moderate, an MSE matrix "smaller" than that associated to LS can be obtained.

3.4.2 Equivalent Degrees of Freedom

For a given regularization matrix P, we have seen (also deriving the structure of the MSE) that the regularization parameter γ controls the influence of the regularization: as γ varies from 0 to ∞, the influence of the regularization $\theta^T P^{-1}\theta$ becomes stronger. In particular, when $\gamma = 0$ there is no regularization and $\hat{\theta}^R$ reduces to $\hat{\theta}^{LS}$. When $\gamma = \infty$ the regularization term $\gamma\theta^T P^{-1}\theta$ overwhelms the data fit $\|Y - \Phi\theta\|_2^2$ and one has $\hat{\theta}^R = 0$.

Often, it is more convenient to exploit a normalized measure of the influence of the regularization instead of considering directly the value of γ. For this goal, we introduce the so-called influence or hat matrix:

$$H = \Phi P \Phi^T (\Phi P \Phi^T + \gamma I_N)^{-1}. \tag{3.63}$$

Such matrix is important since it connects the measured output Y with the predicted output $\hat{Y} = \Phi\hat{\theta}^R$, i.e., one has

$$\hat{Y} = \Phi\hat{\theta}^R = HY. \tag{3.64}$$

It is also important since its trace is indeed a normalized measure of the influence of the regularization. To see this, let $A = \Phi P \Phi^T$ and consider its SVD

$$A = U D U^T,$$

where $UU^T = I$ and D is a diagonal matrix with nonnegative entries d_i^2. Then,

$$H = U D U^T (U D U^T + \gamma I_N U U^T)^{-1} = U D (D + \gamma I_N)^{-1} U^T.$$

Since U is orthogonal, one has $\text{trace}(UMU^T) = \text{trace}(M)$, so that

$$\text{trace}(H) = \text{trace}(D(D + \gamma I_N)^{-1}) = \sum_{i=1}^{n} \frac{d_i^2}{d_i^2 + \gamma}.$$

The above equation implies that $\text{trace}(H)$ is a monotonically decreasing function of γ. It attains its maximum at $\gamma = 0$ and infimum as $\gamma \to \infty$. In particular, for $\gamma = 0$ one has $\hat{\theta}^R = \hat{\theta}^{LS}$ and the hat matrix H becomes $H = \Phi(\Phi^T\Phi)^{-1}\Phi^T$, implying that $\text{trace}(H) = n$ if Φ is full rank. For $\gamma \to \infty$ one instead has $\text{trace}(H) \to 0$. Therefore, it holds that $0 < \text{trace}(H) \leq n$. Hence, since n is the dimension of θ, i.e., the number of parameters in the linear regression model, $\text{trace}(H)$ can be seen as the

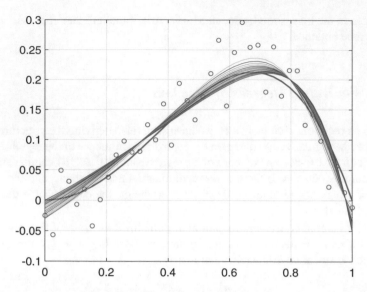

Fig. 3.7 Polynomial regression: true function $g(x)$ (blue line) and ridge regression estimates obtained with 16 different values of the regularization parameter

counterpart of the number of parameters to be estimated in the LS context. In other words, in the regularized framework trace(H) plays the role of the model order. It thus becomes natural to call it the equivalent degrees of freedom for the ReLS-Q estimate $\hat{\theta}^R$, e.g., [21, Sect. 7.6], [4, p. 559]:

$$\mathrm{dof}(\hat{\theta}^R) = \mathrm{trace}(H). \tag{3.65}$$

The notation $\mathrm{dof}(\gamma)$ will be also used in the book in place of $\mathrm{dof}(\hat{\theta}^R)$ to stress the dependence of the equivalent degrees of freedom on the regularization parameter.

Example 3.6 (*Polynomial regression: ridge regression*) As shown in Fig. 3.6, the regression matrix Φ built in the polynomial regression Example (3.26) and (3.27) is ill-conditioned for large n. Here, we consider the case $n = 16$ (corresponding to a polynomial order 15) which leads to $\mathrm{cond}(\Phi) = 1.49 \times 10^{11}$. To illustrate how ridge regression (3.61) can face the ill-conditioning, let $\gamma = \gamma_i, i = 1, \ldots, 16$, with $\gamma_1 = 0.01$ and $\gamma_{16} = 0.31$ and $\gamma_2, \ldots, \gamma_{15}$ evenly spaced between γ_1 and γ_{16}. For each γ_i, we then compute the corresponding ridge regression estimate (3.61) and plot the 16 estimates $\hat{g}(x) = \phi(x)^T \hat{\theta}^R$ in Fig. 3.7. The fits (3.28) are shown in Fig. 3.8 as a function of γ. One can see that $\gamma = 0.11$ gives the best performance obtaining a fit around 89%. Interestingly, such fit is larger than the best result obtained by LS through optimal tuning of the discrete model order, see Fig. 3.4. The base 10 logarithm of the condition number of $\Phi^T \Phi + \gamma I_n$, as a function of γ, is displayed in Fig. 3.9. One can see that the matrix is much better conditioned now. Figure 3.10 plots the equivalent degrees of freedom of $\hat{\theta}^R$. Even if $n = 16$, the actual model complexity in terms of equivalent degrees of freedom is much smaller, around 4 for

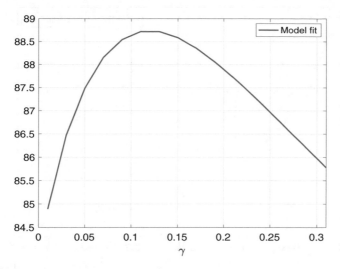

Fig. 3.8 Polynomial regression: profile of the ridge regression fit (3.28) as a function of γ. Large fit values are associated to estimates close to the true function

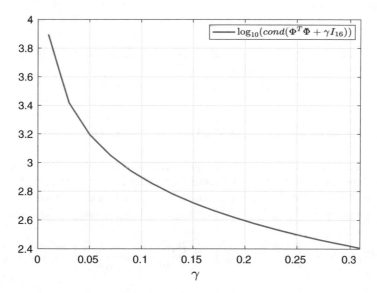

Fig. 3.9 Polynomial regression: profile of the base 10 logarithm of the condition number of $\Phi^T \Phi + \gamma I_n$ as a function of γ

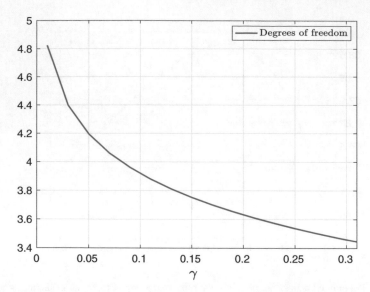

Fig. 3.10 Polynomial regression: profile of the equivalent degrees of freedom (3.65) as a function of γ using ridge regression

the tested values of γ. Finally, the estimates of any component of θ obtained using the different values of γ are shown in Fig. 3.11.

□

3.4.2.1   Regularization Design: The Optimal Regularizer

A natural question is how to design a regularization matrix P and select γ to obtain a "good" model estimate. From a "classic" or "frequentist" point of view, rational choices are those that make the MSE matrix (3.62d) small in some sense, as discussed below. For our purposes, it is useful to rewrite the MSE matrix (3.62d) as follows:

$$
\mathrm{MSE}(\hat{\theta}^{\mathrm{R}}, \theta_0) = \sigma^2 \left(\frac{P\Phi^T\Phi}{\gamma} + I_n \right)^{-1} \left(\frac{P\Phi^T\Phi P}{\gamma^2} + \frac{\theta_0\theta_0^T}{\sigma^2} \right) \left(\frac{\Phi^T\Phi P}{\gamma} + I_n \right)^{-1}.
$$
(3.66)

Then, it is useful to first introduce the following lemma.

Lemma 3.1 (based on [9]) *Consider the matrix*

$$
M(Q) = (QR + I)^{-1}(QRQ + Z)(RQ + I)^{-1},
$$

where Q, R and Z are positive semidefinite matrices. Then for all Q

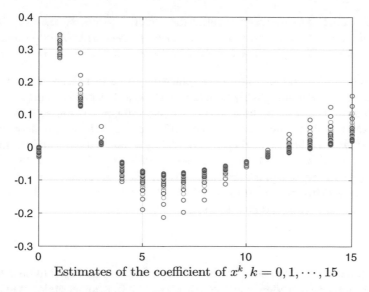

Fig. 3.11 Polynomial regression: profile of the estimates of each component forming the ridge regression estimate (3.61). For each value $k = 0, \ldots, 15$ on the x-axis the plot reports the estimates of the coefficient of the monomial x^k obtained by using different values of the regularization parameter γ

$$M(Z) \preceq M(Q), \tag{3.67}$$

which means that $M(Q) - M(Z)$ is positive semidefinite.

The proof consists of straightforward calculations and can be found in Sect. 3.8.2.

Using (3.66) and Lemma 3.1, the question which P and γ give the best MSE of $\hat{\theta}^R$ has a clear answer: the equation $\sigma^2 P = \gamma \theta_0 \theta_0^T$ needs to be satisfied. Thus, the following result holds.

Proposition 3.1 (Optimal regularization for a given θ_0, based on [9]) *Letting $\gamma = \sigma^2$, the regularization matrix*

$$P = \theta_0 \theta_0^T \tag{3.68}$$

minimizes the MSE matrix (3.66) in the sense of (3.67).

Note that the MSE matrix (3.66) is linear in $\theta_0 \theta_0^T$. This means that if we compute $\hat{\theta}^R$ with the same P for a collection of true systems θ_0, the average MSE over that collection will be given by (3.66) with $\theta_0 \theta_0^T$ replaced by its average over the collection. In particular, if θ_0 is a random vector with $\mathscr{E}(\theta_0 \theta_0^T) = \Pi$, we obtain the following result.

Proposition 3.2 (Optimal regularization for a random system θ_0, based on [9]) *Consider (3.62d) with $\gamma = \sigma^2$. Then, the best average (expected) MSE for a random true system θ_0 with $\mathcal{E}(\theta_0 \theta_0^T) = \Pi$ is obtained by the regularization matrix $P = \Pi$.*

Propositions 3.1 and 3.2 thus give a somewhat preliminary answer to our design problem. Since the best regularization matrix $P = \theta_0 \theta_0^T$ depends on the true system θ_0, such formula cannot be used in practice. Nevertheless, it suggests to choose a regularization matrix which mimics the behaviour of $\theta_0 \theta_0^T$. Using prior knowledge on the true system θ_0, this can be done by postulating a parametrized family of matrices $P(\eta)$ with $\eta \in \Gamma \subset \mathbb{R}^m$, where η is the so-called *hyperparameter* vector, Γ is the set where η can vary and m is the dimension of η. Thus, the choice of a parametrized regularization matrix is similar to model structure selection in system identification. The nature of the optimal regularizer suggests also to set

$$\gamma = \sigma^2. \tag{3.69}$$

However, the noise variance σ^2 is in general unknown and needs to be estimated from the data. One can adopt equations (3.22) or (3.23). Another option is to include σ^2 in η and then estimate it together with the other hyperparameters.

3.5 Regularization Tuning for Quadratic Penalties

3.5.1 Mean Squared Error and Expected Validation Error

Now, assume that a parametrized family of regularization matrices $P(\eta)$ has been defined. The vector η is in general unknown and has to be tuned by using the available measurements. The ReLS-Q estimate $\hat{\theta}^R(\eta)$ in (3.58) depends on η and the estimation strategy depends on the measure used to quantify its quality. We will consider the following two criteria:

- minimizing the MSE;
- minimizing the expected validation error (EVE).

3.5.1.1 Minimizing the MSE

Still adopting a "classic" or "frequentist" point of view, a rational choice of η is one that makes the MSE matrix (3.62d) small in some sense. For ease of estimation, a scalar measure is often exploited. In [25, Chap. 12], it is suggested to use a weighting matrix Q and $\text{trace}(\text{MSE}(\hat{\theta}^R(\eta), \theta_0)Q)$ as a quality measure of $\hat{\theta}^R(\eta)$, where Q reflects the intended use of the model $\hat{\theta}^R(\eta)$. Then an estimate of η, say $\hat{\eta}$, is obtained as follows:

$$\hat{\eta} = \underset{\eta \in \Gamma}{\arg\min} \, \text{trace}(\text{MSE}(\hat{\theta}^{\text{R}}(\eta), \theta_0)Q). \tag{3.70}$$

Note that (3.70) depends on the true system θ_0 that is unknown and thus cannot be used. In practice, we need to first find a "good" estimate, say $\hat{\theta}$, of the true system θ_0 and then to replace θ_0 in (3.70) with $\hat{\theta}$. Then, hopefully, a "good" estimate is given by

$$\hat{\eta} = \underset{\eta \in \Gamma}{\arg\min} \, \text{trace}(\text{MSE}(\hat{\theta}^{\text{R}}(\eta), \hat{\theta})Q). \tag{3.71}$$

Different choices of $\hat{\theta}$ and Q lead to different estimators (3.71). Examples are obtained setting $\hat{\theta}$ to the LS estimate or to the ridge regression estimate of θ_0, while the choice $Q = I_n$ is often used. In any case, the major difficulty underlying the idea of "minimizing the MSE" for hyperparameters tuning lies in whether or not $\hat{\theta}$ is a "good" estimate of θ_0, which is actually our fundamental problem.

3.5.1.2 Minimizing the EVE

An alternative quality measure of $\hat{\theta}^{\text{R}}(\eta)$ is related to model prediction capability on independent validation data and is characterized by the expected validation error (EVE).

To define it, we need to introduce the training/estimation data and the validation data. The training data is used for estimating the model and is contained in the set $\mathscr{D}_{\text{T}}$. The validation data are used to assess model prediction capability and are in the set $\mathscr{D}_{\text{V}}$.

Now, let $\hat{\theta}^{\text{R}}(\eta)$ denote a general ReLS-Q estimate parametrized by the vector η and obtained using only the training data $\mathscr{D}_{\text{T}}$. Let $y_{\text{v}} \in \mathbb{R}$, $\phi_{\text{v}} \in \mathbb{R}^n$ be a validation sample pair. These objects could both be random, e.g., y_{v} can be affected by noise and the regressor could be defined by a stochastic system input. The validation error $\text{EVE}_{\mathscr{D}_{\text{T}}}(\eta)$ is then given by

$$\text{EVE}_{\mathscr{D}_{\text{T}}}(\eta) = \mathscr{E}[(y_{\text{v}} - \phi_{\text{v}}^T \hat{\theta}^{\text{R}}(\eta))^2 | \mathscr{D}_{\text{T}}]. \tag{3.72}$$

In the above equation, the expectation $\mathscr{E}$ is computed w.r.t. the joint distribution of y_{v} and ϕ_{v} conditioned on the training data $\mathscr{D}_{\text{T}}$. If $\phi_{\text{v}} \in \mathbb{R}^n$ is deterministic and, as usual, y_{v} is affected by a noise independent by those entering the training set, the mean is taken just w.r.t. such noise, with $\mathscr{D}_{\text{T}}$ which influences only $\hat{\theta}^{\text{R}}$. In any case, the result is a function of the training set. Now, we can see $\mathscr{D}_{\text{T}}$ as random and then the EVE is

$$\text{EVE}(\eta) \triangleq \mathscr{E}[\text{EVE}_{\mathscr{D}_{\text{T}}}(\eta)], \tag{3.73}$$

where the expectation $\mathscr{E}$ is over the training set. Note that the final result is function of the true θ_0 which determines the probability distributions of the training and validation data.

The $\mathrm{EVE}(\eta)$ measures the prediction capability of the model $\hat{\theta}^R(\eta)$ before seeing any training or validation data: the smaller the $\mathrm{EVE}(\eta)$, the better the expected model prediction capability. Therefore, it is natural to estimate η as follows:

$$\hat{\eta} = \arg\min_{\eta \in \Gamma} \ \mathrm{EVE}(\eta). \tag{3.74}$$

However, as said, the above objective depends on the unknown vector θ_0 so that estimation of θ is not possible in practice. The problem is analogous to that encountered when trying to tune η by minimizing the MSE

Remark 3.2 Interestingly, the idea of "minimizing the MSE" and the idea of "minimizing the EVE" are connected. To see this, we assume for simplicity that the regressors ϕ_i, $i = 1, \ldots, N$ in the training data and ϕ_v in the validation data are deterministic. Then it can be shown that

$$EVE(\eta) = \mathscr{E}[(y_v - \phi_v^T \hat{\theta}^R(\eta))^2] = \sigma^2 + \phi_v^T MSE(\hat{\theta}^R(\eta), \theta_0)\phi_v, \tag{3.75}$$

where the expectation $\mathscr{E}$ is over everything that is random, and $MSE(\hat{\theta}^R(\eta), \theta_0)$ is the MSE matrix of $\hat{\theta}^R(\eta)$ defined in (3.62d). Clearly, (3.75) shows that minimizing $EVE(\eta)$ with respect to η is equivalent to minimizing $\mathrm{trace}(MSE(\hat{\theta}^R(\eta), \theta_0)Q)$ with respect to η when $Q = \phi_v \phi_v^T$.

To overcome the fact that the EVE depends on the unknown θ_0, we could first find a "good" estimate of $\mathrm{EVE}(\eta)$ using the available data and then determine the hyperparameter vector by minimizing it. There are two ways to achieve this goal: by efficient sample reuse of the data and by considering the in-sample EVE instead. More details will be provided in the next two subsections.

3.5.2 Efficient Sample Reuse

One way to estimate $\mathrm{EVE}(\eta)$ by exploiting efficient sample reuse includes cross-validation (CV) [41] and its variants already mentioned in Sects. 2.6.3 and 3.2.2 when discussing model order selection.

3.5.2.1 Hold Out Cross-Validation

The simplest CV is the so-called hold out CV (HOCV), which is widely used to select the model order for the classical PEM/ML. The HOCV can also be used to estimate the hyperparameter $\eta \in \Gamma$ for the ReLS-Q method.

The idea of hold out CV is to first split the given data into two parts: the training data $\mathcal{D}_T$ and the validation data $\mathcal{D}_V$. The prediction capability is measured in terms of the validation error. The model that gives the smallest validation error will be selected. More specifically, the HOCV takes the following three steps:

(1) Split the given data into two parts: $\mathcal{D}_T$ and $\mathcal{D}_V$.
(2) Estimate the model $\hat{\theta}^R(\eta)$ based on $\mathcal{D}_T$ for different values of $\eta \in \Gamma$.
(3) Calculate the validation error for $\hat{\theta}^R(\eta)$ over the validation data $\mathcal{D}_V$:

$$\mathrm{CV}(\eta) = \sum_{(y_v, \phi_v) \in \mathcal{D}_v} (y_v - \phi_v^T \hat{\theta}^R(\eta))^2,$$

where the summation is over all pairs of (y_v, ϕ_v) in the validation data $\mathcal{D}_v$. Then, select the value of η that minimizes $\mathrm{CV}(\eta)$:

$$\hat{\eta} = \arg\min_{\eta \in \Gamma} \mathrm{CV}(\eta). \tag{3.76}$$

It is also possible to change the role of the training and validation sets in order to perform a second validation step: the model is estimated on the previous validation set and the validation error is computed on the previous training set. Finally, the final validation error is obtained by averaging the two validation errors.

3.5.2.2 k-Fold Cross-Validation

The HOCV with swapped sets is a special case of the more general k-fold CV with $k = 2$, e.g., [24]. If the data set size is small, the HOCV may perform poorly. In fact, the training data may not be sufficiently rich to build good models and a validation set of small size may give a too uncertain validation error. In this case, the k-fold CV with $k > 2$ could be used.

The idea of k-fold CV is to first split the data into k parts of equal size. For every $\eta \in \Gamma$, the following procedure is repeated k times. At the ith run with $i = 1, 2, \ldots, k$:

(1) Retain the ith part as the validation data $\mathcal{D}_{V,i}$, and use the remaining $k - 1$ parts as the training data $\mathcal{D}_{T,-i}$.
(2) Estimate $\hat{\theta}^R(\eta)$ based on the training data $\mathcal{D}_{T,-i}$ and then calculate the validation error over the validation data $\mathcal{D}_{V,i}$

$$\text{CV}_{-i}(\eta) = \sum_{(y_v, \phi_v) \in \mathscr{D}_{V,i}} (y_v - \phi_v^T \hat{\theta}^R(\eta))^2,$$

where the summation is over all pairs of (y_v, ϕ_v) in the validation data $\mathscr{D}_{V,i}$.

Finally, the k validation errors $\text{CV}_{-i}(\eta)$ so obtained are summed to obtain the following total validation error for η:

$$\text{CV}(\eta) = \sum_{i=1}^{k} \text{CV}_{-i}(\eta),$$

and the estimate of η is finally given by

$$\hat{\eta} = \arg \min_{\eta \in \Gamma} \text{CV}(\eta). \tag{3.77}$$

3.5.2.3 Predicted Residual Error Sum of Squares and Variants

The computation of the k-fold CV is often expensive and an exception is the leave-one-out CV (LOOCV) where the validation set includes only one validation pair. When the square loss function is used, the total validation error admits a closed-form expression and the LOOCV is also known as the predicted residual error sum of squares (PRESS), e.g., [2].

First, recall the linear regression model (3.13) and the corresponding data $y_i \in \mathbb{R}$ and $\phi_i \in \mathbb{R}^n$ for $i = 1, \dots, N$. Then the ReLS-Q estimate is

$$\begin{aligned}
\hat{\theta}^R &= \arg \min_{\theta} ||Y - \Phi\theta||^2 + \sigma^2 \theta^T P^{-1}(\eta)\theta \\
&= \left(\Phi^T \Phi + \sigma^2 P^{-1}(\eta)\right)^{-1} \Phi^T Y \\
&= \left(\sum_{i=1}^{N} \phi_i \phi_i^T + \sigma^2 P^{-1}(\eta)\right)^{-1} \sum_{i=1}^{N} \phi_i y_i,
\end{aligned} \tag{3.78}$$

where we have set $\gamma = \sigma^2$ following (3.69). For the kth measured output y_k, the corresponding predicted output $\hat{y}_k$ and residual r_k are, respectively,

$$\hat{y}_k = \phi_k^T \left(\sum_{i=1}^{N} \phi_i \phi_i^T + \sigma^2 P^{-1}(\eta)\right)^{-1} \sum_{i=1}^{N} \phi_i y_i, \tag{3.79a}$$

$$r_k = y_k - \hat{y}_k. \tag{3.79b}$$

Then, PRESS selects the value of $\eta \in \Gamma$ that minimizes the sum of squares of the validation errors. One can prove that this corresponds to the following problem:

$$\text{PRESS:} \quad \hat{\eta} = \arg\min_{\eta \in \Gamma} \sum_{k=1}^{N} \frac{r_k^2}{(1 - \phi_k^T M^{-1} \phi_k)^2}, \tag{3.80}$$

where r_k are defined by (3.79) while

$$M = \sum_{i=1}^{N} \phi_i \phi_i^T + \sigma^2 P^{-1}(\eta). \tag{3.81}$$

The derivation of (3.80) can be found in Sect. 3.8.3. It is worth noting that the denominator in (3.80) is strictly related to the diagonal entries of the hat matrix H defined in (3.63). In fact,

$$\phi_k^T M^{-1} \phi_k = H_{kk}$$

so that

$$\text{PRESS:} \quad \hat{\eta} = \arg\min_{\eta \in \Gamma} \sum_{k=1}^{N} \frac{r_k^2}{(1 - H_{kk})^2}.$$

Hence, interestingly, one can conclude that PRESS evaluation requires to compute just the ReLS-Q estimate exploiting the full data set (instead of solving N problems, one for each missing measurement in the training set).

One method that is closely related with PRESS is the so-called generalized cross-validation (GCV), e.g., [18]. GCV is obtained by replacing in (3.80) the factors H_{kk} by their average, i.e., $\text{trace}(H)/N$:

$$\text{GCV:} \quad \hat{\eta} = \arg\min_{\eta \in \Gamma} \frac{1}{(1 - \text{trace}(H)/N)^2} \sum_{k=1}^{N} r_k^2. \tag{3.82}$$

Recalling (3.65), the term $\text{trace}(H)$ defines the degrees of freedom of $\hat{\theta}^R$. Hence, the GCV criterion can be rewritten as follows:

$$\text{GCV:} \quad \hat{\eta} = \arg\min_{\eta \in \Gamma} \frac{1}{(1 - \text{dof}(\theta^R)/N)^2} \sum_{k=1}^{N} r_k^2.$$

3.5.3 Expected In-Sample Validation Error

In the definition of the validation error $\text{EVE}_{\mathscr{D}_T}$ (3.72), reported for convenience also below

$$\text{EVE}_{\mathscr{D}_T}(\eta) = \mathscr{E}[(y_v - \phi_v^T \hat{\theta}^R(\eta))^2 | \mathscr{D}_T],$$

we assumed that the conditional expectation $\mathscr{E}$ is over the independent validation sample pair $y_v \in \mathbb{R}, \phi_v \in \mathbb{R}^n$, which are drawn randomly from their joint distribution. The computation of the validation error (3.72) could become easier if independent validation sample pairs $y_v \in \mathbb{R}, \phi_v \in \mathbb{R}^n$ are generated in a particular way.

For linear regression problems, it is convenient to assume that the same *deterministic* regressors ϕ_i, $i = 1, 2, \ldots, N$, are used for generating both the training data and the validation data. To be specific, still using θ_0 to denote the true parameter vector, we recall from (3.6), that the training output samples are

$$y_i = \phi_i^T \theta_0 + e_i, \quad i = 1, \ldots, N. \tag{3.83}$$

In this case, the training set is

$$\mathscr{D}_T = \{(y_i, \phi_i) \mid y_i \in \mathbb{R}, \phi_i \in \mathbb{R}^n \text{ satisfying } (3.83), \ i = 1, \ldots, N\}. \tag{3.84}$$

Using the same regressors ϕ_i, consider a set of validation output samples $y_{v,i}$ as follows:

$$y_{v,i} = \phi_i^T \theta_0 + e_{v,i}, \quad i = 1, \ldots, N, \tag{3.85}$$

where θ_0 is the true parameter vector, with the noises e_i and $e_{v,i}$ assumed identically and independently distributed. The validation error is now denoted by $\text{EVE}_{\text{in}\mathscr{D}_T}(\eta)$, computed as follows:

$$\text{EVE}_{\text{in}\mathscr{D}_T}(\eta) = \frac{1}{N} \sum_{i=1}^{N} \mathscr{E}[(y_{v,i} - \phi_i^T \hat{\theta}^R(\eta))^2 \mid \mathscr{D}_T], \tag{3.86}$$

and called in-sample validation error [21, p. 228]. Note that, similarly to what discussed after (3.72), the expectation $\mathscr{E}$ in (3.86) is computed w.r.t. the joint distribution of the couples $y_{v,i}, \phi_i$ conditioned on the training data $\mathscr{D}_T$. Thus, the result is function of the training set. As done in (3.73), we can remove such dependence by computing the expected in-sample validation error as

$$\text{EVE}_{\text{in}}(\eta) = \mathscr{E}[\text{EVE}_{\text{in}\mathscr{D}_T}(\eta)], \tag{3.87}$$

with expectation taken over the joint distribution of the training data. In what follows, we will see how to build an unbiased estimator of $\text{EVE}_{\text{in}}(\eta)$ using the training data (3.84), and how to exploit it for hyperparameters tuning.

3.5.3.1 Expectation of the Sum of Squared Residuals, Optimism and Degrees of Freedom

To estimate $\text{EVE}_{\text{in}}(\eta)$, consider the sum of squared residuals

$$\overline{\mathrm{err}}(\eta)_{\mathscr{D}_\mathrm{T}} = \frac{1}{N} \sum_{i=1}^{N} (y_i - \phi_i^T \hat{\theta}^\mathrm{R}(\eta))^2, \tag{3.88}$$

which is function only of the training set. Its expectation w.r.t. the training data (3.84) is

$$\overline{\mathrm{err}}(\eta) = \mathscr{E}\left(\frac{1}{N} \sum_{i=1}^{N} (y_i - \phi_i^T \hat{\theta}^\mathrm{R}(\eta))^2 \right). \tag{3.89}$$

One expects $\mathrm{EVE}_{\mathrm{in}}(\eta)$ to be not smaller than $\overline{\mathrm{err}}(\eta)$ because this latter quantity exploits the same data to fit the model and to assess the error. This intuition is indeed true as shown in the following theorem whose proof is in Sect. 3.8.4.

Theorem 3.7 *Consider the linear regression model (3.13) with the training data (3.84), the validation data (3.85) and the ReLS-Q estimate (3.58). Then it holds that*

$$\overline{err}(\eta) \le \mathrm{EVE}_{\mathrm{in}}(\eta). \tag{3.90}$$

Theorem 3.7 shows that the expectation of the sum of squares of the residuals is an overly optimistic estimator of the expected in-sample validation error $\mathrm{EVE}_{\mathrm{in}}(\eta)$. The difference between $\mathrm{EVE}_{\mathrm{in}}(\eta)$ and $\overline{\mathrm{err}}(\eta)$ is called the optimism in statistics. In particular, one has, see, e.g., [21, p. 229]:

$$\mathrm{EVE}_{\mathrm{in}}(\eta) = \overline{\mathrm{err}}(\eta) + \mathrm{optimism}(\eta), \tag{3.91}$$

where rewriting (3.83) as

$$Y = \Phi\theta_0 + E, \tag{3.92}$$

and defining the output prediction as

$$\hat{Y}(\eta) = \Phi\hat{\theta}^\mathrm{R}(\eta),$$

it holds that

$$\mathrm{optimism}(\eta) = 2\frac{1}{N}\mathrm{trace}(\mathrm{Cov}(Y, \hat{Y}(\eta))) \ge 0. \tag{3.93}$$

Combining arguments contained in the proof of Theorem 3.7 reported in the appendix to this chapter, see, in particular, (3.164), with the definition of equivalent degrees of freedom in (3.65), one obtains that

$$\mathrm{trace}(\mathrm{Cov}(Y, \hat{Y}(\eta))) = \sigma^2\mathrm{dof}(\hat{\theta}^\mathrm{R}(\eta)). \tag{3.94}$$

This thus reveals the deep connection between the optimism and the equivalent degrees of freedom.

3.5.3.2   An Unbiased Estimator of the Expected In-Sample Validation Error

Exploiting (3.94), we can now rewrite (3.91) as

$$\text{EVE}_{\text{in}}(\eta) = \overline{\text{err}}(\eta) + 2\sigma^2 \frac{\text{dof}(\hat{\theta}^{\text{R}}(\eta))}{N}. \tag{3.95}$$

Interestingly, on the left-hand side of (3.95), $\text{EVE}_{\text{in}}(\eta)$, by definition (3.87), is the mean of a random variable which depends on both the training data (3.84) and the validation data (3.85). Instead, on the right-hand side of (3.95), $\overline{\text{err}}(\eta)$ is the expectation of a random variable which depends only on the training data. Hence, an unbiased estimator $\widehat{\text{EVE}}_{\text{in}}(\eta)$ of $\text{EVE}_{\text{in}}(\eta)$ is obtained just replacing $\overline{\text{err}}(\eta)$ with $\overline{\text{err}}(\eta)_{\mathscr{D}_{\text{T}}}$ reported in (3.88). One thus obtains

$$
\begin{aligned}
\widehat{\text{EVE}}_{\text{in}}(\eta) &= \overline{\text{err}}(\eta)_{\mathscr{D}_{\text{T}}} + 2\sigma^2 \frac{\text{dof}(\hat{\theta}^{\text{R}}(\eta))}{N} \\
&= \frac{1}{N} \|Y - \Phi\hat{\theta}^{\text{R}}(\eta)\|_2^2 + 2\sigma^2 \frac{\text{dof}(\hat{\theta}^{\text{R}}(\eta))}{N}.
\end{aligned} \tag{3.96}
$$

So, after observing the training data (3.84), the hyperparameter η can be estimated as follows:

$$\hat{\eta} = \arg\min_{\eta \in \Gamma} \frac{1}{N} \|Y - \Phi\hat{\theta}^{\text{R}}(\eta)\|_2^2 + 2\sigma^2 \frac{\text{dof}(\hat{\theta}^{\text{R}}(\eta))}{N}. \tag{3.97}$$

The hyperparameter estimation criterion (3.97) has different names in statistics: it is known as the CP statistics, e.g., [27] and Stein's unbiased risk estimator (SURE), e.g., [40].

Interestingly, as it will be clear from the proof of Theorem 3.7, the above formula (3.97) still provides an unbiased prediction risk estimator also if we replace $\Phi\theta_0$ in (3.92) with a generic vector μ s.t. $Y = \mu + E$. Hence, one does not need to assume the existence of the true θ_0 and of a regression matrix which describes the linear input–output relation. A variant of the expected in-sample validation error is also discussed in Sect. 3.8.5.

3.5.3.3 Excess Degrees of Freedom*

In the previous subsection, we have discussed how to construct an unbiased estimator of the expected in-sample validation error, see (3.96), and how to use it for
hyperparameters tuning, see (3.97). Irrespective of the particular method adopted
for hyperparameter estimation, the estimate $\hat{\eta}$ of η depends on the data Y, with the
regression matrix Φ here assumed deterministic and known. We stress this by writing

$$\hat{\eta} = \hat{\eta}(Y).$$

Accordingly, the ReLS-Q estimate (3.58) with η replaced by $\hat{\eta}(Y)$ becomes

$$\hat{\theta}^{R}(\hat{\eta}(Y)) = (\Phi^{T}\Phi + \sigma^{2}P^{-1}(\hat{\eta}(Y)))^{-1}\Phi^{T}Y. \tag{3.98}$$

Since $\hat{\eta}$ is a random vector, to design a true unbiased estimator of the expected in-
sample validation error of $\hat{\theta}^{R}(\hat{\eta}(Y))$ one should not use (3.96) since it assumes the
hyperparameter η constant.

In what follows, we will derive an unbiased estimator of the expected in-sample
validation error of $\hat{\theta}^{R}(\hat{\eta}(Y))$. Such an estimator will thus be able to account also for
the price of estimating model complexity (the degrees of freedom) from data. To this
goal, we need the following version of Stein's Lemma [40], a simplified version of
which was already introduced in Chap. 1.

Lemma 3.2 (Stein's Lemma, adapted from [40]) *Consider the following additive
measurement model:*

$$x = \mu + \varepsilon, \quad x, \mu, \varepsilon \in \mathbb{R}^{p},$$

*where μ is an unknown constant vector and $\varepsilon \sim N(0, \Sigma)$. Let $\hat{\mu}(x)$ be an estimator
of μ based on the data x such that $Cov(\hat{\mu}(x), x)$ and $\mathscr{E}(\frac{\partial\hat{\mu}(x)}{\partial x})$ exist. Then*

$$Cov(\hat{\mu}(x), x) = \mathscr{E}\left(\frac{\partial\hat{\mu}(x)}{\partial x}\right)\Sigma.$$

Let

$$Y_{v} = \begin{bmatrix} y_{v,1} \\ y_{v,2} \\ \vdots \\ y_{v,N} \end{bmatrix}, \quad E_{v} = \begin{bmatrix} e_{v,1} \\ e_{v,2} \\ \vdots \\ e_{v,N} \end{bmatrix}, \tag{3.99}$$

so that (3.85) can be rewritten as

$$Y_{v} = \Phi\theta_{0} + E_{v}. \tag{3.100}$$

Now, let us consider the measurements model (3.92) and the validation data (3.100), assuming also that

$$E \sim N(0, \sigma^2 I_N), \ E_v \sim N(0, \sigma^2 I_N). \tag{3.101}$$

Then, using the correspondences

$$x = Y, \mu = \Phi\theta_0, \hat{\mu}(x) = \Phi\hat{\theta}^R(\hat{\eta}(Y)), \tilde{x} = Y_v, \varepsilon = V, \tilde{\varepsilon} = E_v, \Sigma = \sigma^2 I_N$$

$$f(Y, \hat{\eta}) = \Phi\hat{\theta}^R(\hat{\eta}(Y)) = \Phi(\Phi^T\Phi + \sigma^2 P^{-1}(\hat{\eta}(Y)))^{-1}\Phi^T Y,$$

together with (3.161) in the appendix to this chapter, one can prove that

$$\underbrace{\mathscr{E}\left[\frac{1}{N}\mathscr{E}[\|Y_v - \Phi\hat{\theta}^R(\hat{\eta}(Y))\|_2^2 | \mathscr{D}_T]\right]}_{\text{EVE}_{in}(\eta)} - \underbrace{\mathscr{E}\left[\frac{1}{N}\|Y - \Phi\hat{\theta}^R(\hat{\eta}(Y))\|_2^2\right]}_{\overline{\text{err}}(\eta)}$$

$$= 2\frac{1}{N}\operatorname{trace}(\operatorname{Cov}(Y, \Phi\hat{\theta}^R(\hat{\eta}(Y)))).$$

Using Stein's Lemma, one has

$$\operatorname{Cov}(Y, \Phi\hat{\theta}^R(\hat{\eta}(Y))) = \sigma^2 \mathscr{E}[\frac{df(Y, \hat{\eta})}{dY}]$$

$$= \sigma^2 \mathscr{E}[\frac{\partial f(Y, \hat{\eta})}{\partial Y}] + \sigma^2 \mathscr{E}[\frac{\partial f(Y, \hat{\eta})}{\partial \hat{\eta}}\frac{\partial \hat{\eta}}{\partial Y}]$$

$$= \sigma^2 \mathscr{E}[\Phi(\Phi^T\Phi + \sigma^2 P^{-1}(\hat{\eta}(Y)))^{-1}\Phi^T] + \sigma^2 \mathscr{E}[\frac{\partial f(Y, \hat{\eta})}{\partial \hat{\eta}}\frac{\partial \hat{\eta}}{\partial Y}].$$

Therefore, it holds that

$$\text{EVE}_{in} = \overline{\text{err}}(\eta) + 2\sigma^2 \frac{1}{N}\mathscr{E}[\operatorname{trace}(\Phi P(\hat{\eta}(Y)))\Phi^T(\Phi P(\hat{\eta}(Y)))\Phi^T + \sigma^2 I_N)^{-1})]$$

$$+ 2\sigma^2 \frac{1}{N}\operatorname{trace}(\mathscr{E}[\frac{\partial f(Y, \hat{\eta})}{\partial \hat{\eta}}\frac{\partial \hat{\eta}}{\partial Y}])$$

$$= \overline{\text{err}}(\eta) + 2\sigma^2 \frac{\operatorname{dof}(\hat{\theta}^R(\hat{\eta}(Y)))}{N} + 2\sigma^2 \frac{1}{N}\operatorname{trace}(\mathscr{E}[\frac{\partial f(Y, \hat{\eta})}{\partial \hat{\eta}}\frac{\partial \hat{\eta}}{\partial Y}]). \tag{3.102}$$

If $\hat{\eta} = \hat{\eta}(Y)$ were independent of Y, the above objective would coincide with the SURE score reported in (3.97). The difference is instead the presence of the term $2\sigma^2 \frac{1}{N}\operatorname{trace}(\mathscr{E}[\frac{\partial f(Y, \hat{\eta})}{\partial \hat{\eta}}\frac{\partial \hat{\eta}}{\partial Y}])$. It represents the extra optimism induced by the estimation of η and is due to the randomness of the data Y entering the hyperparameter estimator. The term $\operatorname{trace}(\mathscr{E}[\frac{\partial f(Y, \hat{\eta})}{\partial \hat{\eta}}\frac{\partial \hat{\eta}}{\partial Y}])$ is called the excess degrees of freedom [33] and denoted by

$$\operatorname{exdof}(\hat{\theta}^R(\hat{\eta}(Y))) = \operatorname{trace}(\mathscr{E}[\frac{\partial f(Y, \hat{\eta})}{\partial \hat{\eta}}\frac{\partial \hat{\eta}}{\partial Y}]). \tag{3.103}$$

From (3.102), we readily obtain an unbiased estimator of EVE_{in} as follows:

$$\widehat{\text{EVE}_{\text{in}}} = \overline{\text{err}}(\eta)_{\mathscr{D}_{\mathrm{T}}} + 2\sigma^2 \frac{\text{dof}(\hat{\theta}^{\mathrm{R}}(\hat{\eta}(Y)))}{N} + 2\sigma^2 \frac{\widehat{\text{exdof}(\hat{Y}(\hat{\eta}))}}{N}$$

$$= \frac{1}{N} \|Y - \Phi\hat{\theta}^{\mathrm{R}}(\hat{\eta}(Y))\|_2^2 + 2\sigma^2 \frac{\text{dof}(\hat{\theta}^{\mathrm{R}}(\hat{\eta}(Y)))}{N}$$

$$+ 2\sigma^2 \frac{1}{N} \text{trace}(\frac{\partial f(Y, \hat{\eta})}{\partial \hat{\eta}} \frac{\partial \hat{\eta}}{\partial Y}), \tag{3.104}$$

where $\widehat{\text{exdof}(\hat{Y}(\hat{\eta}))}$ is an unbiased estimator of $\text{exdof}(\hat{Y}(\hat{\eta}))$. As discussed in [33], (3.104) can be used to compare different regularized estimators also in terms of the different complexity of the hyperparameters tuning strategies that they adopt.

3.6   Regularized Least Squares with Other Types of Regularizers ★

The general ReLS criterion assumes the following form

$$\hat{\theta}^{\mathrm{R}} = \arg \min_{\theta} \|Y - \Phi\theta\|_2^2 + \gamma J(\theta).$$

The different choices of the regularization term $J(\theta)$ depend on the prior knowledge regarding θ_0. Having discussed the quadratic penalty, we will now consider two other important choices for $J(\theta)$ given by the ℓ_1- or nuclear norm.

3.6.1 ℓ_1-Norm Regularization

ReLS with ℓ_1-norm regularization leads to

$$\hat{\theta}^{\mathrm{R}} = \arg \min_{\theta} \|Y - \Phi\theta\|_2^2 + \gamma \|\theta\|_1, \tag{3.105}$$

where $\|\theta\|_1$ represents the ℓ_1-norm of θ, i.e., $\|\theta\|_1 = \sum_{i=1}^n |\theta_i|$ with θ_i being the ith element of θ. The problem (3.105) is also known as the least absolute shrinkage and selection operator (LASSO) [42] and is equivalently defined as follows:

$$\arg \min_{\theta} \|Y - \Phi\theta\|_2^2, \text{ subj. to } \|\theta\|_1 \leq \beta, \tag{3.106}$$

where $\beta \geq 0$ is a tuning parameter connected with γ that controls the sparsity of θ.

3.6.1.1 Computation of Sparse Solutions

LASSO (3.105) has been widely used for finding sparse solutions. In signal processing, such problem has wide applications in compressive sensing for finding sparse signal representations from redundant dictionaries. In machine learning and statistics, the problem has also been applied extensively for variable selection where the aim is to select a subset of relevant variables to use in model construction.

Recall that a vector $\theta \in \mathbb{R}^n$ is said to be sparse if $\|\theta\|_0 \ll n$, where $\|\theta\|_0$ is the ℓ_0 norm of θ which counts the number of nonzero elements of θ. For linear regression models, sparse estimation requires to find a sparse θ able to well fit the data, i.e., such that $\|Y - \Phi\theta\|_2^2$ is small. More formally, the problem is defined as follows:

$$\min_{\theta} \|\theta\|_0, \text{ subj. to } \|Y - \Phi\theta\|_2^2 \leq \varepsilon, \tag{3.107}$$

where $Y \in \mathbb{R}^N, \theta \in \mathbb{R}^n$ with $n > N, \Phi \in \mathbb{R}^{N \times n}$ assumed of full rank, i.e., $\text{rank}(\Phi) = N$, and $\varepsilon \geq 0$ is a tuning parameter that controls the data fit.

The problem (3.107) is known to be NP-hard, e.g., [31]. It is combinatorial and finding its solution requires an exhaustive search. Hence, one needs approximated methods. The most popular technique relies on a convex relaxation of (3.107) obtained by replacing the ℓ_0-norm with the ℓ_1-norm:

$$\min_{\theta} \|\theta\|_1, \text{ subj. to } \|Y - \Phi\theta\|_2^2 \leq \varepsilon. \tag{3.108}$$

By using the method of Language multipliers, it can be shown that the convex relation (3.108) is equivalent to LASSO (3.105).

A natural question is whether or not the solution of LASSO (3.105) can be sparse. The answer is affirmative. For illustration, we first show this feature when the regression matrix Φ is orthogonal and assuming $N = n$.

3.6.1.2 LASSO Using an Orthogonal Regression Matrix

Let us consider (3.105) with orthogonal regression matrix Φ, i.e., $\Phi^T\Phi = \Phi\Phi^T = I_n$. Then (3.105) is rearranged as follows:

$$\begin{aligned} \hat{\theta}^R &= \arg\min_{\theta} \|(\Phi^T\Phi)^{-1}\Phi^T(Y - \Phi\theta)\|_2^2 + \gamma\|\theta\|_1 \\ &= \arg\min_{\theta} \|\hat{\theta}^{LS} - \theta\|_2^2 + \gamma\|\theta\|_1 \\ &= \arg\min_{\theta} \sum_{i=1}^{n} (\hat{\theta}_i^{LS} - \theta_i)^2 + \gamma|\theta_i|, \end{aligned} \tag{3.109}$$

where $\hat{\theta}_i^{LS}$ is the ith element of $\hat{\theta}^{LS}$.

To derive the optimal solution $\hat{\theta}^R$, we first recall the definition of subderivative and subdifferential of a convex function $f : X \to \mathbb{R}$ with X being an open interval. The subderivative of a convex function $f : X \to \mathbb{R}$ at a point x_0 in the open interval X is a real number a such that

$$f(x) - f(x_0) \geq a(x - x_0)$$

for all x in X. It can be shown that there exist b and c with $b \leq c$ such that the set of subderivatives at x_0 for a convex function is a nonempty closed interval $[b, \ c]$, where b and c are the one-sided limits defined as follows:

$$b = \lim_{x \to x_0^-} \frac{f(x) - f(x_0)}{x - x_0}, \quad c = \lim_{x \to x_0^+} \frac{f(x) - f(x_0)}{x - x_0}.$$

The closed interval $[b, \ c]$ is called the subdifferential of $f(x)$ at the point x_0.

Then, considering (3.109), $\hat{\theta}^R$ is an optimal solution if

$$-2(\hat{\theta}_i^{LS} - \hat{\theta}_i^R) + \gamma \partial |\hat{\theta}_i^R| = 0, \ i = 1, 2, \ldots, n, \tag{3.110}$$

where $\hat{\theta}_i^R$ is the ith element of $\hat{\theta}^R$ and $\partial |\hat{\theta}_i^R|$ represents the subdifferential of $|\hat{\theta}_i^R|$ which is equal to

$$\partial |\hat{\theta}_i^R| = \begin{cases} \{\text{sign}(\hat{\theta}_i^R)\} & \hat{\theta}_i^R \neq 0 \\ [-1, \ 1] & \hat{\theta}_i^R = 0 \end{cases}, \ i = 1, 2, \ldots, n. \tag{3.111}$$

Using (3.110) and (3.111), we obtain the following explicit solution of LASSO for orthogonal Φ:

$$\hat{\theta}_i^R = \text{sign}(\hat{\theta}_i^{LS}) \min \left\{ 0, |\hat{\theta}_i^{LS}| - \frac{\gamma}{2} \right\}, \ i = 1, 2, \ldots, n. \tag{3.112}$$

From (3.112) one can see that the solution of LASSO will be sparse if many absolute values of the elements of $\hat{\theta}^{LS}$ are smaller than $\gamma/2$. So, γ can be used to tune the sparsity of θ. It can also be seen that the nonzero elements of the solution of LASSO are biased and that, compared with the LS solution, they are shrunk towards zero (translated towards zero by a constant factor $\gamma/2$).

3.6.1.3 LASSO Using a Generic Regression Matrix: Geometric Interpretation

For a generic non-orthogonal Φ, LASSO in general has no explicit solutions. To understand why it can still induce sparse solutions, we can use the geometric interpretation of LASSO in the form of (3.106) with $\theta \in \mathbb{R}^2$. In Fig. 3.12, one can see that for the first case coloured in blue (resp., the third case coloured in brown), if

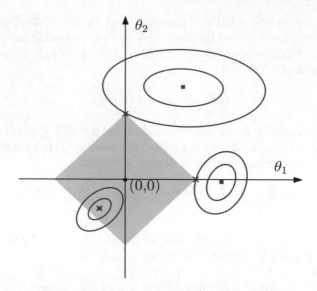

Fig. 3.12 Geometric interpretation of the solution of LASSO in the form (3.106) with non-orthogonal Φ and $\theta = [\theta_1 \ \theta_2]^T \in \mathbb{R}^2$. First, the large grey square represents the constraint $\|\theta\|_1 \leq \beta$. Then, three cases are considered here and coloured in blue, red and brown, respectively. For each case, the tiny square represents the least squares estimate $\hat{\theta}^{LS}$, the elliptical contours represent the level curves of $\|Y - \Phi\theta\|_2^2$ centred at $\hat{\theta}^{LS}$ and the cross represents the solution of LASSO (3.106). For the first case coloured in blue, the cross happens at the top corner of the large grey square and implies that the θ_1-element of the solution of LASSO (3.106) is zero. For the second case coloured in red, the cross and the tiny square coincide and imply that the least square estimate $\hat{\theta}^{LS}$ is also the solution of LASSO (3.106) whose two components are both nonzero. For the third case coloured in brown, the cross happens at the right corner of the large grey square and implies that the θ_2-element of the solution of LASSO (3.106) is zero

the elliptical contour is rotated slightly about the axis perpendicular to the paper and through the blue (resp., brown) cross, the optimal solution of (3.106) will still have a zero θ_1-element (resp., θ_2-element). This explains why LASSO can often induce sparse solutions with a suitable choice of the regularization parameter.

Finally, since the cost function of LASSO (3.105) is a convex function of θ, many standard convex optimization software packages are available to obtain numerical solutions of LASSO very efficiently, such as YALMIP [26], CVX [19], CVXOPT [3], CVXPY [11].

Example 3.7 (*Polynomial regression-LASSO*) We revisit the polynomial regression Examples (3.26) and (3.27) with LASSO (3.105). In particular, we set the model order to $n = 16$, with the regression matrix Φ built according to (3.4) and (3.12). Moreover, we let $\gamma = \gamma_i, i = 1, \ldots, 16$ with $\gamma_1 = 0.01$, $\gamma_{16} = 0.31$ and $\gamma_2, \ldots, \gamma_{15}$ evenly spaced between γ_1 and γ_{16}. For each $\gamma = \gamma_i$, we compute the corresponding solution of the LASSO (3.105). In particular, the estimates $\hat{g}(x) = \phi(x)^T \hat{\theta}^R$ for $x = x_i$, with $i = 1, \ldots, 40$, are plotted in Fig. 3.13.

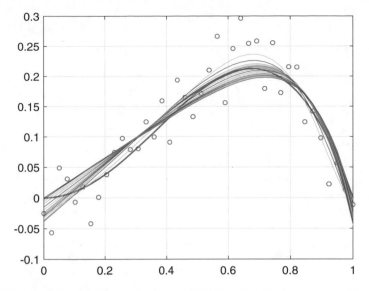

Fig. 3.13 Polynomial regression: true function $g(x)$ (blue) and LASSO estimates (thin) for different values of the regularization parameter γ

The model fits (3.28) obtained for different γ are shown in Fig. 3.14. One can see that $\gamma = 0.15$ gives the best result.

Finally, the LASSO estimates of the components of θ obtained using the different values of γ are shown in Fig. 3.15. It is evident that the LASSO estimate (3.105) is sparse. Comparing it with the ridge regression estimates reported in Fig. 3.11, one can conclude that LASSO may give a simpler model, i.e., depending only on a limited number of components of θ.   □

3.6.1.4 Sparsity Inducing Regularizers Beyond the ℓ_1-Norm

We have seen that the ℓ_1-norm plays a key role for sparse estimation. However, as shown in [34], there are many other sparsity inducing regularizers. Let l be any concave and nondecreasing function on $[0, \infty)$, three examples being reported in the top panel of Fig. 3.16. Then, other penalties which promote sparsity assume the form $J(\theta) = \sum_{i=1}^n l(\theta_i^2)$ and are given by

$$l(\eta) = \eta^{\frac{p}{2}}, \ p \in (0, 2) \qquad \overset{\eta = \theta_i^2}{\Longrightarrow} \qquad J(\theta) = \sum_{i=1}^n |\theta_i|^p, \ p \in (0, 2),$$

$$l(\eta) = \log(|\eta|^{\frac{1}{2}} + \varepsilon), \ \varepsilon > 0 \qquad \overset{\eta = \theta_i^2}{\Longrightarrow} \qquad J(\theta) = \sum_{i=1}^n \log(|\theta_i| + \varepsilon).$$

$$(3.113)$$

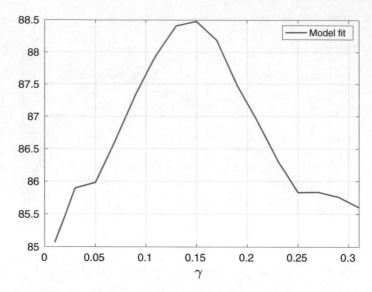

Fig. 3.14 Polynomial regression: profile of the model fit (3.28) obtained by LASSO as a function of the regularization parameter γ

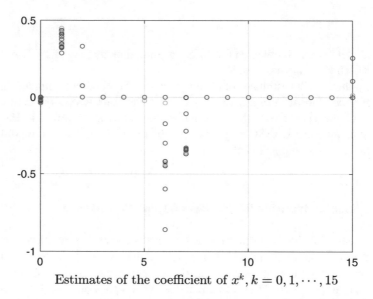

Estimates of the coefficient of $x^k, k = 0, 1, \cdots, 15$

Fig. 3.15 Polynomial regression: profile of the estimates of each component forming the LASSO estimate (3.105). For each value $k = 0, \ldots, 15$ on the x-axis the plot reports the estimates of the coefficient of the monomial x^k obtained by using different values of the regularization parameter γ

Fig. 3.16 The top panel shows profiles of $l(\theta_i)$ given by $\theta_i^{0.05}$, $\log(\theta_i^{0.5} + 1)$ and $\theta_i^{0.5}$ with θ_i ranging over $[0, \ 1]$. The bottom panel displays profiles of sparsity inducing penalties $l(\theta_i^2)$ given by $|\theta_i|^{0.1}$, $\log(|\theta_i| + 1)$ and $|\theta_i|$ with θ_i ranging over $[-1, \ 1]$

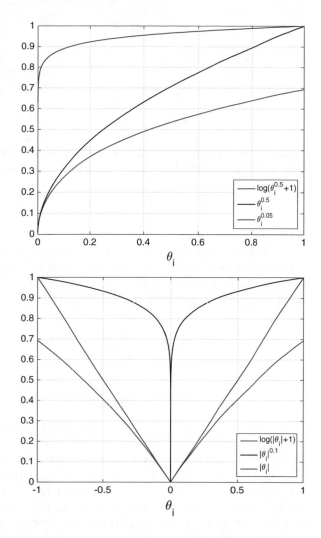

Some of them are displayed in the bottom panel of Fig. 3.16. The use of nonconvex penalties may increase the sparsity in the solution but the drawback is that optimization problems possibly exposed to local minima must be handled.

3.6.1.5   Presence of Outliers and Robust Regression

In practical applications, it may happen that the measurement outputs y_i so far described by the model

$$y_i = \phi_i^T \theta_0 + e_i, \quad i = 1, \ldots, N$$

may be contaminated by outliers which represent unexpected noise model deviations. They can be due to the failure of some sensors or to mistakes in the setting of the experiment. In this case, data can actually be generated by the following system:

$$y_i = \phi_i^T \theta_0 + e_i + v_{0,i}, \quad i = 1, \ldots, N, \tag{3.114}$$

where the e_i form a white noise with mean zero and variance σ^2 while the $v_{0,i}$ represents the outliers which are assumed to be zero most of time. Hence, the vector

$$V_0 = \begin{bmatrix} v_{0,1} & v_{0,2} & \cdots & v_{0,N} \end{bmatrix}^T$$

is assumed to be sparse.

When data come from (3.114), straightforward application of the LS method may lead to a poor estimate $\hat{\theta}^{LS}$ of θ_0. For illustration, let us consider an extreme case by assuming $v_{0,i} = 0$ for $i = 1, 2, \ldots, N - 1$ while the $|\phi_i^T \theta_0 + e_i|$ for $i = 1, \ldots, N$ are all negligible compared to $|v_{0,N}|$. LS leads to

$$\hat{\theta}^{LS} = \arg\min_\theta \sum_{i=1}^{N} (y_i - \phi_i^T \theta)^2$$

$$= \arg\min_\theta \sum_{i=1}^{N-1} (\phi_i^T \theta_0 + e_i - \phi_i^T \theta)^2$$

$$+ (\phi_N^T \theta_0 + e_N + v_{0,N} - \phi_N^T \theta)^2.$$

The first $N - 1$ terms in the above cost function are the same encountered in absence of outliers while the last term is different due to $v_{0,N}$. The $|\phi_i^T \theta_0 + e_i|, i = 1, \ldots, N$ are negligible compared to $|v_{0,N}|$, a phenomenon then further amplified by the quadratic criterion here adopted. To make the last term as small as possible, $\hat{\theta}^{LS}$ will mainly tend to fit only $v_{0,N}$. Hence, the terms $\phi_i^T \theta_0 + e_i$ which carry information on the true system will be little regarded. This will lead to a poor estimate of θ_0.

Many robust regression methods are available hinging on loss functions less sensitive to outliers than the square loss. An example is Huber estimation

$$\hat{\theta}^{Huber} = \arg\min_\theta \sum_{i=1}^{N} l^{Huber}(y_i - \phi_i^T \theta) \tag{3.115}$$

where the Huber loss function l^{Huber} is defined as follows:

$$l^{Huber}(x) = \begin{cases} x^2 & |x| < \frac{\gamma}{2} \\ \gamma|x| - \frac{1}{4}\gamma^2 & |x| \geq \frac{\gamma}{2} \end{cases}. \tag{3.116}$$

In (3.116), the parameter $\gamma > 0$ is a tuning parameter whose role will become clear shortly. The Huber loss function (3.116) is less sensitive to outliers because it grows linearly for $|x| \geq \gamma/2$. Note that a limit case of the Huber loss is the ℓ_1-norm obtained with γ which tends to zero.

3.6.1.6 An Equivalence Between ℓ_1-Norm Regularization and Huber Estimation

Let

$$\tilde{y}_i = y_i - \phi_i^T \theta, \quad i = 1, 2, \ldots, N,$$
$$\tilde{Y} = \begin{bmatrix} \tilde{y}_1 & \tilde{y}_2 & \ldots & \tilde{y}_N \end{bmatrix}^T.$$

Consider the ℓ_1-norm regularization given by

$$\arg\min_{\theta, V_0} \sum_{i=1}^{N} (y_i - \phi_i^T \theta - v_{0,i})^2 + \gamma |v_{0,i}| \tag{3.117}$$

whose peculiarity is to require joint optimization w.r.t. the parameter vector θ and the outliers $v_{0,i}$ contained in V_0. Interestingly, (3.117) is actually equivalent to Huber estimation (3.115), i.e., they have the same optimal solution. To show this, one needs just to prove that

$$\sum_{i=1}^{N} l^{\text{Huber}}(y_i - \phi_i^T \theta) = \min_{V} \|\tilde{Y} - V_0\|_2^2 + \gamma \|V_0\|_1. \tag{3.118}$$

The right-hand side of (3.118) corresponds to LASSO (3.105) with an orthogonal regression matrix given by the identity. It thus follows from (3.112) that the components of the optimal solution $\hat{V}_0^R$ admit the following closed-form expression:

$$\hat{v}_{0,i}^R = \text{sign}(\tilde{y}_i) \min \left\{ 0, |\tilde{y}_i| - \frac{\gamma}{2} \right\}, \quad i = 1, 2, \ldots, N. \tag{3.119}$$

Now we replace V_0 in the cost function of the right-hand side of (3.118) with $\hat{V}_0^R$ and it is straightforward to check that the following identify holds:

$$\sum_{i=1}^{N} l^{\text{Huber}}(y_i - \phi_i^T \theta) = \|\tilde{Y} - \hat{V}_0^R\|_2^2 + \gamma \|\hat{V}_0^R\|_1. \tag{3.120}$$

Therefore, (3.117) is indeed equivalent to the Huber estimation (3.115).

3.6.2 Nuclear Norm Regularization

So far the output Y, the parameter θ and the noise E in (3.13) have been assumed to be vectors. In what follows, we allow them to be matrices and consider the following linear regression model:

$$Y = \Phi\theta_0 + E, \quad Y \in \mathbb{R}^{N \times m}, \quad \Phi \in \mathbb{R}^{N \times n}, \quad \theta_0 \in \mathbb{R}^{n \times m}, \quad E \in \mathbb{R}^{N \times m}. \quad (3.121)$$

The ReLS with nuclear norm regularization takes the following form:

$$\hat{\theta}^R = \arg\min_{\theta} \|Y - \Phi\theta\|_F^2 + \gamma \|h(\theta)\|_*, \quad (3.122)$$

where $\| \cdot \|_F$ is the Frobenius norm of a matrix, $h(\theta)$ is a matrix that is affine in θ and $\|h(\theta)\|_*$ is the nuclear norm of the matrix $h(\theta)$, see also Sect. 3.8.1, the appendix to this chapter, for a brief review of matrix and vector norms.

3.6.2.1 Nuclear Norm Regularization for Matrix Rank Minimization

Matrix rank minimization problems (RMP) are a class of optimization problems that involve minimizing the rank of a matrix subject to convex constraints. They are often encountered in signal processing, image processing and statistics. For example, a typical statistical problem is to obtain a low-rank covariance matrix able to describe some available data and/or consistent with some prior assumptions. Formally, the RMP is defined as follows:

$$\text{RMP:} \quad \begin{array}{c} \min_X \text{rank}(X) \\ \text{subj. to } X \in \mathfrak{C} \subset \mathbb{R}^{n \times m}, \end{array} \quad (3.123)$$

with X belonging to a convex set $\mathfrak{C}$ while $\text{rank}(X)$ describes the order (complexity) of the underlying model.

In general, the RMP (3.123) is NP-hard and thus there is need for approximated methods. Several heuristic methods have been proposed, such as the nuclear norm heuristic [14] and the log-det heuristic [15]. In particular, for a convex set $\mathfrak{C}$ the convex envelope of a function $f : \mathfrak{C} \to \mathbb{R}$ is defined as the largest convex function g such that $g(x) \le f(x)$ for every $x \in \mathfrak{C}$, e.g., [22]. For a nonconvex f, solving

$$\min_{x \in \mathfrak{C}} f(x) \quad (3.124)$$

may be difficult. In this case, if it is possible to derive the convex envelope g of f, then

$$\min_{x \in \mathfrak{C}} g(x) \quad (3.125)$$

turns out a convex approximation of (3.124) and, in particular, the minimum of (3.125) can represent a lower bound of that of (3.124). Moreover, if necessary, the minimizing argument of (3.125) can be chosen as the initial point for a more complicated nonconvex local search aiming to solve (3.124).

As shown in Theorem 1 of [13, Chap. 5], the convex envelope of the rank function $\text{rank}(X)$ with $X \in \mathfrak{C} = \{X \mid \|X\|_2 \leq 1, X \in \mathbb{R}^{n \times m}\}$ is the nuclear norm of X, i.e., $\|X\|_*$. As a result, the nuclear norm heuristic to solve the RMP (3.123) is obtained by replacing the rank of X with the nuclear norm of X, i.e.,

$$\text{Nuclear norm heuristic:} \quad \begin{aligned} &\min_X \|X\|_* \\ &\text{subj. to } X \in \mathfrak{C} \subset \mathbb{R}^{n \times m}. \end{aligned} \tag{3.126}$$

Without loss of generality, we assume that $X \in \mathfrak{C} = \{X \mid \|X\|_2 \leq M, X \in \mathbb{R}^{n \times m}\}$ for some $M > 0$. Then, from the definition of the convex envelope, for $X \in \mathfrak{C}$ we have

$$\left\| \frac{X}{M} \right\|_* \leq \text{rank}\left(\frac{X}{M}\right) \quad \Longrightarrow \quad \frac{1}{M}\|X\|_* \leq \text{rank}(X).$$

In addition

$$\frac{1}{M}\|X^{\text{copt}}\|_* \leq \text{rank}(X^{\text{opt}}) \leq \text{rank}(X^{\text{copt}}), \tag{3.127}$$

where X^{opt} and X^{copt} denote the optimal solution of the RMP (3.123) and that of the nuclear norm heuristic (3.126), respectively. The inequalities in (3.127) thus provide an upper and lower bound for the optimal solution of the RMP (3.123).

As shown in [13, Chap. 5], the nuclear norm heuristic (3.126) can be equivalently formulated as a semidefinite program (SDP):

$$\begin{aligned} &\min_{X,Y,Z} \text{trace } Y + \text{trace } Z \\ &\text{subj. to } \begin{bmatrix} Y & X \\ X^T & Z \end{bmatrix} \geq 0, \ X \in \mathfrak{C}, \end{aligned} \tag{3.128}$$

where $Y \in \mathbb{R}^{n \times n}$, $Z \in \mathbb{R}^{m \times m}$ and both Y and Z are symmetric. The SDP problem (3.128) can be solved by interior point methods. For this purpose, some convex optimization software packages which can be used include YALMIP [26], CVX [19], CVXOPT [3] and CVXPY [11].

3.6.2.2 Application in Covariance Matrix Estimation with Low-Rank Structure

Now we go back to the linear regression model (3.121) and the ReLS with nuclear norm regularization (3.122). Consider the problem of covariance matrix estimation with low-rank structure, e.g., [38]. In particular, in (3.121), we take $N = m = n$, let Y be a sample covariance matrix, $\Phi = I_n$, and θ_0 be a positive semidefinite matrix which has low-rank structure. Moreover, in (3.122), we take $h(\theta) = \theta$. We can then obtain a matrix estimate $\hat{\theta}^R$ with low-rank structure using ReLS with nuclear norm regularization as follows:

$$\hat{\theta}^R = \arg\min_\theta \|Y - \theta\|_F^2 + \gamma \|\theta\|_*, \tag{3.129}$$

for a suitable choice of $\gamma > 0$. An example is reported below.

Example 3.8 (*Covariance matrix estimation problem*) First, we construct a block-diagonal rank-deficient covariance matrix θ_0 that has 4 blocks denoted by $A_i \in \mathbb{R}^{n_i \times n_i}$ with $n_1 = 20$, $n_2 = 10$, $n_3 = 5$ and $n_4 = 15$. Using *blkdiag* to represent a block-diagonal matrix, one thus has $\theta_0 = blkdiag(A_1, A_2, A_3, A_4)$. Each A_i is generated by summing up $v_{i,j} v_{i,j}^T$, $j = 1, \ldots, n_i - 2$, where the $v_{i,j}$ are n_i-dimensional vectors with components independent and uniformly distributed on $[-1, 1]$. It comes that $rank(\theta_0) = 42$ since the rank of each ith block is $n_i - 2$. Then we draw 20000 samples x_i from the Gaussian distribution $\mathcal{N}(0, \theta_0)$. The available measurements are $z_i = x_i + e_i$ where the e_i are independent and distributed as $\mathcal{N}(0, 0.6)$. Using the z_i we calculate the sample covariance Y as follows:

$$Y = \frac{1}{20000} \sum_{i=1}^{20000} (z_i - \bar{z})(z_i - \bar{z})^T, \quad \bar{z} = \frac{1}{20000} \sum_{i=1}^{20000} z_i. \tag{3.130}$$

We solve the ReLS problem (3.129) with the data Y defined above and γ in the set $\{0.1411, 0.1414, 0.1419, 0.1423, 0.1427\}$, obtaining different estimates $\hat{\theta}^R$ of the covariance matrix.

The top panel of Fig. 3.17 shows the base 10 logarithm of the 50 estimated singular values. Each profile is obtained with a different regularization parameter. Such results show that, seeing the tiny singular values as null, a suitable value of the regularization parameter, like $\gamma = 0.1427$, leads to $rank(\hat{\theta}^R) = 42$. Note in fact that the green curve, which is associated to such γ, has a jump towards zero when passing from 42 to 43 on the x-axis. The influence of the nuclear norm regularization is also visible in the bottom panel which shows the profile of the relative error of $\hat{\theta}^R$ as a function of γ. When γ is small, e.g., $\gamma = 0.1411$, the influence is invisible, $\hat{\theta}^R$ is almost the same as the sample covariance Y and $rank(\hat{\theta}^R) = 50$. When γ becomes larger, the regularization influence becomes more visible, making $\hat{\theta}^R$ closer to the true covariance θ_0.   □

Fig. 3.17 Covariance estimation with low-rank structure. Panel **a** shows the base 10 logarithm of the 50 singular values of the estimated covariance matrix $\hat{\theta}^R$ with different values of γ. Panel **b** shows the profile of the relative error of the estimated covariance matrix $\hat{\theta}^R$ as a function of γ

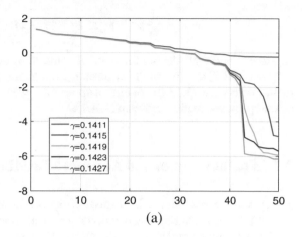

(a)

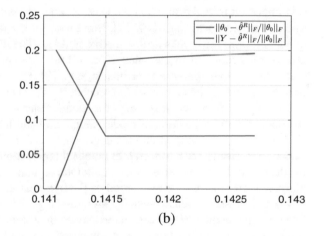

(b)

3.6.2.3   Vector Case: ℓ_1-Norm Regularization

The nuclear norm heuristic and inequalities (3.127) also justify the use of the ℓ_1-norm regularization (3.108) for the problem of finding sparse solutions (3.107).

For the vector case, i.e., $\theta \in \mathbb{R}^{n \times m}$ with $m = 1$, we can take X and $\mathfrak{C}$ in the previous section to be $X = \theta$ and $\mathfrak{C} = \{\theta \in \mathbb{R}^n \mid \|Y - \Phi\theta\|_2^2 \le \varepsilon\}$. Then it is easy to see that the ℓ_1-norm is the convex envelope of the ℓ_0-norm for $\|\theta\|_\infty \le 1$, i.e.,

$$\|\theta\|_1 \le \|\theta\|_0, \ \text{for } \|\theta\|_\infty \le 1.$$

Then, the RMP (3.123) and the nuclear norm heuristic (3.126) become the problem of finding sparse solutions (3.107) and the ℓ_1-norm regularization (3.108), respectively. Similar to what is done to obtain (3.127), we assume that $\|\theta\|_\infty \le M$ for some $M > 0$. If $\|\theta\|_\infty \le M$, one has

$$\frac{1}{M}\|\theta^{\mathrm{copt}}\|_1 \leq \|\theta^{\mathrm{opt}}\|_0 \leq \|\theta^{\mathrm{copt}}\|_0, \tag{3.131}$$

where θ^{opt} and θ^{copt} denote the optimal solution of the problem of finding sparse solution (3.107) and that of the ℓ_1-norm regularization (3.108), respectively. Similar to the matrix case, (3.131) provides an upper and lower bound for the optimal solution of the sparse estimation problem (3.107).

3.7 Further Topics and Advanced Reading

The systematic treatment of the regression theory is available in many textbooks, e.g., [12, 35]. The noise variance estimation is a critical issue in practical applications and has been discussed in details in [48]. When the regression matrix is ill-conditioned, it is important to make sure that the least squares estimate is calculated in an accurate and efficient way, e.g., [10, 17]. Moreover, for the regularized least squares in quadratic form, the regularization matrix could also be ill-conditioned. In this case, extra care is required in the calculation of both the regularized least squares estimate and the hyperparameter estimates, e.g., [8]. For given data, the quality of a model depends on the control of its complexity, which can be described by different measures in different contexts, e.g., the model order and the equivalent degrees of freedom. A good exposition of model complexity and its selection can be found in [21]. It is worth to mention that the degrees of freedom for LASSO have also been defined and discussed in [43, 51]. In practical applications, there are two key issues for the regularized least squares with quadratic regularization: the design of the regularization matrix and the estimation of the hyperparameter. While the latter issue has been discussed extensively in the literature, e.g., [21, 36, 46, 47], there are much fewer results on the former issue in the context of system identification, as discussed in [7]. The asymptotic properties of some widely used hyperparameter estimators, such as the maximum marginal likelihood estimator, Stein's unbiased risk estimator, generalized cross-validation, etc., have been reported in [29, 30]. LASSO and its variants have been extremely popular in practical applications, as described in [16, 28, 32, 50]. The nuclear norm heuristic to solve matrix rank minimization problems has wide applications in practical applications, see, e.g., [5, 6, 14, 15, 37]. Beyond the Huber loss function [23], the square loss function can be replaced also by other convex functions like the Vapnik loss function [45] as discussed later on in Chap. 6.

3.8 Appendix

3.8.1 Fundamentals of Linear Algebra

In this section, we review some fundamentals of linear algebra used in this chapter.

3.8.1.1   QR Factorization and Singular Value Decomposition

We begin with giving the definitions of QR factorization and SVD, which are very important decompositions used for many purposes other than solving LS problems.

For any $\Phi \in \mathbb{R}^{N \times n}$ with $N \geq n$, Φ can be decomposed as follows:

$$\Phi = QR, \tag{3.132}$$

where $Q \in \mathbb{R}^{N \times N}$ is orthogonal, i.e., $Q^T Q = Q Q^T = I_N$, and $R \in \mathbb{R}^{N \times n}$ is upper triangular. Further assume that Φ has full rank. Then Φ can be decomposed as follows:

$$\Phi = Q_1 R_1 \tag{3.133}$$

where $Q_1 = Q(:, 1 : n)$ and $R_1 = R(1 : n, 1 : n)$ with $Q(:, 1 : n)$ being the matrix consisting of the first n columns of Q and $R(1 : n, 1 : n)$ being the matrix consisting of the first n rows and n columns of R. The factorizations (3.132) and (3.133) are called the full and thin QR factorization, respectively. In particular, when R_1 has positive diagonal entries, the thin QR factorization (3.133) is unique.

We start providing the "economy size" definition of the SVD. For any $\Phi \in \mathbb{R}^{N \times n}$ with $N \geq n$, Φ can be decomposed as follows:

$$\Phi = U \Lambda V^T, \tag{3.134}$$

where $U \in \mathbb{R}^{N \times n}$ satisfies $U^T U = I_N$, $\Lambda = \mathrm{diag}(\sigma_1, \sigma_2, \ldots, \sigma_n)$ with $\sigma_1 \geq \sigma_2 \geq \cdots \geq \sigma_n \geq 0$, and $V \in \mathbb{R}^{n \times n}$ is orthogonal. The factorization (3.134) is called the singular value decomposition (SVD) of Φ and the σ_i, $i = 1, \ldots, n$ are called the singular values of Φ.

The SVD admits also the "full size" formulation, as given in (3.29). One has that (3.134) still holds but U is an orthogonal $N \times N$ matrix and Λ is a rectangular $N \times n$ diagonal matrix, while V is still an orthogonal $n \times n$ matrix. In this second formulation, V and U can be associated to orthonormal change of coordinates in the domain and codomain of Φ such that, in the new coordinates, the linear operator is diagonal.

3.8.1.2   Vector and Matrix Norms

Important vector norms are the ℓ_1, ℓ_2 and ℓ_∞ norms. For a given vector $\theta \in \mathbb{R}^n$, they are denoted by $\|\theta\|_1$, $\|\theta\|_2$ and $\|\theta\|_\infty$, respectively, and are defined as follows:

$$\|\theta\|_1 = \sum_{i=1}^{n} |\theta_i|, \tag{3.135}$$

$$\|\theta\|_2 = \sqrt{\sum_{i=1}^{n} \theta_i^2}, \tag{3.136}$$

$$\|\theta\|_\infty = \max\{|\theta_1|, |\theta_2|, \ldots, |\theta_n|\}, \tag{3.137}$$

where the ℓ_2 norm is also known as the Euclidean norm.

Important matrix norms are the nuclear norm, the Frobenius norm and the spectral norm. For a given matrix $\Phi \in \mathbb{R}^{N \times n}$ with $N \geq n$, these three matrix norms are denoted by $\|\Phi\|_*$, $\|\Phi\|_F$ and $\|\Phi\|_2$, respectively, and are defined as follows:

$$\|\Phi\|_* = \sum_{i=1}^{n} \sigma_i(\Phi), \tag{3.138}$$

$$\|\Phi\|_F = \sqrt{\sum_{i=1}^{N} \sum_{j=1}^{n} \Phi_{i,j}^2} = \sqrt{\sum_{i=1}^{n} \sigma_i^2(\Phi)}, \tag{3.139}$$

$$\|\Phi\|_2 = \sigma_{\max}(\Phi), \tag{3.140}$$

where $\sigma_i(\Phi)$ represents the ith largest singular value of Φ, $\sigma_{\max}(\Phi) = \sigma_1(\Phi)$ and $\Phi_{i,j}$ is the (i, j)th element of Φ.

Now, we report some properties of the vector and matrix norms. The ith largest singular value of Φ is equal to the square root of the ith largest eigenvalue of $\Phi^T \Phi$, or equivalently $\Phi \Phi^T$. If Φ is square and positive semidefinite, then the nuclear norm of Φ is equal to the trace of Φ, i.e., $\|\Phi\|_* = \text{trace}(\Phi)$. For matrices $A, B \in \mathbb{R}^{N \times n}$, we can define the inner product on $\mathbb{R}^{N \times n} \times \mathbb{R}^{N \times n}$ as $\langle A, B \rangle = \text{trace}(A^T B) = \sum_{i=1}^{N} \sum_{j=1}^{n} A_{i,j} B_{i,j}$. So the Frobenius norm is the norm associated with this inner product. The spectral norm is defined as the induced 2-norm, i.e., for $\Phi \in \mathbb{R}^{N \times n}$,

$$\|\Phi\|_2 = \underset{\theta \neq 0}{\text{maximize}} \frac{\|\Phi\theta\|_2}{\|\theta\|_2} = \underset{\|\theta\|_2=1}{\text{maximize}} \|\Phi\theta\|_2. \tag{3.141}$$

To show that (3.141) is equal to (3.140), note that $\max_{\|\theta\|_2=1} \|\Phi\theta\|_2$ is equivalent to $\max_{\|\theta\|_2^2=1} \|\Phi\theta\|_2^2$, which is further equivalent to

$$\max_{\theta} \|\Phi\theta\|_2^2 + \lambda(1 - \|\theta\|_2^2) = \max_{\theta} \theta^T \Phi^T \Phi\theta + \lambda(1 - \theta^T \theta), \tag{3.142}$$

where λ is the Lagrange multiplier. Checking the optimality condition of (3.142) yields that the optimal solution will satisfy

$$\Phi^T \Phi\theta - \lambda\theta = 0, \quad \theta^T \theta = 1.$$

The above equation implies that λ is an eigenvalue of $\Phi^T \Phi$, and moreover,

$$\theta^T \Phi^T \Phi \theta = \lambda \theta^T \theta = \lambda. \tag{3.143}$$

As a result, we have

$$\max_{\|\theta\|_2=1} \|\Phi\theta\|_2 = (\max_{\|\theta\|_2^2=1} \theta^T \Phi^T \Phi \theta)^{\frac{1}{2}} = (\max_{\|\theta\|_2^2=1} \lambda)^{\frac{1}{2}} = (\lambda_{\max})^{\frac{1}{2}},$$

where $\lambda_{\max}$ is the largest eigenvalue of $\Phi^T \Phi$ that is equal to $\sigma_{\max}^2(\Phi)$. Thus (3.141) is indeed equal to (3.140).

The aforementioned three matrix norms, the nuclear norm, the Frobenius norm and the spectral norm, can be seen as natural extensions of the three vector norms: the ℓ_1, ℓ_2 and ℓ_∞ norms,, respectively. In particular, if we construct an n-dimensional vector with the n singular values of Φ as its elements, then the three matrix norms $\|\Phi\|_*$, $\|\Phi\|_F$ and $\|\Phi\|_2$ correspond to the ℓ_1, ℓ_2 and ℓ_∞ norms of the constructed vector, respectively. Moreover, for any given norm $\| \cdot \|$ on $\mathbb{R}^{N \times n}$, there exists a dual norm $\| \cdot \|_d$ of $\| \cdot \|$ defined as

$$\|A\|_d = \sup\{\text{trace}(A^T B) | B \in \mathbb{R}^{N \times n}, \|B\| \leq 1\}. \tag{3.144}$$

For the vector norms, the dual norm of the ℓ_1 norm is the ℓ_∞ norm and the dual norm of the ℓ_2 norm is the ℓ_2 norm. The properties for the vector norms extend to the matrix norms we have defined: the dual norm of the nuclear norm is the spectral norm, see, e.g., [37], and the dual norm of the Frobenius norm is itself.

3.8.1.3 Matrix Inversion Lemma, Based on [49]

The matrix inversion lemma is also known as Sherman–Morrison–Woodbury formula and refers to the following identity:

$$(A + UCV)^{-1} = A^{-1} - A^{-1}U(C^{-1} + VA^{-1}U)^{-1}VA^{-1}, \tag{3.145}$$

where A and C are square $n \times n$ and $m \times m$ matrices.

3.8.2 Proof of Lemma 3.1

Define $W = -(QR + I_n)^{-1}$ and $W_0 = -(ZR + I_n)^{-1}$. Then (3.67) can be rewritten as

$$W(QRQ + Z)W^T \geq W_0(ZRZ + Z)W_0^T. \tag{3.146}$$

Note that

$$I_n + W = -WQR, \qquad I_n + W_0 = -W_0 ZR \qquad (3.147)$$

thus (3.67) can be further rewritten as

$$(I_n + W)R^{-1}(I_n + W)^T + WZW^T$$
$$\geq (I_n + W_0)R^{-1}(I_n + W_0)^T + W_0 Z W_0^T. \qquad (3.148)$$

In the following, we show that

$$(I_n + W)R^{-1}(I_n + W)^T + WZW^T$$
$$- (I_n + W_0)R^{-1}(I_n + W_0)^T - W_0 Z W_0^T$$
$$= (W - W_0)(R^{-1} + Z)(W - W_0)^T. \qquad (3.149)$$

Simple calculation shows that (3.149) is equivalent to

$$(I_n + W_0)R^{-1}W^T + WR^{-1}(I_n + W_0^T)$$
$$- (I + W_0)R^{-1}W_0^T - W_0 R^{-1}(I_n + W_0^T)$$
$$= 2W_0 Z W_0^T - W_0 Z W^T - WZW_0^T. \qquad (3.150)$$

It follows from the second equation of (3.147) that

$$(I_n + W_0)R^{-1} = -W_0 Z. \qquad (3.151)$$

Now inserting (3.151) into the left-hand side of (3.150) shows that (3.150) and thus (3.149) holds. Moreover, since $(W - W_0)(R^{-1} + Z)(W - W_0)^T$ in (3.149) is positive semidefinite, Eq. (3.148) holds as well, which in turn implies (3.67) holds. This completes the proof.

3.8.3 Derivation of Predicted Residual Error Sum of Squares (PRESS)

For the case when the kth measured output y_k, $k = 1, \ldots, N$, is not used, the corresponding ReLS-Q estimate becomes

$$\hat{\theta}_{-k}^{R} = \left(\sum_{i=1, i \neq k}^{N} \phi_i \phi_i^T + \sigma^2 P^{-1}(\eta) \right)^{-1} \sum_{i=1, i \neq k}^{N} \phi_i y_i. \qquad (3.152)$$

For the kth measured output y_k, $k = 1, \ldots, N$, the corresponding predicted output $\hat{y}_{-k}$ and validation error r_{-k} are

$$\hat{y}_{-k} = \phi_k^T \left(\sum_{i=1, i \neq k}^{N} \phi_i \phi_i^T + \sigma^2 P^{-1}(\eta) \right)^{-1} \sum_{i=1, i \neq k}^{N} \phi_i y_i, \tag{3.153a}$$

$$r_{-k} = y_k - \hat{y}_{-k}. \tag{3.153b}$$

With M defined in (3.81) and by Woodbury matrix identity, e.g., [10, 17], we have

$$\left(\sum_{i=1, i \neq k}^{N} \phi_i \phi_i^T + \sigma^2 P^{-1}(\eta) \right)^{-1} = (M - \phi_k \phi_k^T)^{-1}$$

$$= M^{-1} - \frac{M^{-1} \phi_k \phi_k^T M^{-1}}{-1 + \phi_k^x T M^{-1} \phi_k}. \tag{3.154}$$

Then we have

$$r_{-k} = y_k - \phi_k^T M^{-1} \sum_{i=1, i \neq k}^{N} \phi_i y_i + \phi_k^T \frac{M^{-1} \phi_k \phi_k^T M^{-1}}{-1 + \phi_k^T M^{-1} \phi_k} \sum_{i=1, i \neq k}^{N} \phi_i y_i$$

$$= r_k + \phi_k^T M^{-1} \phi_k y_k + \phi_k^T \frac{M^{-1} \phi_k \phi_k^T M^{-1}}{-1 + \phi_k^T M^{-1} \phi_k} \sum_{i=1, i \neq k}^{N} \phi_i y_i$$

$$- r_k + \phi_k^T M^{-1} \phi_k \left(y_k + \frac{\phi_k^T M^{-1}}{-1 + \phi_k^T M^{-1} \phi_k} \sum_{i=1, i \neq k}^{N} \phi_i y_i \right)$$

$$= r_k + \frac{\phi_k^T M^{-1} \phi_k}{-1 + \phi_k^T M^{-1} \phi_k} \tag{3.155}$$

$$\times \left(-y_k + \phi_k^T M^{-1} \phi_k y_k + \phi_k^T M^{-1} \sum_{i=1, i \neq k}^{N} \phi_i y_i \right)$$

$$= r_k - \frac{\phi_k^T M^{-1} \phi_k}{-1 + \phi_k^T M^{-1} \phi_k} r_k$$

$$= r_k \frac{1}{1 - \phi_k^T M^{-1} \phi_k},$$

which shows that r_{-k} is actually obtained by scaling r_k with a factor $1/(1 - \phi_k^T M^{-1} \phi_k)$. Accordingly, we have the sum of squares of the validation errors

$$\sum_{k=1}^{N} r_{-k}^2 = \sum_{k=1}^{N} \frac{r_k^2}{(1 - \phi_k^T M^{-1} \phi_k)^2}. \tag{3.156}$$

Then the PRESS (3.80) is obtained by minimizing (3.156) with respect to $\eta \in \Gamma$.

3.8.4   Proof of Theorem 3.7

Using (3.92) and (3.100), it is easy to see that proving (3.90) is equivalent to show that

$$\underbrace{\mathscr{E}\left[\frac{1}{N}\|Y - \Phi\hat{\theta}^{R}(\eta)\|_{2}^{2}\right]}_{\overline{\mathrm{err}}(\eta)} \leq \underbrace{\mathscr{E}\left[\frac{1}{N}\mathscr{E}[\|Y_{v} - \Phi\hat{\theta}^{R}(\eta)\|_{2}^{2}|\mathscr{D}_{\mathrm{T}}]\right]}_{\mathrm{EVE}_{\mathrm{in}}(\eta)} \qquad (3.157)$$

and to prove the above inequality we need the following lemma.

Lemma 3.3 *Consider the following additive measurement model:*

$$x = \mu + \varepsilon, \; x, \mu, \varepsilon \in \mathbb{R}^{p}, \qquad (3.158)$$

where μ is an unknown constant vector and ε is a random variable with zero-mean and covariance matrix $\mathscr{E}(\varepsilon\varepsilon^{T}) = \Sigma$. Let $\hat{\mu}(x)$ be an estimator of μ based on the data x and let $\tilde{x}$ be new data generated from

$$\tilde{x} = \mu + \tilde{\varepsilon}, \; \tilde{x} \in \mathbb{R}^{p}, \qquad (3.159)$$

where $\tilde{\varepsilon}$ is a random variable uncorrelated with ε and has zero-mean and covariance matrix $\mathscr{E}(\tilde{\varepsilon}\tilde{\varepsilon}^{T}) = \Sigma$. Then it holds that

$$\mathscr{E}(\|\tilde{x} - \hat{\mu}(x)\|_{2}^{2}) = \mathscr{E}(\|\mu - \hat{\mu}(x)\|_{2}^{2}) + \mathrm{trace}(\Sigma) \qquad (3.160)$$

$$= \mathscr{E}(\|x - \hat{\mu}(x)\|_{2}^{2}) + 2\,\mathrm{trace}(\mathrm{Cov}(\hat{\mu}(x), x)), \qquad (3.161)$$

where the expectation is over both ε and $\tilde{\varepsilon}$.

Proof Firstly, we consider (3.160). We have

$$\mathscr{E}(\|\tilde{x} - \hat{\mu}(x)\|_{2}^{2}) = \mathscr{E}(\|\tilde{x} - \mu + \mu - \hat{\mu}(x)\|_{2}^{2})$$
$$= \mathscr{E}(\|\mu - \hat{\mu}(x)\|_{2}^{2}) + \mathscr{E}(\|\tilde{x} - \mu\|_{2}^{2}) + 2\mathscr{E}[(\tilde{x} - \mu)^{T}(\mu - \hat{\mu}(x))]$$
$$= \mathscr{E}(\|\mu - \hat{\mu}(x)\|_{2}^{2}) + \mathscr{E}(\|\tilde{\varepsilon}\|_{2}^{2}),$$

which shows that (3.160) is true.

Secondly, we consider (3.161). Similarly, we have

$$\mathscr{E}(\|\tilde{x} - \hat{\mu}(x)\|_{2}^{2}) = \mathscr{E}(\|\tilde{x} - x + x - \hat{\mu}(x)\|_{2}^{2})$$
$$= \mathscr{E}(\|x - \hat{\mu}(x)\|_{2}^{2}) + \mathscr{E}(\|\tilde{x} - x\|_{2}^{2}) + 2\mathscr{E}[(\tilde{x} - x)^{T}(x - \hat{\mu}(x))]$$
$$= \mathscr{E}(\|x - \hat{\mu}(x)\|_{2}^{2}) + \mathscr{E}(\|\tilde{\varepsilon} - \varepsilon\|_{2}^{2}) + 2\mathscr{E}[(\tilde{\varepsilon} - \varepsilon)^{T}(\varepsilon + \mu - \hat{\mu}(x))]$$
$$= \mathscr{E}(\|x - \hat{\mu}(x)\|_{2}^{2}) + 2\,\mathrm{trace}(\Sigma) - 2\mathscr{E}[\varepsilon^{T}(\varepsilon + \mu - \hat{\mu}(x))]$$
$$= \mathscr{E}(\|x - \hat{\mu}(x)\|_{2}^{2}) + 2\mathscr{E}[\varepsilon^{T}\hat{\mu}(x)],$$

which implies that (3.161) is true.   □

Now we prove (3.157) by applying Lemma 3.3. Let

$$x = Y, \mu = \Phi\theta_0, \hat{\mu}(x) = \Phi\hat{\theta}^R, \tilde{x} = Y_v, \varepsilon = E, \tilde{\varepsilon} = E_{\text{test}}, \Sigma = \sigma^2 I_N, \quad (3.162)$$

and then it follows from (3.161) that

$$\underbrace{\mathcal{E}\left[\frac{1}{N}\mathcal{E}[\|Y_v - \Phi\hat{\theta}^R(\eta)\|_2^2|\mathscr{D}_{\text{T}}]\right]}_{\text{EVE}_{\text{in}}(\eta)} - \underbrace{\mathcal{E}\left[\frac{1}{N}\|Y - \Phi\hat{\theta}^R(\eta)\|_2^2\right]}_{\overline{\text{err}}(\eta)}$$

$$= 2\frac{1}{N}\,\text{trace}(\text{Cov}(Y, \Phi\hat{\theta}^R(\eta))). \quad (3.163)$$

Next we show that the right-hand side of (3.163) is nonnegative. For the ReLS-Q problem (3.58a) with the ReLS-Q estimate (3.58b), the predicted output $\hat{Y}(\eta)$ of Y is

$$\hat{Y}(\eta) = \Phi\hat{\theta}^R(\eta) = \Phi P\Phi^T(\Phi P\Phi^T + \sigma^2 I_N)^{-1}Y.$$

Then we have

$$\text{Cov}(Y, \Phi\hat{\theta}^R(\eta)) = \text{Cov}(Y, \hat{Y}(\eta))$$
$$= \mathcal{E}(Y - \mathcal{E}(Y))(\hat{Y} - \mathcal{E}(\hat{Y}(\eta)))^T$$
$$= \mathcal{E}(Y - \mathcal{E}(Y))(Y - \mathcal{E}(Y))^T \Phi P\Phi^T(\Phi P\Phi^T + \sigma^2 I_N)^{-1}$$
$$= \sigma^2 \Phi P\Phi^T(\Phi P\Phi^T + \sigma^2 I_N)^{-1} = \sigma^2 H, \quad (3.164)$$

where H is the hat matrix defined in (3.63). One has

$$\text{trace}(\text{Cov}(Y, \Phi\hat{\theta}^R(\eta))) = \sigma^2\,\text{trace}(H) \geq 0.$$

Therefore, the right-hand side of (3.163) is nonnegative and thus (3.90) holds true completing the proof of Theorem 3.7.

3.8.5 A Variant of the Expected In-Sample Validation Error and Its Unbiased Estimator

It is possible to derive variants of the expected in-sample validation error and its unbiased estimator by modifying (3.92) and (3.100).

Assume that Φ is full rank, i.e., rank(Φ) = n. Then, multiplying both sides of (3.92) and (3.100) with $(\Phi^T\Phi)^{-1}\Phi^T$ yields

$$(\Phi^T\Phi)^{-1}\Phi^T Y = \theta_0 + (\Phi^T\Phi)^{-1}\Phi^T E, \quad (3.165)$$
$$(\Phi^T\Phi)^{-1}\Phi^T Y_v = \theta_0 + (\Phi^T\Phi)^{-1}\Phi^T E_v, \quad (3.166)$$

which will be our new "true system" and new "validation data", respectively.

Different from (3.162), we now take

$$x = (\Phi^T \Phi)^{-1} \Phi^T Y, \mu = \theta_0, \hat{\mu}(x) = \hat{\theta}^R(\eta), \tilde{x} = (\Phi^T \Phi)^{-1} \Phi^T Y_v,$$
$$\varepsilon = (\Phi^T \Phi)^{-1} \Phi^T E, \tilde{\varepsilon} = (\Phi^T \Phi)^{-1} \Phi^T E_v, \Sigma = \sigma^2 (\Phi^T \Phi)^{-1}. \qquad (3.167)$$

Note that $\hat{\theta}^{LS} = (\Phi^T \Phi)^{-1} \Phi^T Y$ and then it follows from (3.160) and (3.161) that

$$\mathscr{E}(\|(\Phi^T \Phi)^{-1} \Phi^T Y_v - \hat{\theta}^R(\eta)\|_2^2) = \mathscr{E}(\|\hat{\theta}^R(\eta) - \theta_0\|_2^2) + \sigma^2 \operatorname{trace}((\Phi^T \Phi)^{-1})$$
$$= \mathscr{E}(\|\hat{\theta}^{LS} - \hat{\theta}^R(\eta)\|_2^2) + 2 \operatorname{trace}(\operatorname{Cov}(\hat{\theta}^R(\eta), \hat{\theta}^{LS})).$$

From the above two equations, we have

$$\mathscr{E}(\|\hat{\theta}^R(\eta) - \theta_0\|_2^2) = \mathscr{E}(\|\hat{\theta}^{LS} - \hat{\theta}^R(\eta)\|_2^2)$$
$$+ 2 \operatorname{trace}(\operatorname{Cov}(\hat{\theta}^R(\eta), \hat{\theta}^{LS})) - \sigma^2 \operatorname{trace}((\Phi^T \Phi)^{-1}).$$

Further note that

$$\hat{\theta}^R(\eta) = (\Phi^T \Phi + \sigma^2 P^{-1}(\eta))^{-1} \Phi^T Y = (\Phi^T \Phi + \sigma^2 P^{-1}(\eta))^{-1} \Phi^T \Phi \hat{\theta}^{LS},$$
$$\operatorname{Cov}(\hat{\theta}^{LS}, \hat{\theta}^{LS}) = \sigma^2 (\Phi^T \Phi)^{-1},$$

then we have

$$\mathscr{E}(\|\hat{\theta}^R(\eta) - \theta_0\|_2^2) = \mathscr{E}(\|\hat{\theta}^{LS} - \hat{\theta}^R(\eta)\|_2^2)$$
$$+ 2\sigma^2 \operatorname{trace}((\Phi^T \Phi + \sigma^2 P^{-1}(\eta))^{-1} - 0.5(\Phi^T \Phi)^{-1}).$$
$$(3.168)$$

Note that $\mathscr{E}(\|\hat{\theta}^R(\eta) - \theta_0\|_2^2)$ is equal to $\operatorname{trace}(\operatorname{MSE}(\hat{\theta}^R(\eta), \theta_0))$, then we denote it by mse_η and we readily obtain an unbiased estimator of mse_η as follows:

$$\widehat{\operatorname{mse}_\eta} = \|\hat{\theta}^{LS} - \hat{\theta}^R(\eta)\|_2^2 + 2\sigma^2 \operatorname{trace}((\Phi^T \Phi + \sigma^2 P^{-1}(\eta))^{-1} - 0.5(\Phi^T \Phi)^{-1}).$$
$$(3.169)$$

Now given the training data (3.84), the corresponding estimate $\widehat{\operatorname{mse}_\eta}$ of mse_η can be used to estimate the hyperparameter η: we should take the value of $\eta \in \Gamma$ that minimizes (3.169), i.e.,

$$\hat{\eta} = \arg \min_{\eta \in \Gamma} \|\hat{\theta}^{LS} - \hat{\theta}^R(\eta)\|_2^2 + 2\sigma^2 \operatorname{trace}((\Phi^T \Phi + \sigma^2 P^{-1}(\eta))^{-1} - 0.5(\Phi^T \Phi)^{-1}).$$
$$(3.170)$$

The criterion (3.170) is known as the SURE of the expected in-sample validation error for the true system (3.165) and the validation data (3.166), e.g., [33, 40].

References

1. Akaike H (1974) A new look at the statistical model identification. IEEE Trans Autom Control AC–19:716–723
2. Allen DM (1974) The relationship between variable selection and data augmentation and a method for prediction. Technometrics 16(1):125–127
3. Andersen MS, Dahl J, Vandenberghe L (2012) CVXOPT: a Python package for convex optimization, version 1.1.5. http://abel.ee.ucla.edu/cvxopt
4. Bishop CM (2006) Pattern recognition and machine learning. Springer, New York
5. Candès EJ, Recht B (2009) Exact matrix completion via convex optimization. Found Comput Math 9(6):717
6. Candès EJ, Tao T (2010) The power of convex relaxation: near-optimal matrix completion. IEEE Trans Inf Theory 56(5):2053–2080
7. Chen T (2018) On kernel design for regularized LTI system identification. Automatica 90:109–122
8. Chen T, Ljung L (2013) Implementation of algorithms for tuning parameters in regularized least squares problems in system identification. Automatica 49:2213–2220
9. Chen T, Ohlsson H, Ljung L (2012) On the estimation of transfer functions, regularizations and Gaussian processes - revisited. Automatica 48:1525–1535
10. Demmel JW (1997) Applied numerical linear algebra. SIAM, Philadelphia
11. Diamond S, Boyd S (2016) CVXPY: a Python-embedded modeling language for convex optimization. J Mach Learn Res 17:1–5
12. Draper NR, Smith H (1981) Applied regression analysis, 2nd edn. Wiley, New York
13. Fazel M (2002) Matrix rank minimization with applications. PhD thesis, Department of Electrical Engineering, Stanford University
14. Fazel M, Hindi H, Boyd SP (2001) A rank minimization heuristic with application to minimum order system approximation. In: Proceedings of the 2001 American control conference, pp 4734–4739
15. Fazel M, Hindi H, Boyd SP (2003) Log-det heuristic for matrix rank minimization with applications to Hankel and Euclidean distance matrices. In: Proceedings of the 2003 American control conference, vol 3, pp 2156–2162
16. Friedman J, Hastie T, Tibshirani R (2008) Sparse inverse covariance estimation with the graphical Lasso. Biostatistics 9(3):432–441
17. Golub GH, Van Loan CF (2013) Matrix computations, 4th edn. The Johns Hopkins University Press, Baltimore
18. Golub GH, Heath M, Wahba G (1979) Generalized cross-validation as a method for choosing a good ridge parameter. Technometrics 21(2):215–223
19. Grant M, Boyd S, Ye Y (2009) MATLAB software for disciplined convex programming
20. Grenander U, Szegö G (1956) Toeplitz forms and their applications, vol 321. University of California Press
21. Hastie T, Tibshirani R, Friedman J (2001) The elements of statistical learning. Springer, Berlin
22. Hiriart-Urruty JB, Lemaréchal C (1993) Convex analysis and minimization algorithms II: advanced theory and bundle methods
23. Huber PJ (1981) Robust statistics. Wiley, New York
24. Kohavi R (1995) A study of cross-validation and bootstrap for accuracy estimation and model selection. In: Proceedings of the 14th international joint conference on artificial intelligence, San Francisco, CA, USA, pp 1137–1143
25. Ljung L (1999) System identification - theory for the user, 2nd edn. Prentice-Hall, Upper Saddle River

26. Lofberg J (2004) YALMIP: a toolbox for modeling and optimization in MATLAB. In: 2004 IEEE international symposium on computer aided control systems design, pp 284–289
27. Mallows CL (1973) Some comments on CP. Technometrics 15(4):661–675
28. Meinshausen N, Buhlmann P (2006) High-dimensional graphs and variable selection with the Lasso. Ann Stat 34(3):1436–1462
29. Mu B, Chen T, Ljung L (2018) On asymptotic properties of hyperparameter estimators for kernel-based regularization methods. Automatica 94:381–395
30. Mu B, Chen T, Ljung L (2018) Asymptotic properties of hyperparameter estimators by using cross-validations for regularized system identification. In: Proceedings of the 57th IEEE conference on decision and control, pp 644–649
31. Natarajan BK (1995) Sparse approximate solutions to linear systems. SIAM J Comput 24(2):227–234
32. Park T, Casella G (2008) The Bayesian Lasso. J Am Stat Assoc 103(482):681–686
33. Pillonetto G, Chiuso A (2015) Tuning complexity in regularized kernel-based regression and linear system identification: the robustness of the marginal likelihood estimator. Automatica 58:106–117
34. Rao BD, Engan K, Cotter SF, Palmer J, Kreutz-Delgado K (2003) Subset selection in noise based on diversity measure minimization. IEEE Trans Signal Process 51(3):760–770
35. Rao CR (1973) Linear statistical inference and its applications. Wiley, New York
36. Rasmussen CE, Williams CKI (2006) Gaussian processes for machine learning. The MIT Press, Cambridge
37. Recht B, Fazel M, Parrilo P (2010) Guaranteed minimum-rank solutions of linear matrix equations via nuclear norm minimization. SIAM Rev 52(3):471–501
38. Richard E, Savalle P, Vayatis N (2012) Estimation of simultaneously sparse and low rank matrices. In: The 29th international conference on machine learning (ICML)
39. Rissanen J (1978) Modelling by shortest data description. Automatica 14:465–471
40. Stein C (1981) Estimation of the mean of a multivariate normal distribution. Ann Stat 9:1135–1151
41. Stone M (1974) Cross-validatory choice and assessment of statistical predictions. J R Stat Soc Ser B Stat Methodol 111–147
42. Tibshirani R (1996) Regression shrinkage and selection via the Lasso. J R Stat Soc Ser B Stat Methodol 58:267–288
43. Tibshirani R, Taylor J (2012) Degrees of freedom in Lasso problems. Ann Stat 40(2):1198–1232
44. Toeplitz O (1911) Zur theorie der quadratischen und bilinearen formen von unendlichvielen veränderlichen. Math Ann 70(3):351–376
45. Vapnik V (1998) Statistical learning theory. Wiley, New York
46. Wahba G (1990) Spline models for observational data. SIAM, Philadelphia
47. Wahba G (1999) Support vector machines, reproducing kernel Hilbert spaces, and the randomized GACV. In: Scholkopf B, Burges C, Smola A (eds) Advances in kernel methods - support vector learning. MIT Press, Cambridge, pp 69–88
48. Wolter KM (2007) Introduction to variance estimation, 2nd edn. Springer, Berlin
49. Woodbury MA (1950) Inverting modified matrices. Memorandum Rept. 42. Princeton University, Princeton
50. Zou H (2006) The adaptive Lasso and its oracle properties. J Am Stat Assoc 101(476):1418–1429
51. Zou H, Hastie T, Tibshirani R (2007) On the degrees of freedom of the Lasso. Ann Stat 35(5):2173–2192

Chapter 4
Bayesian Interpretation
of Regularization

Abstract In the previous chapter, it has been shown that the regularization approach is particularly useful when information contained in the data is not sufficient to obtain a precise estimate of the unknown parameter vector and standard methods, such as least squares, yield poor solutions. The fact itself that an estimate is regarded as poor suggests the existence of some form of prior knowledge on the degree of acceptability of candidate solutions. It is this knowledge that guides the choice of the regularization penalty that is added as a corrective term to the usual sum of squared residuals. In the previous chapters, this design process has been described in a deterministic setting where only the measurement noises are random. In this chapter, we will see that an alternative formalization of prior information is obtained if a subjective/Bayesian estimation paradigm is adopted. The major difference is that the parameters, rather than being regarded as deterministic, are now treated as a random vector. This stochastic setting permits the definition of new powerful tools for both priors selection, e.g., through the maximum entropy principle, and for regularization parameters tuning, e.g., through the empirical Bayes approach and its connection with the concept of equivalent degrees of freedom.

4.1 Preliminaries

We have seen that the regularization approach can be used to effectively solve estimation problems that are otherwise ill-conditioned. In particular, a penalty is added as a corrective term to the usual sum of squared residuals. In this way, between two candidate solutions achieving the same squared loss, the regularizer is chosen such as to penalize candidate solutions that depart from our prior knowledge on some features of the unknown parameter vector.

It is worth noting that the regularization approach lies within a frequentist paradigm in which the observed data, affected by noise, are random variables, but the unknown parameter vector is deterministic in nature. For linear-in-parameter

G. Pillonetto et al., *Regularized System Identification*, Communications and Control Engineering, https://doi.org/10.1007/978-3-030-95860-2_4

models, regularization yields an estimate that, though biased, may be preferable to the unbiased least squares estimate in view of the smaller variance. In particular, the tuning of the regularization parameter aims at an advantageous solution of the bias-variance dilemma. By trading an excessive variance for some bias, a smaller mean squared error may be achieved, as exemplified by the James–Stein estimator. An alternative formalization of prior information is obtained if a subjective/Bayesian estimation paradigm is adopted. The major difference is that the parameters, rather than being regarded as deterministic, are now treated as a random vector.

In order to introduce the Bayesian paradigm, it can be useful to start with a simple example in which the parameters do depend on the result of a random experiment. Consider a metabolism model for which the parameter vector θ can take only two possible values, θ_h and θ_d, associated with healthy and diabetic patients, respectively. The model specifies $p(Y|\theta)$, where Y are observations collected from a randomly chosen patient with 90% probability of being healthy and 10% probability of being diabetic. In this simple case, model identification amounts to deciding between θ_h and θ_d. It is also clear that θ is a discrete random variable with $p(\theta = \theta_h) = 0.9$ and $p(\theta = \theta_d) = 0.1$. These probabilities summarize the prior information about the unknown parameter, before any observation is collected. Once the data Y become available, the Bayes formula can be used to compute the posterior probability

$$p(\theta_h|Y) = \frac{p(Y|\theta_h)p(\theta_h)}{p(Y)} = \frac{p(Y|\theta_h)p(\theta_h)}{p(Y|\theta_h)p(\theta_h) + p(Y|\theta_d)p(\theta_d)}. \tag{4.1}$$

Of course, $p(\theta_d|Y) = 1 - p(\theta_h|Y)$. In particular, if the data Y are consistent with diabetes symptoms, it may well happen that $p(\theta_d|Y) > 0.5$, in which case $\theta = \theta_d$ would be taken as the final estimate.

In the previous example, the prior probability distribution assigned to θ reflects a real experiment that is the random choice of a patient from a population where 90% of subjects are healthy, which implies a prejudice in favour of $\theta = \theta_h$. In other words, the prior distribution ranks the candidate parameters according to the available a priori knowledge. If we look at the numerator of (4.1), we see that it combines a priori information with the data through the product of the *prior probability* $p(\theta_h)$ and the *likelihood* $p(Y|\theta_h)$. In the example, the population was a binary one (either healthy or diabetic), but we can imagine more complex populations allowing for several countable or even uncountable possible values of θ.

In the actual Bayesian paradigm a further step is made: the parameters θ are assigned a prior probability $p(\theta_h)$, even if there does not exist an underlying experiment that draws the model from a population of possible models. According to the subjective definition of probability, $p(\theta = \bar{\theta})$ represents the (subjective) degree of belief that θ is going to take the value $\bar{\theta}$. In particular, in analogy with the regularization penalty, it is possible to rank the possible values of θ, assigning a low probability to values whose occurrence is deemed unlikely. In our context, the intrinsically subjective nature of the prior probability, a controversial issue in the confrontation between the frequentist and Bayesian paradigms, is specular to the subjective choice of the regularization penalty: rather than expressing the preference for some solutions through the choice of a proper penalty, the preference is formulated by means a prior distribution.

As shown in the following, many formulas and results can be indifferently derived adopting either the regularization or the Bayesian paradigm. However, the Bayesian approach has its pros. In particular, the tuning of the regularization parameter, rather than being addressed on an ad hoc basis, can be formulated as a statistical estimation problem. Moreover, the Bayesian paradigm offers a very natural way to asses uncertainty intervals, whereas the regularization paradigm has a harder time assessing the amount of bias in the estimate. Among the cons, one may mention the need for a deeper probabilistic background in order to gain a full comprehension of all aspects.

Throughout the chapter we will mainly focus on the linear Gaussian case, but the approach is more general and some hints at generalizations will be provided. In addition, we will use θ to denote the stochastic vector that has generated the data, in contrast with the deterministic θ_0 used in the classical setting discussed in the previous chapter.

4.2 Incorporating Prior Knowledge via Bayesian Estimation

We consider the problem of estimating a parameter vector $\theta \in \mathbb{R}^n$, based on the observation vector $Y \in \mathbb{R}^N$. The two ingredients of Bayesian estimation are the prior distribution of θ, also known by short as *prior*, and the conditional distribution of Y given θ. As already observed, the basic assumption is that the parameter vector θ is not completely unknown, but rather some prior knowledge is available that is formulated in terms of *subjective probability*, specified as a probability density function:

$$\mathrm{p}(\theta) : \mathbb{R}^n \mapsto \mathbb{R}.$$

The density function $\mathrm{p}(\theta)$ is chosen by the user so as to assign a low probability to values whose occurrence is deemed unlikely. For instance, if θ is a scalar parameter whose value is believed to lie more or less around 30, hardly smaller than 20 and hardly larger than 40, this prior knowledge can be embedded in a Gaussian density with $\mathscr{E}\theta = \mu_\theta = 30$ and standard deviation $\sigma_\theta = 5$:

$$\theta \sim \mathscr{N}(30, 25).$$

In fact, under this distribution, $\mathrm{p}(|\theta - \mu_\theta| > 2\sigma_\theta) = \mathrm{p}(|\theta - 30| > 10) < 0.05$. Although not impossible, it is considered unlikely that values of θ too distant from 30 are going to occur. A natural question is how and when our prior knowledge is sufficient to specify a distribution. This crucial issue calls for the notion and role of *hyperparameters*, see Sect. 4.2.4, and for the possible use of the *maximum entropy principle* as a way to obtain an entire probability distribution from partial knowledge relative to its moments, see Sect. 4.6.

The second ingredient is the conditional distribution of Y given θ that, when considered as a function of θ, is also known as *likelihood*:

$$L(\theta|Y) = p(Y|\theta) = \frac{p(Y, \theta)}{p(\theta)},$$

where $p(Y, \theta)$ is the joint probability distribution of the random vectors Y and θ. The likelihood is usually obtained from some mathematical model of the data. Consider, for instance, the simple model

$$Y_i = \theta\sqrt{i} + e_i, \qquad i = 1, \ldots, N,$$

where $e_i \sim \mathcal{N}(0, \sigma^2)$ are independent and identically distributed measurement errors, with known variance σ^2. Conditional on θ, i.e., assuming that θ is known, Y_i is Gaussian with

$$\mathscr{E}[Y_i|\theta] = \theta\sqrt{i}, \qquad \mathrm{Var}(Y_i|\theta) = \sigma^2$$

so that, in view of independence, the likelihood is

$$L(\theta|Y) = p(Y|\theta) = \prod_{i=1}^{N} p(Y_i|\theta), \qquad p(Y_i|\theta) = \mathcal{N}(\theta\sqrt{i}, \sigma^2).$$

When both the prior distribution $p(\theta)$ and the likelihood $p(Y|\theta)$ have been specified, the Bayes formula yields the *posterior distribution*

$$p(\theta|Y) = \frac{p(Y|\theta)p(\theta)}{p(Y)}.$$

We have seen that all our prior knowledge was embedded in the prior. In a similar way, all the knowledge obtained by the combination of prior information with the new information brought by the observations is now embedded in the posterior distribution $p(\theta|Y)$, denoted by short as *posterior*.

Although all the relevant information is encapsulated within the posterior, a *point estimate* is often required for practical or communication purposes. The *Maximum A Posteriori (MAP) estimate* is the value that maximizes the posterior:

$$\theta^{\mathrm{MAP}} = \arg\max_{\theta} p(Y|\theta). \tag{4.2}$$

Its interpretation is simple, as it represents the most likely value, once the prior knowledge has been updated taking into account the observations. Alternatively, the mean squared error

$$\text{MSE}(\hat{\theta}) = \mathscr{E}\left[\left(\hat{\theta} - \theta\right)^2 | Y\right]$$

can be used as a criterion to select the point estimate $\hat{\theta}$. Above, $\mathscr{E}(\cdot|Y)$ denotes the expected value taken with respect to the posterior distribution $p(\theta|Y)$. The following classical result from estimation theory (whose proof is in Sect. 4.13.1) then holds.

Theorem 4.1 *The minimizer of the MSE*

$$\theta^{\text{B}} = \arg\min_{\hat{\theta}} \text{MSE}(\hat{\theta})$$

is known as Bayes estimate *and can be shown to be equal to the conditional mean:*

$$\theta^{\text{B}} = \mathscr{E}[\theta|Y].$$

A third point estimate is the conditional median used especially in view of its statistical robustness when the posterior is obtained numerically via stochastic simulation algorithms, see Sect. 4.10.

When, in addition to a point estimate, an assessment of the uncertainty is needed, it can be derived from the posterior through the computation of a properly defined *credible region* $C_\gamma \in \mathbb{R}^n$ such that

$$\Pr(\theta \in C_\gamma|Y) = \gamma. \tag{4.3}$$

For example, C_γ could be taken as the smallest region such that (4.3) holds, a choice that goes under the name of highest posterior density region.

4.2.1 Multivariate Gaussian Variables

In this subsection, some basic properties and definitions of multivariate Gaussian variables are recalled. This review is instrumental to the derivation of the Bayesian estimator when observations and parameters are jointly Gaussian, see Sect. 4.2.2. In turn, this will pave the way to the analysis of the linear model under additive Gaussian measurement errors, see Sect. 4.2.3.

A random vector $Z = [Z_1 \ldots Z_m]^T$ is said to be distributed according to a nondegenerate m-variate Gaussian distribution if its joint probability density function is of the type

$$p(z_1, \ldots, z_m) = \frac{1}{\sqrt{(2\pi)^m \det V}} \exp^{-\frac{1}{2}(z-\mu)^T V^{-1}(z-\mu)}, \tag{4.4}$$

where V is a symmetric positive definite matrix and μ is some vector in $\mathbb{R}^m$.

It can be shown that

$$\mathscr{E}(Z) = \mu, \quad \text{Var}(Z) = V.$$

Then, the notation

$$Z \sim \mathcal{N}(\mu, V)$$

(already used before in the scalar case) indicates that Z is a multivariate Gaussian (Normal) random vector with mean μ and variance matrix V.

Property 4.1 *If $Z \sim \mathcal{N}(\mu, V)$ and $Y = AZ$, where $A \in \mathbb{R}^{n \times m}, n \leq m$, is a full-rank deterministic matrix, then*

$$Y \sim \mathcal{N}(A\mu, AVA^T).$$

In particular, it follows that the marginal distributions of the entries of Z are Gaussian:

$$Z_i \sim \mathcal{N}(\mu_i, V_{ii}).$$

Property 4.2 *Assuming $Z \sim \mathcal{N}(\mu, V)$, let $X = [Z_1 \ldots Z_n]^T, Y = [Z_{n+1} \ldots Z_m]^T$, where $1 \leq n < m$, and partition μ and V accordingly:*

$$\mu = \begin{bmatrix} \mu_X \\ \mu_Y \end{bmatrix}, \quad \begin{bmatrix} V_{XX} & V_{XY} \\ V_{YX} & V_{YY} \end{bmatrix}.$$

Then, $\mathrm{p}(X|Y = y)$ is a multivariate Gaussian density function with

$$\mathscr{E}(X|Y = y) = \mu_X + V_{XY} V_{YY}^{-1}(y - \mu_Y)$$
$$\mathrm{Var}(X|Y = y) = V_{XX} - V_{XY} V_{YY}^{-1} V_{YX}$$

and we can write

$$(X|Y = y) \sim \mathcal{N}\left(\mu_X + V_{XY} V_{YY}^{-1}(y - \mu_Y), V_{XX} - V_{XY} V_{YY}^{-1} V_{YX}\right),$$

where $X|Y = y$ stands for the random vector X conditional on $Y = y$.

4.2.2   The Gaussian Case

Let us consider the case in which the observation vector $Y \in \mathbb{R}^N$ and the unknown vector $\theta \in \mathbb{R}^n$ are jointly Gaussian:

$$\begin{bmatrix} \theta \\ Y \end{bmatrix} \sim \mathcal{N}\left(\begin{bmatrix} \mu_\theta \\ \mu_Y \end{bmatrix}, \begin{bmatrix} \Sigma_\theta & \Sigma_{\theta Y} \\ \Sigma_{Y\theta} & \Sigma_Y \end{bmatrix}\right), \quad \Sigma_Y > 0. \tag{4.5}$$

The key idea behind Bayesian estimation is referring to the posterior distribution of θ given Y as representative of the state of knowledge about the unknown vector. It follows from Property 4.2 that such posterior is Gaussian as well:

$$\theta|Y \sim \mathcal{N}\left(\mu_\theta + \Sigma_{\theta Y}\Sigma_Y^{-1}(Y - \mu_Y), \Sigma_\theta - \Sigma_{\theta Y}\Sigma_Y^{-1}\Sigma_{Y\theta}\right). \qquad (4.6)$$

In view of Gaussianity, θ^{MAP} coincides with the conditional expectation $\mathscr{E}(\theta|Y)$:

$$\theta^{\mathrm{B}} = \theta^{\mathrm{MAP}} = \mathscr{E}(\theta|Y) = \mu_\theta + \Sigma_{\theta Y}\Sigma_Y^{-1}(Y - \mu_Y). \qquad (4.7)$$

The reliability of the estimate can be assessed by the posterior variance

$$\Sigma_{\theta|Y} = \mathrm{Var}(\theta|Y) = \Sigma_\theta - \Sigma_{\theta Y}\Sigma_Y^{-1}\Sigma_{Y\theta}$$

based on which the so-called credible intervals can be derived as explained below.

The posterior variance of θ_i is the i-th diagonal entry of the posterior covariance matrix:

$$\sigma_{\theta_i|Y}^2 = \left[\Sigma_{\theta|Y}\right]_{ii}.$$

Observing that $\theta_i|Y \sim \mathcal{N}(\theta_i^{\mathrm{B}}, \sigma_{\theta_i|Y}^2)$, it follows that

$$\mathrm{Pr}\left(\theta_i^{\mathrm{B}} - 1.96\sigma_{\theta_i|Y} \le \theta_i \le \theta_i^{\mathrm{B}} - 1.96\sigma_{\theta_i|Y}|Y\right) = 0.95 \qquad (4.8)$$

so that $[\theta_i^{\mathrm{B}} - 1.96\sigma_{\theta_i|Y}, \theta_i^{\mathrm{B}} + 1.96\sigma_{\theta_i|Y}]$ is the 95%-credible interval for the parameter θ_i, given the observation vector Y. If two or more parameters are jointly considered, the notion of credible region can be obtained in a similar way. In the Gaussian case, such regions are suitable (hyper)-ellipsoids centred in θ^{B}.

4.2.3 The Linear Gaussian Model

The Bayesian approach can be applied to the estimation of the standard linear model in matrix form

$$Y = \Phi\theta + E, \quad E \sim \mathcal{N}(0, \Sigma_E), \quad \Sigma_E > 0 \qquad (4.9)$$

in which $Y \in \mathbb{R}^N$ and the parameter vector θ is no more regarded as a deterministic quantity, but as a random vector independent of E. In particular, we assume that some prior information is available which is embedded in a Gaussian prior distribution

$$\theta \sim \mathcal{N}(\mu_\theta, \Sigma_\theta), \quad \Sigma_\theta > 0.$$

Since Y is the linear combination of the jointly Gaussian vectors θ and E, the vectors Y and θ are jointly Gaussian as well. Hereafter, positive definiteness of Σ_θ is assumed if not stated otherwise. The singular case, see Remark 4.1, amounts to assuming perfect knowledge of some linear combination of the unknown parameters or, equivalently, to constrain the estimated vector θ to belong to a prescribed subspace. The ability to incorporate this type of constraint is not unique to the Bayesian approach. In the context of the deterministic regularization, an example is given by the optimal regularization matrix $P = \theta_0 \theta_0^T$, derived in Sect. 3.4.2.1.

In order to obtain the Bayes estimate according to (4.7), we need to compute $\mu_Y = \mathscr{E}(Y)$, $\Sigma_{\theta Y} = \text{Cov}(\theta, Y)$, and $\Sigma_Y = \text{Var}(Y)$:

$$\mu_Y = \mathscr{E}(Y) = \Phi \mu_\theta$$
$$\text{Var}(Y) = \text{Var}(\Phi\theta) + \text{Var}(E) = \Phi \Sigma_\theta \Phi^T + \Sigma_E$$
$$\text{Cov}(\theta, Y) = \text{Cov}(\theta, \Phi\theta) + \text{Cov}(\theta, E) = \Sigma_\theta \Phi^T.$$

Then, we can apply (4.7) to obtain

$$\theta^B = \mu_\theta + \Sigma_\theta \Phi^T (\Phi \Sigma_\theta \Phi^T + \Sigma_E)^{-1}(Y - \Phi\mu_\theta) \tag{4.10}$$
$$\text{Var}(\theta|Y) = \Sigma_\theta - \Sigma_\theta \Phi^T (\Phi \Sigma_\theta \Phi^T + \Sigma_E)^{-1} \Phi \Sigma_\theta. \tag{4.11}$$

The proofs of the following two classical results are reported in Sects. 4.13.2 and 4.13.3.

Theorem 4.2 (Orthogonality property)

$$\mathscr{E}\left[(\theta^B - \theta)Y^T\right] = 0. \tag{4.12}$$

The following lemma, whose proof is in Sect. 4.13.3, is useful in order to obtain an alternative expression that proves more convenient, especially when $n \ll N$.

Lemma 4.1 *It holds that*

$$\Sigma_\theta \Phi^T (\Phi \Sigma_\theta \Phi^T + \Sigma_E)^{-1} = (\Phi^T \Sigma_E^{-1} \Phi + \Sigma_\theta^{-1})^{-1} \Phi^T \Sigma_E^{-1}.$$

By applying the previous lemma, the alternative expression of the Bayes estimate is obtained

$$\theta^B = (\Phi^T \Sigma_E^{-1} \Phi + \Sigma_\theta^{-1})^{-1}(\Phi^T \Sigma_E^{-1} Y + \Sigma_\theta^{-1} \mu_\theta) \tag{4.13}$$
$$\text{Var}(\theta|Y) = (\Phi^T \Sigma_E^{-1} \Phi + \Sigma_\theta^{-1})^{-1}. \tag{4.14}$$

As already noted, the Bayes estimate coincides with θ^{MAP}, the maximum of the posterior density:

$$p(\theta|Y) \propto p(Y|\theta)p(\theta).$$

Recall that, in view of the assumed linear model (4.9),

$$Y|\theta \sim \mathscr{N}(\Phi\theta, \Sigma_E)$$

and note that

$$\log p(\theta) = c_1 - \frac{1}{2}(\theta - \mu_\theta)^T \Sigma_\theta^{-1}(\theta - \mu_\theta) \tag{4.15}$$

$$\log p(Y|\theta) = c_2 - \frac{1}{2}(Y - \Phi\theta)^T \Sigma_E^{-1}(Y - \Phi\theta), \tag{4.16}$$

where c_1 and c_2 are constants we are not concerned with. Therefore, the maximization of the posterior density can be written as

$$
\begin{aligned}
\theta^{\text{MAP}} &= \arg\max_\theta \log p(Y|\theta) + \log p(\theta) \\
&= \arg\max_\theta (Y - \Phi\theta)^T \Sigma_E^{-1}(Y - \Phi\theta) + (\theta - \mu_\theta)^T \Sigma_\theta^{-1}(\theta - \mu_\theta)
\end{aligned}
$$

whose solution is easily shown to be given by (4.13). This shows that, under Gaussianity assumptions, the Bayes estimate of the linear model can be seen as a regularized least squares estimator with quadratic regularization term (ReLS-Q), see Sect. 3.4. In particular, if

$$\Sigma_E = \sigma^2 I_N, \quad \mu_\theta = 0, \tag{4.17}$$

the Bayes and MAP estimators,

$$\theta^{\text{B}} = \theta^{\text{MAP}} = \arg\min_\theta \|Y - \Phi\theta\|^2 + \theta^T P^{-1}\theta, \tag{4.18}$$

coincide with the ReLS estimator with regularization matrix $P = \Sigma_\theta/\sigma^2$. Under the further assumption $\Sigma_\theta = \lambda I_n$, the MAP estimator coincides with a ridge regression estimator with $\gamma = \sigma^2/\lambda$.

Remark 4.1 When $\Sigma_\theta = P$, where $P = P^T \geq 0$ is singular, one can still use (4.10) to obtain the Bayes estimate, while (4.13) and the quadratic problem (4.18) are no more valid due to the nonexistence of Σ_θ^{-1}. Nevertheless, by replicating the derivation in Remark 3.1, it is still possible to interpret the Bayes estimate as the solution of a constrained quadratic problem. In particular, under (4.17), we have that

$$\theta^{\text{B}} = \arg\min_\theta \|Y - \Phi\theta\|_2^2 + \theta^T P^+\theta \tag{4.19}$$

$$\text{subj. to} \quad U_2^T \theta = 0, \tag{4.20}$$

where U_2 was defined in Remark 3.1, as part of the singular value decomposition of P. The result can be interpreted as follows. A singular variance matrix means that we have perfect knowledge on some linear combination of the parameter vector. In particular,

$$
\begin{aligned}
\text{Var}\left[U_2^T \theta\right] &= U_2^T \text{Var}(\theta) U_2 \\
&= U_2^T \begin{bmatrix} U_1 & U_2 \end{bmatrix} \begin{bmatrix} \Lambda_P & 0 \\ 0 & 0 \end{bmatrix} \begin{bmatrix} U_1 & U_2 \end{bmatrix}^T U_2 = 0,
\end{aligned}
$$

where, with reference to the SVD of P, we have exploited the fact that $U_2^T U_1 = 0$. As a consequence,

$$\Pr(U_2^T \theta = U_2 \mu_\theta) = 1,$$

thus justifying the presence of the equality constraints in the quadratic problem (4.19)–(4.20), where $\mu_\theta = 0$ is assumed. Recalling the orthogonality of U_1 and U_2, we have that $U_2^T \theta = 0$ implies that $\theta \in \text{Range}(U_1) = \text{Range}(P)$. Therefore, the constrained quadratic problem (4.19)–(4.20) can also be equivalently reformulated as

$$\theta^B = \underset{\theta \,\in\, Range(P)}{\arg\min} \quad \|Y - \varPhi\theta\|^2 + \theta^T P^+ \theta. \tag{4.21}$$

One can also assess that the solution of this problem can be written as

$$\theta^B = P\varPhi^T (\varPhi P \varPhi^T + \Sigma_E)^+ Y,$$

an expression which does not require invertibility of any matrix.

In conclusion, the Bayes estimate always exists and is unique. In any case, it can be written as (4.7) with Σ_Y^{-1} replaced by its pseudoinverse.

The Bayesian interpretation of deterministic regularization can be exploited to obtain a guideline for the selection of the regularization matrix. The simplest case is when some statistics, e.g., based on samples coming from past problems, is available for the parameter vector θ. Then, the Bayesian interpretation suggests to select the covariance matrix of θ, divided by the error variance σ^2, as regularization matrix. If examples from the past are not available, one may rely on prior knowledge, telling that some entries of θ have smaller variance than others or that some correlation exists between the entries.

4.2.4 Hierarchical Bayes: Hyperparameters

In the cases in which prior information on the parameters is not sufficient to specify a prior, it is common to resort to hierarchical Bayesian models. Instead of fixing the prior, a family of priors is considered, parametrized by one or more *hyperparameters*. As an example, consider the case in which prior knowledge could be formalized in terms of zero-mean independent and equally distributed parameters whose absolute value is not too large. In absence of more precise information on their size, we could adopt the following prior:

$$\theta \sim \mathcal{N}(0, \lambda I_N),$$

where the scalar λ, called hyperparameter, enters the game as a further unknown quantity. More in general, the prior distribution $p(\theta|\alpha)$ may depend on a hyperparameter vector α. One may also want to consider a hyperparameter vector β entering the definition of the likelihood $p(Y|\theta, \beta)$. The most common example is when the measurement variance σ^2 is not known and is therefore treated as a hyperparameter. In the following, the vector of all hyperparameters will be denoted by

$$\eta = \left[\alpha^T \beta^T\right]^T.$$

For a given η, we will denote by $\theta^{\mathrm{MAP}}(\eta)$ and $\theta^B(\eta)$ the corresponding MAP and Bayes estimates:

$$\theta^{\mathrm{MAP}}(\eta) = \arg \max_{\theta} \mathrm{p}(\theta|Y, \eta) \tag{4.22}$$

$$\theta^B(\eta) = \mathscr{E}(\theta|Y, \eta) = \int \theta \mathrm{p}(\theta|Y, \eta)d\theta, \tag{4.23}$$

where

$$\mathrm{p}(\theta|Y, \eta) = \frac{\mathrm{p}(Y|\theta, \beta)\mathrm{p}(\theta|\alpha)}{\int \mathrm{p}(Y|\theta, \beta)\mathrm{p}(\theta|\alpha)d\theta}. \tag{4.24}$$

4.3 Bayesian Interpretation of the James–Stein Estimator

In this section, we show that the James–Stein estimator can be seen as a particular Bayesian estimator. As seen, in Eq. (1.2), the measurements model is

$$Y = \theta + E, \quad E \sim \mathscr{N}(0, \sigma^2 I_N). \tag{4.25}$$

In a Bayesian setting, the parameter vector is regarded as a random vector, whose distribution reflects our state of knowledge. In particular, we assume

$$\theta \sim \mathscr{N}(0, \lambda I_N), \tag{4.26}$$

where λ plays the role of hyperparameter. It follows that θ and Y are zero-mean jointly Gaussian variables with

$$\Sigma_{\theta Y} = \mathscr{E}(\theta Y^T) = \mathscr{E}(\theta\theta^T) = \lambda I_N, \quad \Sigma_Y = \mathscr{E}(YY^T) = (\lambda + \sigma^2)I_N. \tag{4.27}$$

According to (4.7), the Bayes estimate is given by the conditional expectation

$$\mathscr{E}(\theta|Y) = \Sigma_{\theta Y}\Sigma_Y^{-1}Y = \frac{\lambda}{\lambda + \sigma^2}Y = \left(1 - r_{\mathrm{Bayes}}\right)Y, \tag{4.28}$$

where

$$r_{\mathrm{Bayes}} = \frac{\sigma^2}{\lambda + \sigma^2}. \tag{4.29}$$

It is apparent that the estimator (4.28) has the same structure as James–Stein's one, with r replaced by r_{Bayes}.

Since Y and θ are jointly Gaussian, $\mathscr{E}(\theta|Y) = \theta^{\text{MAP}}$, where

$$\theta^{\text{MAP}} = \arg\min_{\theta} \frac{\|Y - \theta\|^2}{\sigma^2} + \frac{\|\theta\|^2}{\lambda} = \arg\min_{\theta} \|Y - \theta\|^2 + \frac{\sigma^2}{\lambda}\|\theta\|^2$$

which highlights the fact that $\mathscr{E}(\theta|Y)$ is the solution of a regularized least squares problem, controlled by the regularization parameter σ^2/λ.

If the variances λ and σ^2 could be assigned on the basis of prior knowledge, the similarity would be only formal. Let us make a step forward, considering the case in which the variance σ^2 is given, while λ is estimated from the data. The basic idea is that the hyperparameter λ could be tuned based on the observed vector Y and plugged into (4.29) to obtain an estimate of r_{Bayes}. Alternatively, one may focus directly on finding a sensible estimate of r_{Bayes}. In this respect, we are going to show that Stein's r is an unbiased estimate of r_{Bayes} under the Gaussian model (4.25) and (4.26) [6]. For this purpose, we will exploit a property of the inverse chi-square variable.

Definition 4.1 *(chi-square random variable)* The sum of the squares of n standard Gaussian independent random variables is a nonnegative valued random variable known as *chi-square variable* with n degrees of freedom:

$$\chi_n^2 = \sum_{i=1}^{n} X_i^2, \quad X_i \sim \mathscr{N}(0, 1).$$

Its mean and expectation are

$$\mathscr{E}\left(\chi_n^2\right) = n, \quad \text{Var}\left(\chi_n^2\right) = 2n.$$

The inverse of a chi-square variable is called *inverse chi-square*. For $n > 2$, its mean is

$$\mathscr{E}\left[\frac{1}{\chi_n^2}\right] = \frac{1}{n-2}. \tag{4.30}$$

Now, assume $N > 2$ and observe that

$$\frac{\|Y\|^2}{\lambda + \sigma^2} = \frac{\sum_i^n Y_i^2}{\lambda + \sigma^2} \sim \chi_N^2.$$

Recalling that the expectation of the inverse chi-square is equal to $1/(N-2)$, we have that

$$\mathscr{E}\left[\frac{\lambda + \sigma^2}{\|Y\|^2}\right] = \mathscr{E}\left[\frac{1}{\chi_N^2}\right] = \frac{1}{(N-2)}.$$

Therefore,

$$\mathcal{E}(r) = \mathcal{E}\left[\frac{(N-2)\sigma^2}{\|Y\|^2}\right] = \frac{\sigma^2}{\lambda + \sigma^2} = r_{Bayes}.$$

This means that James–Stein's shrinking coefficient r can be seen as an unbiased estimator of the shrinking coefficient r_{Bayes} appearing in the formula of the posterior expectation.

The example is instructive under several respects. First, it shows that, under suitable probabilistic assumptions, the typical structure of regularized estimators can be justified through Bayesian arguments. The second point has to do with the tuning of the regularization parameters. In the empirical Bayes approach, see Sect. 4.4, there is a preliminary step in which a point estimate of hyperparameters is obtained by standard estimation methods. Then, this point estimate is plugged into the expression of the Bayesian estimator. Although a full Bayesian approach would call for the joint estimation of parameters and hyperparameters, the two-step empirical Bayes approach not only conjugates simplicity and effectiveness but provides a probabilistic underpinning to regularized identification methods.

4.4 Full and Empirical Bayes Approaches

When the prior, and possibly the likelihood, include hyperparameters, Bayesian estimation becomes more complex and gives rise to alternative approaches. In principle, we want to obtain the posterior distribution

$$p(\theta|Y) = \frac{p(Y|\theta)p(\theta)}{p(Y)}.$$

However, if a hierarchical Bayesian model is adopted, we do not know $p(\theta)$, but only $p(\theta|\eta)$. At the cost of assigning a prior $p(\eta)$ also to the hyperparameters, the prior $p(\theta)$ can be obtained by marginalization of the joint probability density:

$$p(\theta) = \int p(\theta, \eta)d\eta = \int p(\theta|\eta)p(\eta)d\eta.$$

In general, this integral has to be computed numerically, e.g., by Monte Carlo methods. This leads to *full Bayesian* methods that compute the desired $p(\theta|Y)$ regarding both parameters and hyperparameters as random variables. Some remarks on these methods will be given in Sect. 4.10.

The justification for a simpler computational scheme stems from the following reformulation of the posterior:

$$p(\theta|Y) = \int p(\theta, \eta|Y)d\eta = \int p(\theta|\eta, Y)p(\eta|Y)d\eta. \tag{4.31}$$

Observe that

$$p(\eta|Y) \propto p(Y|\eta)p(\eta), \tag{4.32}$$

where $L(\eta|Y) = p(Y|\eta)$ is the likelihood of the hyperparameter vector η. It is also called *marginal likelihood* because it is obtained from the marginalization with respect to θ of the joint density $p(Y, \theta|\eta)$:

$$L(\eta|Y) = \int p(Y, \theta|\eta)d\theta = \int p(Y|\theta, \eta)p(\theta|\eta)d\theta. \tag{4.33}$$

If data are sufficiently informative, the marginal likelihood has good chances to be unimodal and sharply peaked in a neighbourhood of the maximum likelihood estimate

$$\eta^{ML} = \arg \max_{\eta} p(Y|\eta).$$

When this happens and $p(\eta)$ is rather uninformative (as it should be), from (4.32) it follows that $p(\eta|Y)$ is peaked as well. Then, as long as the properties of $p(\theta|\eta, Y)$ do not change rapidly with η near η^{ML}, the integral (4.31) can be approximated as

$$p(\theta|Y) \simeq p(\theta|\eta^{ML}, Y) = \frac{p(Y|\theta, \eta^{ML})p(\theta|\eta^{ML})}{p(Y|\eta^{ML})}.$$

In practice, this suggests to compute the posterior using the prior $p^*(\theta) = p(\theta|\eta^{ML})$ associated with the maximum likelihood estimate of hyperparameters. More in general, Empirical Bayes (EB) methods adopt a two-stage scheme. In the first step, a point estimate η^* is computed which is then kept fixed in the second step, when the posterior of the parameters is obtained, based on the prior $p^*(\theta) = p(\theta|\eta^*)$.

Among the advantages of the approach one may mention its simplicity, especially when there are few hyperparameters and the posterior $p(\theta|Y, \eta^{ML})$ is easily obtained as in the jointly Gaussian case. Moreover, the tuning of η admits an intuitive interpretation as the counterpart of model order selection in classic parametric estimation methods. The main drawback is that the EB method fails to propagate the uncertainty of the point estimate η^*.

Under the linear Gaussian model (4.9), the integral (4.33) admits a closed-form solution. In fact, since

$$Y \sim \mathcal{N}(\Phi\mu_\theta(\eta), \Sigma(\eta)), \quad \Sigma(\eta) = \Phi\Sigma_\theta(\eta)\Phi^T + \Sigma_E(\eta),$$

we have

$$\log L(\eta|Y) = -\frac{1}{2}\log(2\pi \det(\Sigma)) - \frac{1}{2}(Y - \Phi\mu_\theta)^T \Sigma^{-1}(Y - \Phi\mu_\theta), \tag{4.34}$$

where in the right-hand side dependence on η has been omitted for simplicity.

Therefore, application of Empirical Bayes estimation to the linear model (4.9) would consist of the following two steps:

Step 1:

$$\eta^* = \eta^{\mathrm{ML}} = \arg\max_{\eta} L(\eta|Y).$$

Step 2: Let $\mu_\theta = \mu_\theta(\eta^*)$, $\Sigma_E = \Sigma(\eta^*)$, $\Sigma_\theta = \Sigma_\theta(\eta^*)$ and compute the posterior expectation according to Sect. 4.2.3.

When the likelihood and the prior are such that integral (4.33) cannot be computed explicitly, an approximation is needed. In particular, one can resort to the Laplace approximation, which is based on a second-order Taylor expansion of $\log p(Y, \theta|\eta)$ around $\theta^{\mathrm{MAP}}(\eta)$ defined in (4.22), from which an integrable approximation of $p(Y, \theta|\eta)$ appearing in (4.33) is obtained. Note, however, that the Laplace approximation has to be recalculated for each evaluation of $L(\eta|Y)$ occurring during the iterative computation of η^{ML}.

4.5 Improper Priors and the Bias Space

The use of priors is most useful whenever the data alone are not sufficient to provide reliable parameter estimates but there exists some a priori knowledge that can be exploited. It may happen that for some parameters the introduction of a prior is not possible or not desirable, because their estimation can be satisfactorily performed anyway, given the information in the data. This can be accounted for by assuming that such parameters have *improper priors*.

In order to deal with the case where p parameters $\theta^P \in \mathbb{R}^p$ have a proper prior and the remaining $n - p$ parameters $\theta^I \in \mathbb{R}^{n-p}$ have an improper prior, consider the following model:

$$Y = \Phi\theta + E, \quad \Phi = \begin{bmatrix} \Omega & \Psi \end{bmatrix}, \quad \theta = \begin{bmatrix} \theta^P \\ \theta^I \end{bmatrix} \tag{4.35}$$

$$\theta \sim \mathcal{N}(0, \Sigma_\theta), \quad E \sim \mathcal{N}(0, \sigma^2 I_N) \tag{4.36}$$

$$\Sigma_\theta = \begin{bmatrix} \Sigma & 0 \\ 0 & aI_{n-p} \end{bmatrix}, \quad \Sigma > 0. \tag{4.37}$$

The (asymptotically) improper prior for θ^I is obtained by letting $a \to \infty$ so that θ^I has infinite variance, i.e., its density is flat. This amounts to complete lack of prior knowledge for the last $n - p$ entries of the parameter vector θ that, for simplicity, is assumed to be zero mean. The use of improper priors in a Bayesian setting has the same effect as the introduction of a *bias space* in a deterministic regularization setting. Within such a subspace, parameters are immune from regularization, a feature that could be useful to apply regularization only where needed without causing undesired distortions. The following theorem, whose proof is in Sect. 4.13.4, is analogous to a result obtained in [22] to obtain a Bayesian interpretation of smoothing splines. It illustrates the asymptotic behaviour of posterior means and variances as a goes to infinity.

Theorem 4.3 (adapted from [22]) *If* $\text{rank}(\Phi) = n$ *and* $\text{rank}(\Omega) = n - p$, *then*

$$\lim_{a \to \infty} \mathscr{E}(\theta^I | Y) = (\Psi^T M^{-1} \Psi)^{-1} \Psi^T M^{-1} Y$$

$$\lim_{a \to \infty} \mathscr{E}(\theta^P | Y) = \Sigma \Omega^T M^{-1} (I_n - \Psi (\Psi^T M^{-1} \Psi)^{-1} \Psi^T M^{-1}) Y$$

$$M = \Omega \Sigma \Omega^T + \sigma^2 I_N$$

$$\lim_{a \to \infty} \text{Var}(\theta | Y) = \sigma^2 \left(\Phi^T \Phi + \sigma^2 \begin{bmatrix} 0 & 0 \\ 0 & \Sigma^{-1} \end{bmatrix} \right)^{-1}.$$

An interesting benefit of improper priors is the possibility of reducing the number of hyperparameters by treating some of them as unknowns whose prior is improper. Letting the symbol $\mathbf{1}_{n \times 1}$ denotes a column vector of ones, assume, for example, that $\theta \sim \mathscr{N}(\mu \mathbf{1}_{n \times 1}, \Sigma_\theta)$, i.e., all the scalar entries of θ share the same prior mean μ. In most cases, very little is known about μ that could be therefore regarded as a hyperparameter to be tuned by marginal likelihood maximization. It can be then treated as a deterministically known variable, according to the Empirical Bayes approach, see Sect. 4.4. By this choice, however, the hyperparameter is fixed to its point estimate and its uncertainty is not propagated, implying that the uncertainty of θ^B will be underestimated if assessed by (4.14).

Alternatively, μ can be treated as a further random parameter. For this purpose, define $\tilde{\theta} = \theta - \mu$ and consider the model

$$\bar{\theta} = \begin{bmatrix} \tilde{\theta} \\ \mu \end{bmatrix}, \quad \Sigma_{\bar{\theta}} = \begin{bmatrix} \Sigma_\theta & 0 \\ 0 & a \end{bmatrix}$$

$$Y = \bar{\Phi} \bar{\theta} + E, \quad \bar{\Phi} = \begin{bmatrix} \Phi & \Phi \mathbf{1}_{n \times 1} \end{bmatrix}$$

$$\bar{\theta} \sim \mathscr{N}(0, \Sigma_{\bar{\theta}}), \quad E \sim \mathscr{N}(0, \sigma^2 I_N).$$

This formulation decreases the number of hyperparameters, without introducing prejudices (provided we let $a \to \infty$). More importantly, it is now possible to assess the joint uncertainty of the estimates of μ and $\tilde{\theta}$ through the posterior variance $\text{Var}(\bar{\theta} | Y)$.

4.6 Maximum Entropy Priors

A major appeal of the Bayesian paradigm lies in its ability to provide a rational foundation to regularization: one starts from prior knowledge and then proceeds with its formalization in terms of a probabilistic prior, from which the regularization penalty is finally derived. However, there is a stumbling block in the way, because the available prior knowledge is often too vague to avoid arbitrariness in the choice of the prior distribution. As a matter of fact, the derivation of systematic approaches for the selection of prior distributions is a classic topic of Bayesian estimation theory. In this section, the approach based on entropy maximization is briefly reviewed.

The starting point is the observation that, even when prior information is absent or very limited, there are candidate distributions that are obviously preferable, due to symmetry arguments. Assume, for instance, that candidate values for a scalar parameter θ are known to belong to a finite set $\{\theta_i, i = 1, \ldots m\}$ and no further information is available. Then, the only reasonable prior distribution will be $p(\theta = \theta_i) = 1/m$. In fact, assigning unequal probabilities would create an unjustified asymmetry, given that our prior information does not make any distinction between the m possible values of the parameter.

The case of a continuous-valued parameter θ taking values in a finite interval $[a, b]$ can be addressed in a similar way. In this case, a reasonable prior distribution is the uniform one:

$$p(\theta) = \begin{cases} \frac{1}{b-a}, & a \leq \theta \leq b \\ 0, & \text{elsewhere} \end{cases}.$$

In both examples, we might say the chosen distributions are those that reflect the maximum ignorance about the unknown parameter.

The next step is to formalize this notion of maximum ignorance in contexts where some partial information about θ is available. This can be done by means of the notion of entropy of a probability distribution. For a discrete distribution $p(\cdot)$ taking values $p(\theta_i)$ on a numerable set $\{\theta_i\}$, the entropy H is defined as

$$H(p) = -\sum_i p(\theta_i) \log p(\theta_i).$$

Note that the minimum possible entropy $H(p) = 0$ occurs when the probability is concentrated at a unique value $\bar{\theta}$. This is the case of a maximally informative distribution such that $p(\theta = \bar{\theta}) = 1$. Conversely, if the set $\{\theta_i\}$ has cardinality m, the maximum value $H(p) = \log(m)$ is achieved in correspondence of the uniform distribution $p(\theta = \theta_i) = 1/m, \forall i$. In other words, the larger the entropy, the less information is conveyed by the distribution.

For continuous-valued random variables, the notion of *differential entropy* $h(p)$ is introduced:

$$h(p) = -\int_{D_\theta} p(\theta) \log p(\theta) d\theta,$$

where D_θ denotes the support of the distribution. Note that, among distributions with finite support, the maximum possible (differential) entropy is achieved by the uniform distribution.

The principle of *Maximum Entropy* (MaxEnt) states that the admissible distribution with largest entropy is the one that best represents the current state of knowledge.

The admissible distributions are those that satisfy a set of constraints, chosen so as to incorporate all the available prior knowledge. For instance, if the prior knowledge amounts to knowing that $\theta \in [a, b]$, the prior suggested by the MaxEnt principle is the uniform distribution. Other types of constraints are typically expressed as expectations of functions of the parameters θ. In particular, consider a random variable θ, subject to known values η_i of m expectations

$$\mathscr{E}[g_i(\theta)] = \int g_i(\theta)\mathrm{p}(\theta)d\theta = \eta_i, \quad i = \ldots, m. \tag{4.38}$$

Then, we have the following useful result.

Theorem 4.4 (General form of maximum entropy distributions, based on [12]) *Among all the distributions satisfying (4.38), the maximum entropy one is of exponential type*

$$\mathrm{p}(\theta) = A \exp(-\lambda_1 g_1(\theta) - \ldots - \lambda_m g_m(\theta)), \tag{4.39}$$

where λ_i are m constants determined from (4.38) and A is such that

$$A \int_{-\infty}^{+\infty} \exp(-\lambda_1 g_1(\theta) - \ldots - \lambda_m g_m(\theta))d\theta = 1. \tag{4.40}$$

Example 4.5 *(MaxEnt prior from information on expected absolute value)* Assume that prior knowledge is summarized by the expectation $\mathscr{E}|\theta| = \eta$. Then, the MaxEnt prior is the solution of the constrained optimization problem

$$\max_{\mathrm{p}} h(\mathrm{p}) \quad \text{s.t.} \quad \mathscr{E}|\theta| = \eta.$$

Obviously, $m = 1$ and $g_1(\theta) = |\theta|$. In view of (4.39) and (4.40), $\mathrm{p}(\theta)$ is a Laplace distribution:

$$\mathrm{p}(\theta) = 0.5\lambda e^{-\lambda|\theta|}.$$

The value of λ is found by imposing the constraint on the expectation:

$$\int_{-\infty}^{+\infty} 0.5|\theta|\lambda e^{-\lambda|\theta|}d\theta = \eta.$$

Since the constraint on the expectation is satisfied for $\lambda = 1/\eta$, the following Laplace distribution is eventually obtained:

$$\mathrm{p}(\theta) = \frac{e^{-\frac{|\theta|}{\eta}}}{2\eta}.$$

Therefore, starting from a very partial information, such as a guess on the expected absolute value of the parameter, it is possible to completely specify a prior distribu-

tion that: (i) is coherent with the prior knowledge and (ii) does not introduces undue
assumptions because it is the least informative one, so far as entropy is taken as a
measure of informativeness. One could object that it is scarcely realistic to assume
prior knowledge of the expected absolute value of θ. However, if we adopt the empirical Bayes framework, the objection is circumvented by the possibility of treating η
as a hyperparameter that will be estimated from data.

Therefore, prior knowledge may just tell that the expectation of $|\theta|$ is finite,
without specifying a value for this expectation. The MaxEnt principle then suggests
the functional form of the prior that incorporates a hyperparameter η, whose tuning,
e.g., by marginal likelihood maximization, see Sect. 4.4, will be the first step of the
actual estimation algorithm. As it will be seen in the following, this particular prior
is associated with the Bayesian interpretation of the regularization penalty employed
by the so-called Lasso estimator that has been already introduced in a deterministic
regularization setting in Sect. 3.6.1.1.   □

For our purposes, of particular interest are MaxEnt priors satisfying constraints
on the second-order moments. In the scalar case, we have the following classical
result, e.g., see [19].

Proposition 4.1 (based on [12]) *Let θ be a zero-mean random variable with known
variance $\mathscr{E}\theta^2 = \lambda$. Then, the MaxEnt distribution is normal:*

$$\theta \sim \mathscr{N}(0, \lambda).$$

Also in this case, the necessity of specifying λ is not an issue, because the unknown
variance can be regarded as a hyperparameter and tuned by marginal likelihood
maximization. In other words, if the only prior knowledge is that θ has a finite,
yet unknown, variance, the MaxEnt principle suggests the use of a normal prior
parametrized by its variance.

When θ is a vector, a multivariate prior might be derived according to the following
proposition.

Proposition 4.2 (based on [12]) *Let θ be a zero-mean n-dimensional random vector whose entries have known variances $\mathscr{E}\theta_i^2 = \lambda_i, i = 1, \ldots, n$. Then, the MaxEnt
distribution is a multivariate normal with diagonal covariance matrix:*

$$\theta \sim \mathscr{N}(0, \Sigma_\theta), \quad \Sigma_\theta = \text{diag}\{\lambda_i\}.$$

The importance of this result is twofold. First, also in the multivariate case, the
least informative distribution under second moment constraints is of normal type.
Moreover, if the covariances are unknown, it is seen that the MaxEnt principle yields
independent distributions.

A shortcoming of the maximum entropy approach is that the resulting distributions
are not invariant with respect to reparametrizations of the unknown vector. To make
an example, we have already seen that the maximum entropy distribution of θ in a
finite interval [1, 2] is uniform. On the other hand, if the reparametrization $\psi = 1/\theta$

is adopted and the MaxEnt approach is applied to ψ, the resulting prior will be a uniform distribution for ψ in [0.5,1], which corresponds to

$$p(\theta) = \begin{cases} \frac{2}{\theta^2}, & 1 \leq \theta \leq 2 \\ 0, & \text{elsewhere,} \end{cases}$$

which is obviously different from a uniform distribution. A possible way to limit arbitrariness is to specify that, before applying the MaxEnt principle, one should first identify the "object of interest". Indeed, choosing either θ or $1/\theta$ as object of interest is going to yield different MaxEnt priors.

4.7 Model Approximation via Optimal Projection ⋆

Approximate low-order models are commonly used even when there is awareness that the real data are generated by a more complex model. Motivations may range from their use for control design purposes to better interpretability of the phenomena under investigation. Unfortunately, under model misspecification, several nice properties enjoyed by standard estimators are no more valid. In particular, a naive application of the least squares may provide far less than satisfactory results. In this section, it is shown that, within the Bayesian framework, the search for an optimal approximate model can be given a rigorous formulation that admits a projection-based solution.

We assume that the data Y are distributed according to (4.9), which summarizes our state of knowledge. However, rather than resorting to Bayesian estimation of the vector θ, an approximate model, typically of low order, is searched for. For instance, if θ_i were the samples of an impulse response, one might be interested in approximating them by a parametric model:

$$\theta \simeq g(\zeta), \quad g(\zeta) = \begin{bmatrix} g_1(\zeta) & \cdots & g_n(\zeta) \end{bmatrix}^T,$$

where $\zeta = \begin{bmatrix} \zeta_1 & \cdots & \zeta_q \end{bmatrix}^T$ is the unknown parameter vector. For example, in order to approximate the sequence θ_i by means of a single exponential function, it suffices to let $q = 2$ and

$$g_i(\zeta) = \zeta_1 e^{\zeta_2 i},$$

where ζ_1 is the amplitude and ζ_2 is the rate coefficient of the exponential.

A very natural estimator is the least squares one:

$$\zeta^{LS} = \arg \min_{\zeta} \|Y - \Phi g(\zeta)\|^2.$$

Note that ζ^{LS} coincides with the maximum likelihood estimate if the following model is assumed:

$$Y = \Phi g(\zeta) + E, \quad E \sim \mathcal{N}(0, \sigma^2 I_N).$$

In the present context, however, no claim is made that reality conforms to our approximate model. It may well be that the true θ, being more complex than its parsimonious parametric model $g(\zeta)$, is better represented by the model (4.9). Nevertheless, we are interested in finding the best approximation of θ within a set $\mathscr{P} = \{g(\zeta)|\zeta \in \mathbb{R}^q, \}$ of parametric approximations.

Under model (4.9), the optimal approximate model g^* can be defined as the one that minimizes the mean squared error $\mathscr{E}\|\theta - g\|^2$. For a generic model $g = g(\zeta)$, parametrized by the vector $\zeta \in \mathbb{R}^q$, $q \leq n$, we have that

$$g^* = g(\zeta^*), \quad \zeta^* := \arg\min_{\zeta} \mathscr{E}\left[\|\theta - g(\zeta)\|^2|Y\right], \tag{4.41}$$

where the conditional expectation is taken with reference to the probability measure specified by (4.9). The following theorem, whose proof is in Sect. 4.13.5, was first derived in the context of linear system identification [20]. It shows that the optimal approximation is the projection of the Bayes estimate θ^B onto the set $\mathscr{P}$.

Theorem 4.6 (Optimal approximation, based on [20]) *Assume that (4.9) holds. Then,*

$$\zeta^* = \arg\min_{\zeta} \|\theta^B - g(\zeta)\|^2. \tag{4.42}$$

In view of the last theorem, the best approximation $g(\zeta) \in \mathscr{P}$ can be computed by a two-step procedure. First, the Bayes estimate θ^B is obtained and in the second step the optimal $g(\zeta^*)$ is calculated as the solution of the least squares problem (4.42).

An interesting question is whether the obtained approximation is still optimal if the goal is minimizing the error, not with respect to θ, but with respect to the noiseless output $\Phi\theta$. In other words, the goal is finding g^o that minimizes $\|\Phi\theta - \Phi g^o)\|^2$. This can be done by introducing a weighted norm in the cost function:

$$g^o = g(\zeta^o), \quad \zeta^o := \arg\min_{\zeta} \mathscr{E}\left[\|\theta - g(\zeta)\|_W^2|Y\right], \tag{4.43}$$

where $\|x\|_W^2$ stands for $x^T W x$. In particular, if $W = \Phi^T\Phi$, then

$$\|\theta - g(\zeta)\|_W^2 = \|\Phi\theta - \Phi g(\zeta)\|^2.$$

By extending the proof of Theorem 4.6 to the case of a weighted norm, the following projection result is obtained.

Theorem 4.7 (Optimal weighted approximation, based on [20]) *Assume that (4.9) holds. Then,*

$$\zeta^o = \arg\min_{\zeta} \|\theta^B - g(\zeta)\|_W^2. \tag{4.44}$$

The consequence is that different approximations g^o are obtained depending on their prospective use. If the scope is just approximating θ, then $W = I_n$, but, if the scope is predicting the outputs, then $W = \Phi^T\Phi$ and a different result is obtained.

4.8   Equivalent Degrees of Freedom

In this section, the Bayesian estimation problem for the linear model is analysed by means of a diagonalization approach. The purpose is twofold: (i) the equivalent degrees of freedom of the Bayesian estimator are introduced together with their relationship with suitable weighted squared sums of residuals and squared sums of estimated parameters; (ii) it is shown that η^{ML}, the ML estimate of the hyperparameter vector, satisfies meaningful conditions involving the degrees of freedom. Finally, the obtained results are applied to the tuning of the regularization parameter, defined as the ratio between scaling factors for the noise variance Σ_E and the parameter variance Σ_θ. For the sake of simplicity, in this section, we assume $\mu_\theta = 0$.

Let us consider the case when the hyperparameters are just two scaling factors for the covariance matrices Σ_E and Σ_θ, that is,

$$\Sigma_\theta = \lambda K, \quad \lambda > 0 \tag{4.45}$$

$$\Sigma_E = \sigma^2 \Psi, \quad \sigma^2 > 0 \tag{4.46}$$

$$\eta = \left[\lambda \; \sigma^2 \right]^T, \tag{4.47}$$

where K and Ψ are known definite positive matrices. In such a case, it is immediate to see that the Bayes estimate

$$\theta^{\mathrm{B}} = \left(\Phi^T \Psi^{-1} \Phi + \frac{\sigma^2}{\lambda} K^{-1} \right)^{-1} \Phi^T \Psi^{-1} Y$$

depends only on the ratio $\gamma = \sigma^2 / \lambda$, which behaves as a deterministic regularization parameter. This means that only the ratio between the scaling factors is relevant to the computation of a point estimate, although both of them are needed to compute the posterior variance (4.14). When $\Psi = I_N$ and $K = I_n$, the above estimator provides a Bayesian interpretation to the classical ridge regression estimator. In particular, γ can be interpreted as a noise-to-signal ratio and its tuning reformulated as a statistical estimation problem.

Given a positive definite symmetric matrix S, let $S^{1/2} = \left(S^{1/2} \right)^T$ be its symmetric square root, i.e., $S^{1/2} S^{1/2} = S$. Now, consider the singular value decomposition

$$\Psi^{-1/2} \Phi K^{1/2} = U D V^T,$$

where U and V are square matrices such that $U^T U = I_N$ and $V^T V = I_n$ and $D \in \mathbb{R}^{N \times n}$ is a diagonal matrix with diagonal entries $\{d_i\}$, $i = 1, \ldots, n$, see (3.134). Moreover, define

$$\bar{Y} = U^T \Psi^{-1/2} Y$$
$$\bar{E} = U^T \Psi^{-1/2} E$$
$$\bar{\theta} = V^T K^{-1/2} \theta.$$

Observe that

$$\mathscr{E}\left(\bar{E}\bar{E}^T\right) = U^T\Psi^{-1/2}\mathscr{E}EE^T\Psi^{-1/2}U = \sigma^2 U^T U = \sigma^2 I_N.$$

Analogously, $\mathscr{E}\left(\bar{\theta}\bar{\theta}^T\right) = \lambda I_n$. Moreover,

$$\bar{Y} = U^T\Psi^{-1/2}(\Phi\theta + E) = U^T\Psi^{-1/2}\Phi K^{1/2}VV^T K^{-1/2}\theta + \bar{E}$$
$$= U^T U D V^T V\bar{\theta} + \bar{E} = D\bar{\theta} + \bar{E}.$$

In view of these properties, it follows that the original Bayesian estimation problem admits the following diagonal reformulation:

$$\bar{Y} = D\bar{\theta} + \bar{E}, \quad \bar{E} \sim \mathcal{N}(0, \sigma^2 I_N), \quad \bar{\theta} \sim \mathcal{N}(0, \lambda I_n), \qquad (4.48)$$

where $\bar{E}$ and $\bar{\theta}$ are independent of each other.

In view of statistical independence, we have N independent scalar models:

$$\bar{y}_i = d_i\bar{\theta}_i + \bar{v}_i, \quad i = 1, \ldots, n$$
$$\bar{y}_i = \bar{v}_i, \quad i = n+1, \ldots, N,$$

where $\bar{v}_i \sim \mathcal{N}(0, \sigma^2)$, $i = 1, \ldots, N$, and $\bar{\theta}_i \sim \mathcal{N}(0, \lambda)$, $i = 1, \ldots, n$.

By (4.11), it is straightforward to see that the Bayes estimates are

$$\bar{\theta}_i^{\mathrm{B}} = \frac{\lambda d_i \bar{y}_i}{\sigma^2 + \lambda d_i^2} = \frac{d_i \bar{y}_i}{\gamma + d_i^2}, \quad l = 1, \ldots, n$$

or, in matrix form,

$$\bar{\theta}^{\mathrm{B}} = (D^T D + \gamma I_n)^{-1} D^T \bar{Y}.$$

Let the residuals be defined as $\bar{\varepsilon}_i = \bar{y}_i - \bar{d}_i\bar{\theta}_i^{\mathrm{B}}$, $i = 1, \ldots, N$, where

$$\bar{d}_i = \begin{cases} d_i, & 1 \le i \le n \\ 0, & n+1 \le i \le N \end{cases} . \qquad (4.49)$$

Then, we have

$$\bar{\varepsilon}_i = \bar{y}_i - \frac{\bar{d}_i^2 \bar{y}_i}{\gamma + \bar{d}_i^2} = \frac{\gamma \bar{y}_i}{\gamma + \bar{d}_i^2} \qquad (4.50)$$

$$\mathscr{E}\bar{\varepsilon}_i^2 = \frac{\gamma^2 \mathscr{E}\bar{y}_i^2}{(\gamma + \bar{d}_i^2)^2} = \frac{\gamma^2(\bar{d}_i^2\lambda + \sigma^2)}{(\gamma + \bar{d}_i^2)^2} = \frac{\sigma^2\gamma}{\gamma + \bar{d}_i^2} = \sigma^2\left(1 - \frac{\bar{d}_i^2}{\gamma + \bar{d}_i^2}\right) \quad (4.51)$$

or, in matrix form,

$$\bar{\varepsilon} = \gamma (D^T D + \gamma I_N)^{-1} \bar{Y}, \quad \mathcal{E}\|\bar{\varepsilon}\|^2 = \sigma^2 \left(N - \text{trace}(D(D^T D + \gamma I_N)^{-1} D^T) \right).$$
$$(4.52)$$

It is worth noting that the above relationships do not hold for a generic regularization parameter γ, but only when $\gamma = \sigma^2/\lambda$. In the remaining part, we present some results that were first derived in the context of Bayesian deconvolution in [5]. The proof of the following proposition is in Sect. 4.13.6.

Proposition 4.3 (based on [5]) *For a given hyperparameter vector η, let* WRSS *denote the following weighted squared sum of residuals:*

$$\text{WRSS} = (Y - \Phi\theta^{\text{B}})^T \Psi^{-1}(Y - \Phi\theta^{\text{B}}),$$

where $\theta^{\text{B}} = \mathcal{E}[\theta|Y, \eta]$. Then,

$$\mathcal{E}(\text{WRSS}) = \sigma^2(N - \text{trace}(H(\gamma))),$$

where

$$H(\gamma) = \Phi(\Phi^T \Psi^{-1}\Phi + \gamma K^{-1})^{-1}\Phi^T \Psi^{-1}$$

is the so-called hat matrix.

As already noted, see (3.64), when $\Sigma_E = \sigma^2 I_N$, the predicted output $\hat{Y} = \Phi\theta^{\text{B}}$ and the measured output Y are related through the hat matrix:

$$\hat{Y} = H(\gamma)Y.$$

In order to better understand the link between the hat matrix and the degrees of freedom, just consider the standard linear model $Y = \Phi\theta + E$, $\theta \in \mathbb{R}^n$, and the corresponding LS estimate $\theta^{\text{LS}} = (\Phi^T \Phi)^{-1}\Phi^T Y$. The predicted output is $\hat{Y} = H^{\text{LS}}Y$, where $H^{\text{LS}} = \Phi(\Phi^T \Phi)^{-1}\Phi^T$ enjoys the property $\text{trace}(H^{\text{LS}}) = n$.

It is this analogy that justifies the introduction of equivalent degrees of freedom which we already encountered in (3.65) as a function of the regularized estimate θ^{R} described in the deterministic context. Its definition, here derived starting from the stochastic context, is reported below stressing its dependence on the regularization parameter γ.

Definition 4.2 *(equivalent degrees of freedom)* The quantity

$$\text{dof}(\gamma) = \text{trace}(H(\gamma)), \quad 0 \le \text{dof}(\gamma) \le n \qquad (4.53)$$

is called *equivalent degrees of freedom*.

In view of (4.52),

$$\text{dof}(\gamma) = \sum_{i=1}^{n} \frac{d_i^2}{d_i^2 + \gamma}$$

so that $\mathrm{dof}(\gamma)$ is a monotonically decreasing function of γ with $0 \leq \mathrm{dof}(\gamma) \leq n$. The equivalent degrees of freedom provide an easily understandable measure of the flexibility of estimator: for instance, if they are approximately equal to three, the Bayesian estimator has a flexibility comparable to a model with three parameters. For linear-in-parameter models estimated by ordinary or weighted least squares, the degrees of freedom coincide with the rank of the regressor matrix and, therefore, they can take only integer values. The equivalent degrees of freedom of the Bayesian estimator, conversely, are a nonnegative real number controlled by γ.

The next theorem establishes a connection between the degrees of freedom and the ML estimate

$$\eta^{\mathrm{ML}} = \left[\lambda^{\mathrm{ML}} \; \left(\sigma^2\right)^{\mathrm{ML}} \right]^T$$

of the hyperparameter vector. Accordingly, we define

$$\gamma^{\mathrm{ML}} = \frac{\left(\sigma^2\right)^{\mathrm{ML}}}{\lambda^{\mathrm{ML}}}.$$

Moreover, we introduce the following weighted squared sum of estimated parameters:

$$\mathrm{WPSS} = (\theta^{\mathrm{B}})^T K^{-1} \theta^{\mathrm{B}} = \left\| \bar{\theta}^{\mathrm{B}} \right\|^2 = \sum_{i=1}^{n} \frac{d_i^2 \, \bar{y}_i^2}{(\gamma + d_i^2)^2}. \tag{4.54}$$

The proof of the following result is in Sect. 4.13.7.

Theorem 4.8 (based on [5]) *Assume that model (4.9) holds where Σ_θ and Σ_E are as in (4.46)–(4.47). Then, the* ML *estimates of the hyperparameters satisfy the following necessary conditions:*

$$\mathrm{WRSS} = \left(\sigma^2\right)^{\mathrm{ML}} \left(N - \mathrm{dof}(\gamma^{\mathrm{ML}})\right) \tag{4.55}$$

$$\mathrm{WPSS} = \lambda^{\mathrm{ML}} \mathrm{dof}(\gamma^{\mathrm{ML}}). \tag{4.56}$$

By taking the ratio between (4.55) and (4.56), the following proposition is obtained.

Proposition 4.4 (based on [5]) *If λ^{ML} and $\left(\sigma^2\right)^{\mathrm{ML}}$ are nonnull and finite, then*

$$\gamma^{\mathrm{ML}} = \frac{\mathrm{dof}(\gamma^{\mathrm{ML}})}{N - \mathrm{dof}(\gamma^{\mathrm{ML}})} \frac{\mathrm{WRSS}}{\mathrm{WPSS}}. \tag{4.57}$$

This last corollary can be used as a simple and practical tuning procedure as it requires just a line search on the scalar γ. Of course, (4.57) relies on the necessary conditions of Theorem 4.8, so that one has to check if the solution corresponds to a maximum of the likelihood function.

4.9 Bayesian Function Reconstruction

In this section, the Bayesian estimation approach is illustrated through its application to the reconstruction of an unknown function from noisy samples. The observations will be generated by adding pseudorandom noise to a known function $g(x)$, so that the performances of alternative estimators can be directly assessed by comparison with the ground truth. The selected $g(x)$ is the same function (3.26) used in the previous chapter in order to illustrate polynomial regression:

$$g(x) = (\sin(x))^2(1 - x^2), \quad x \in [0, 1]. \tag{4.58}$$

Also the noise model is the same:

$$y_i = g(x_i) + e_i, \quad i = 1, \dots, N. \tag{4.59}$$

We let $N = 40$, $x_1 = 0$, $x_{40} = 1$, and $x_2, \dots, x_{39}$ are evenly spaced points between x_1 and x_{40}. Finally, e_i, $i = 1, \dots, 40$, are i.i.d. Gaussian distributed with mean zero and standard deviation 0.034.

The problem of estimating $\theta_i = g(t_i)$, i.e., the samples of the unknown function, is a particular case of the linear Gaussian model (4.9) with $\Phi = I_N$, that is,

$$Y = \theta + E, \quad E \sim \mathcal{N}(0, \sigma^2 I_N). \tag{4.60}$$

Since Φ is square, in this case, the number n of unknowns coincides with the number N of observations.

The noisy data and the true function are displayed in the top left panel of Fig. 4.1. It is assumed that the available prior knowledge regards the "regularity" of $g(\cdot)$ and the knowledge that $g(0) = 0$. A possible probabilistic translation of this qualitative knowledge is assuming that θ_i is a so-called random walk:

$$\theta_i = \theta_{i-1} + w_i, \quad i = 1, \dots, N, \quad \theta_0 = 0,$$

where $w_i \sim \mathcal{N}(0, \lambda)$ are independent random variables. In fact, under the random walk model, the first difference

$$\theta_i - \theta_{i-1} = w_i$$

has a finite variance, equal to λ. Hence, if we approximate the derivative of $g(\cdot)$ by the first difference $\theta_i - \theta_{i-1}$, this approximation is less than $1.96\sqrt{\lambda}$ with probability 0.95, which guarantees that the profile of the function cannot vary too quickly. Note that, due to the qualitative nature of the prior knowledge, the precise value of λ is unknown, so that it has to be treated as a hyperparameter. Conversely, it is assumed that the true value of σ^2 is known. Summarizing, we have

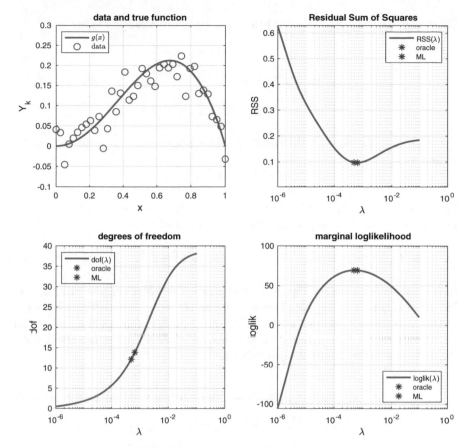

Fig. 4.1 Function reconstruction example. Top left: noisy data and true function. Top right, bottom left and bottom right: Residual sum of squares, i.e., the sum of the squared differences between the function values and their estimates, degrees of freedom and marginal loglikelihood against the hyperparameter λ. The oracle denotes the value that minimizes RSS while ML indicates the maximizer of the marginal likelihood

$$\theta_i = \sum_{j=1}^{i} w_j, \quad i = 1, \ldots, N$$

or, in matrix form,

$$\theta = Fw, \quad F = \begin{bmatrix} 1\,0\,0 \ldots 0 \\ 1\,1\,0 \ldots 0 \\ 1\,1\,1 \ldots 0 \\ \vdots\,\vdots\,\vdots\,\ddots\,\vdots \\ 1\,1\,1 \ldots 1 \end{bmatrix}, \quad w = \begin{bmatrix} w_1 \\ w_2 \\ w_3 \\ \vdots \\ w_N \end{bmatrix}.$$

Observing that $\mathrm{Var}(w) = \lambda I_N$, the prior variance of the parameter vector is

$$\Sigma_\theta = \lambda F F^T = \lambda \begin{bmatrix} 1 & 1 & \cdots & 1 \\ 1 & 2 & \cdots & 2 \\ \vdots & \vdots & \vdots & \vdots \\ 1 & 2 & \cdots & N \end{bmatrix}.$$

For a given λ, the Bayes estimate θ^B is obtained according to (4.10) and can be written as

$$\theta^B = \Sigma_\theta \left(\Sigma_\theta + \sigma^2 I_N \right)^{-1} Y.$$

The corresponding equivalent degrees of freedom, obtained by (4.53), are now thought as a (monotonically nondecreasing) function of λ, i.e.,

$$\mathrm{dof}(\lambda) = \mathrm{trace}\, H(\lambda), \quad H(\lambda) = \Sigma_\theta \left(\Sigma_\theta + \sigma^2 I_N \right)^{-1}, \quad \Sigma_\theta = \lambda F F^T.$$

In the bottom left panel of Fig. 4.1, the degrees of freedom are plotted against λ. For small values of λ they are close to zero and get closer to $N = 40$ as λ goes to infinity. It is a rather general feature that the function $\mathrm{dof}(\lambda)$ is better visualized on a semilog scale. In order to tune the regularization parameter λ, one can resort to the maximization of the marginal loglikelihood:

$$\lambda^{\mathrm{ML}} = \arg\max_\lambda \left\{ -\frac{1}{2} \log(2\pi \det(\Sigma)) - \frac{1}{2} Y^T \Sigma^{-1} Y \right\}$$
$$\Sigma = \Sigma_\theta + \sigma^2 I_N = \lambda F F^T + \sigma^2 I_N.$$

It turns out that $\lambda^{\mathrm{ML}} = 4.92e - 4$, the corresponding degrees of freedom being 12.17. For the sake of comparison, $\lambda = 6.61e - 4$ is the best possible value, i.e., the one provided by an oracle that exploits the knowledge of the true function in order to minimize the sum of the squared reconstruction errors. This latter quantity is function of λ and here denoted by $\mathrm{RSS}(\lambda)$. As seen in the top right panel of Fig. 4.1, marginal likelihood maximization achieves $\mathrm{RSS} = 9.80e - 2$, not much worse than $\mathrm{RSS} = 9.71e - 2$ achieved by the oracle, whose associated degrees of freedom are 13.88. Therefore, in this specific case, the marginal likelihood criterion somehow underestimates the complexity of the model.

In Fig. 4.2, the estimates obtained in correspondence of six different values of λ are displayed. It is apparent that for $\lambda = 1e - 6$ and $\lambda = 1e - 5$ the estimated function is overregularized, while overfitting occurs for $\lambda = 1e - 1$ and $\lambda = 1e - 2$. The two bottom panels display the oracle and ML estimates, the former exhibiting a slightly more regular profile.

Finally, observing that in our case $\Sigma_{\theta Y} = \Sigma_\theta$, we have

$$\Sigma_{\theta|Y} = \mathrm{Var}(\theta|Y) = \Sigma_\theta - \Sigma_\theta \Sigma_Y^{-1} \Sigma_\theta$$

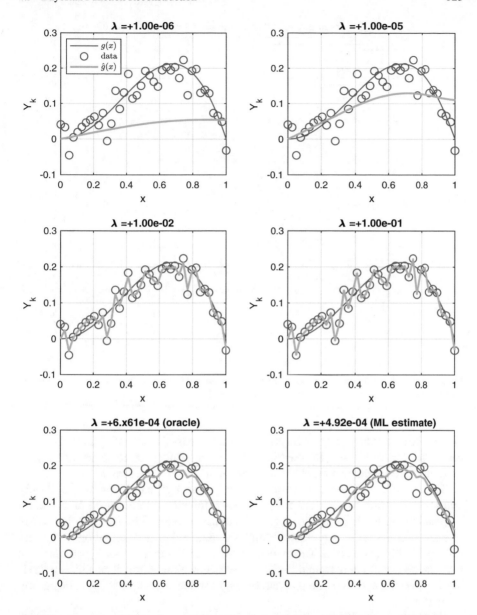

Fig. 4.2 Function reconstruction example. The panels display the Bayes estimates $\hat{g}(x)$ corresponding to six different values of the hyperparameter λ, including the one provided by the oracle and the maximum likelihood one

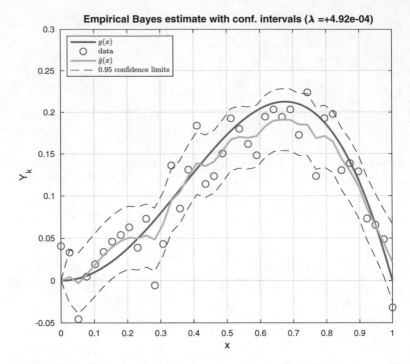

Fig. 4.3 Function reconstruction example. True function and Empirical Bayes estimate $\hat{g}(x)$ based on λ^{ML} together with its 95% Bayesian credible intervals

and we can compute the 95% Bayesian credible intervals, according to (4.8). As it can be seen from Fig. 4.3, the credible limits successfully capture the uncertainty, as demonstrated by the fact that the true function lies within the limits.

This simple example has shown that Bayesian estimation can be effectively employed in order to reconstruct an unknown function without need of assuming a specific parametric structure, e.g., polynomial or other. The key idea is the use of a smoothness prior, expressed through the assumed prior distribution of the first differences of the function. The associated variance λ is treated as a hyperparameter that can be tuned via marginal likelihood maximization. Altogether, this is a flexible Empirical Bayes scheme that can be employed as a general-purpose black-box estimator.

Of interest is also the fact that the considered function could have been the impulse response of a dynamical system. In this respect, the example highlights also the limits of the approach. A first issue, easily fixable, has to do with the insufficient smoothness of the estimate. As seen in Fig. 4.3, the true function is significantly smoother than its estimate. As a matter of fact, it is not difficult to increase the regularity of the Bayes estimate: for instance, it suffices to assume that the samples $\theta_i = g(x_i)$ are an integrated random walk:

$$\theta_i = \theta_{i-1} + \xi_i$$
$$\xi_i = \xi_{i-1} + w_i,$$

where $w_i \sim \mathcal{N}(0, \lambda)$ are again independent and identically distributed. This prior distribution is going to yield smoother profiles. Rather interestingly, the obtained estimate can be seen as the discrete-time counterpart of cubic smoothing splines, a method widely used for the nonparametric reconstruction of unknown functions.

A more serious issue regards extrapolation properties of the estimate that are in turn connected with the type of asymptotic decay shown by stable impulse responses. As it can be seen from Fig. 4.3, oscillations and credible intervals do not tend to dampen as x increases. While it would be easy to compute the Bayes estimate also for values far beyond the observation window, the result would be disappointing. Indeed, coherently with the diffusive nature of random walks, the width of the credible band would diverge, which is unnecessarily conservative when a stable impulse response is reconstructed. It appears that the task of identifying impulse responses calls for prior distributions that are specifically suited to the their features, especially the asymptotic ones. The development of these prior distributions, or equivalently the design of suitable regularization penalties, will be a central topic of the subsequent chapters.

4.10 Markov Chain Monte Carlo Estimation

As already mentioned in Sect. 4.4, in the full Bayesian approach the estimate

$$p(\theta|Y) = \int p(\theta, \eta|Y)d\eta = \int p(\theta|\eta, Y)p(\eta|Y)d\eta$$

requires a marginalization with respect to the hyperparameter vector η. In general, this integral cannot be computed analytically. Nevertheless it can be computed numerically by means of Markov Chain Monte Carlo (MCMC) methods that generate pseudorandom samples drawn from the joint posterior density $p(\theta, \eta|Y)$. The Gibbs sampling (GS) algorithm is the most straightforward and popular MCMC method. Its goal is to simulate a realization of a Markov chain, whose samples, though not independent of each other, form an ergodic process whose stationary distribution coincides with the desired posterior. Hence, provided that the burn-in phase is discarded, the posterior distribution is approximated by the histogram of the samples. In order to generate the samples, at each step a random extraction is made from a proposal distribution. In the Gibbs sampler, the proposal distribution is the so-called full conditional, that is, the probability of a given element of the parameter vector given the data and the current values of all other elements.

For the linear Gaussian model (4.9), a Gibbs sampler may be implemented as follows:

1. Select initializations η^0, θ^0, and let $k = 0$.
2. Draw a sample $\eta^{(k+1)}$ from the full conditional distribution $p(\eta|\theta^{(k)}, Y)$.
3. Draw a sample $\theta^{(k+1)}$ from the full conditional distribution $p(\theta|\eta^{(k+1)}, Y)$.
4. If $k = k_{max}$, end, else $k = k + 1$ and go to step 2.

This stochastic simulation algorithm generates a Markov chain whose stationary distribution coincides with $p(\theta, \eta|Y)$. Therefore, though correlated, the generated samples $\{\theta^{(k)}, \eta^{(k)}\}$ can be used to estimate the (joint and marginal) posterior distributions and also the posterior expectations via the proper sample averages. For example,

$$\frac{1}{N} \sum_{k=1}^{N} \theta^{(k)} \simeq \mathscr{E}(\theta|Y).$$

The choice of the prior distributions $p(\theta|\eta)$ and $p(\eta|Y)$ has a critical influence on the efficiency of the scheme. The priors are called *conjugate*, when for each parameter the prior and the full conditional belong to the same distribution family. This implies that the same random variable generators can be used throughout the simulation.

Consider model (4.9), where Σ_E is known and $\Sigma_\theta = \lambda K$, with λ unknown. Below, we describe a Gibbs sampling scheme for obtaining the posterior distributions of θ and $\eta = \lambda$. For θ, the prior is $\theta|\lambda \sim \mathscr{N}(0, \lambda K)$. A conjugate prior for λ is the inverse Gamma distribution:

$$\frac{1}{\lambda} \sim \Gamma(g_1, g_2), \quad g_1, g_2 > 0.$$

In other words, it is assumed that $1/\lambda$ is distributed as a Gamma random variable, so that

$$p\left(\frac{1}{\lambda}\right) \propto \left(\frac{1}{\lambda}\right)^{g_1 - 1} e^{-\left(\frac{g_2}{\lambda}\right)}.$$

With this choice of the prior, the full conditional of $1/\lambda$ will be distributed as a suitable Gamma variable, $\forall k$. More precisely, it can be shown that, if

$$p\left(\bar{\theta}|\lambda\right) \sim \mathscr{N}(0, \lambda I_N), \quad p\left(\frac{1}{\lambda}\right) \sim \Gamma(g_1, g_2)$$

then

$$p\left(\frac{1}{\lambda}\bigg|\bar{\theta}\right) \sim \Gamma\left(g_1 + \frac{N}{2}, g_2 + \frac{\|\bar{\theta}\|^2}{2}\right). \tag{4.61}$$

Recall that the mean and variance of the Gamma random variable are g_1/g_2 and g_1/g_2^2, respectively. For the prior to be as uninformative as possible, we let g_1 and g_2 decrease to zero. Under these assumptions, the Gibbs sampler unfolds as follows:

1. Initialize λ and θ, e.g., using the empirical Bayes estimates

$$\lambda^{(0)} = \lambda^{ML}, \quad \theta^0 = \theta^B = \mathscr{E}(\theta|\lambda^{ML}, Y)$$

and let $k = 0$.

2. Draw a sample $1/\lambda^{(k+1)}$ from the full conditional distribution

$$\text{p}\left(\frac{1}{\lambda}\bigg|\theta^{(k)}, Y\right) = \text{p}\left(\frac{1}{\lambda}\bigg|\theta^{(k)}\right) = \Gamma\left(\frac{N}{2}, \frac{\theta^{(k)T} K^{-1}\theta^{(k)}}{2}\right). \tag{4.62}$$

3. Draw a sample $\theta^{(k+1)}$ from the full conditional distribution

$$\text{p}\left(\theta|\lambda^{(k+1)}, Y\right) = \mathcal{N}\left(\mathscr{E}(\theta|\lambda^{(k+1)}, Y), \text{Var}(\theta|\lambda^{(k+1)}, Y)\right)$$

whose mean and variance are obtained according to (4.10) or (4.13).

4. If $k = k_{max}$, end, else $k = k + 1$ and go to step 2.

Above, the expression of the full conditional (4.62) is a direct consequence of the conjugacy property (4.61), as it can be seen by letting $\bar{\theta} = K^{-1/2}\theta^{(k)}$, where $K^{-1/2}$ is a symmetric matrix such that $K^{-1/2}K^{-1/2} = K^{-1}$.

When there are other hyperparameters to tune, e.g., the noise variance σ^2, the MCMC scheme can be properly extended. Provided that they exist, conjugate priors ensure an efficient sampling from the proposal distributions that generate the random samples, although a variety of MCMC schemes are available that deal with non-conjugate priors at the cost of an increased computational effort.

The main advantage of MCMC methods is that they implement the full Bayesian framework that is only approximated by the empirical Bayes scheme. In particular, MCMC methods do not neglect the hyperparameter uncertainty which is correctly propagated to the parameter estimate. However, as already discussed in Sect. 4.4, if data are informative enough to ensure a precise estimate of the hyperparameters, the difference between MCMC and empirical Bayes estimates (and associated credible regions) may be of minor importance.

4.11 Model Selection Using Bayes Factors

As discussed in Sect. 2.6.2, one fundamental issue is the selection of the "best" model inside a class of postulated structures. In the classical setting, this can be performed using criteria like AIC (2.34) and BIC (2.36) or adopting a cross-validation strategy. We will now see that the Bayesian approach provides a powerful alternative based on the concept of posterior model probability.

Let $\mathscr{M}^i$ be a model structure parametrized by the vector x^i. In the system identification scenario discussed in Chap. 2, the structures could be ARMAX models of different complexity. Hence, each x^i would correspond to the θ^i parametrizing (2.1) and containing the coefficients of rational transfer functions of different orders. If little knowledge on them were available, poorly informative prior densities could be assigned. Another example concerns the function estimation problem illustrated in Sect. 4.9. Here, x^i could contain the samples θ^i of the unknown function g modelled as a stochastic process. Then, the different structures could represent different covariances of g defined by a random walk or an integrated random walk. Each covariance would then depend on an unknown hyperparameter vector η^i containing the variance of the random walk increments and possibly also of the measurement noise. So, in this case, one would have $x^i = [\theta^i \ \eta^i]$. Here, η^i is a random vector with flat priors typically assigned to the variances to include just nonnegativity information.

Now, suppose that we are given m competitive structures $\mathcal{M}^i$. An important conceptual step is to interpret even them as (discrete) random variables, each having probability $\Pr(\mathcal{M}^i)$ before seeing the data Y. The selection of the best model has then a natural answer: one should select the structure having the largest posterior probability $\Pr(\mathcal{M}^i|Y)$. Using Bayes rule, one has

$$\Pr(\mathcal{M}^i|Y) = \frac{\int p(Y|\mathcal{M}^i, x^i) dx^i \, \Pr(\mathcal{M}^i)}{p(Y)}.$$

A typical choice is to think of the structures as equiprobable, so that $\Pr(\mathcal{M}^i) = 1/m$ for any i. Then, one can select the $\mathcal{M}^i$ maximizing the so-called Bayesian evidence given by

$$p(Y|\mathcal{M}^i) = \int p(Y|\mathcal{M}^i, x^i) dx^i.$$

Note that this corresponds to the marginal likelihood where all the parameter uncertainty connected with the i-th structure has been integrated out. Given two structures $\mathcal{M}^1$ and $\mathcal{M}^2$, the Bayes factor is also defined as follows:

$$B_{12} = \frac{p(Y|\mathcal{M}^1)}{p(Y|\mathcal{M}^2)}.$$

Hence, large values of B_{12} indicate that data strongly support $\mathcal{M}^1$ as opposed to $\mathcal{M}^2$.

For the computation of the Bayesian evidence, the same numerical considerations reported at the end of Sect. 4.4 then hold. In particular, when the evidence cannot be computed explicitly, approximations are needed given by the Laplace approximation. Also the BIC criterion is often adopted. In particular, in the function estimation problem one can integrate out θ. Then, one can evaluate the complexity of the model using the marginal likelihood optimized w.r.t. the hyperparameters η^i, then adding a term which penalizes the dimension of the hyperparameter vector. This will be also discussed later on in Sect. 7.2.1.1.

MCMC can be also used to compute the evidence by simulating from posterior distributions and using the harmonic mean of the likelihood values, see Sect. 4.3 in [14]. A more powerful and complex approach employs MCMC techniques able to jump between models of different dimensions, an approach known in the literature as reversible jump Markov chain Monte Carlo computation [10].

4.12   Further Topics and Advanced Reading

There is an extensive literature debating on the interpretation of probability as a quantification of personal belief and it would be impossible to give a satisfactory account of all the contributions. The reader interested in studying motivations and foundations of *subjective probability* may refer to [4, 16]. One of the merits of Bayesian probability is its efficacy in addressing ill-posed and ill-conditioned problems, including also a wide class of statistical learning problems. The connection between deterministic regularization and Bayesian estimation has been pointed out by several authors in different contexts. Two examples related to spline approximation and neural networks are given by [8, 15].

The choice and tuning of the priors is undoubtedly the crux of any Bayesian approach. It is not a surprise that the tuning of hyperparameters via the *Empirical Bayes* approach emerged early as a practical and effective way to deploy Bayesian methods in real-world contexts, see [6] for its use in the study of the James–Stein estimator. Since the 1980s, thanks to the advent of Markov chain Monte Carlo methods, *full Bayesian* approaches have become a viable alternative, motivating reflections on the pros and cons of the two approaches, see, for instance, [17]. In particular, the connection between Stein's Unbiased Risk Estimator (SURE), equivalent degrees of freedom and the robustness of marginal likelihood hyperparameter tuning has been investigated by [1, 21]. The choice of the prior distributions is somehow more controversial. In the present chapter, we exposed the principles of the maximum entropy approach, mainly following [12], but other approaches have been advocated for finding non-informative priors. A requirement could be invariance with respect to change of coordinates, enjoyed, for instance, by Jeffreys' prior [13].

It not unusual to have parameters that should be left immune from regularization. In the Bayesian approach, this corresponds to the absence of prior information, usually expressed through an infinite variance prior. Although the case could be treated by assigning large variances to some parameters, it is numerically more robust useful to use the exact formulas. Their derivation by a limit argument followed [22].

The idea of deriving approximated parametric models by a suitable projection of the Bayes estimate conforms to Hjalmarsson's advice "always first model as well as possible" [11]. The projection result has been derived in [23] for Gaussian processes and subsequently extended to general distributions in [20].

The equivalent degrees of freedom of a regularized estimator have been studied in the context of smoothing by additive [2] and spline models [3, 9], while a discussion specialized to the case of Bayesian estimation can be found in [5, 17].

Starting by the seminal paper [7], the use of stochastic simulation for computing posterior distributions according to a full Bayesian paradigm has gained a wider and wider adoption, especially when there exist conjugate priors that allow efficient sampling schemes. In particular, this is possible for the linear model discussed in this chapter, whose MCMC estimation is discussed in [18].

4.13 Appendix

4.13.1 Proof of Theorem 4.1

For simplicity, the proof is given in the scalar parameter case. We have that

$$
\begin{aligned}
\frac{d\text{MSE}(\hat{\theta})}{d\hat{\theta}} &= \frac{d}{d\hat{\theta}} \int_{-\infty}^{+\infty} (\hat{\theta} - \theta)^2 \mathrm{p}(\theta|Y) d\theta \\
&= 2\hat{\theta} \int_{-\infty}^{+\infty} \mathrm{p}(\theta|Y) d\theta - 2 \int_{-\infty}^{+\infty} \theta \mathrm{p}(\theta|Y) d\theta \\
&= 2 \left(\hat{\theta} - \mathscr{E}[\theta|Y] \right).
\end{aligned}
$$

Moreover,

$$
\frac{d^2 \text{MSE}(\hat{\theta})}{d\hat{\theta}^2} = 2 \int_{-\infty}^{+\infty} \mathrm{p}(\theta|Y) d\theta = 2 > 0.
$$

Therefore, $\theta^{\text{B}} = E[\theta|Y]$ minimizes $\text{MSE}(\hat{\theta})$.

4.13.2 Proof of Theorem 4.2

Let $X = \theta^{\text{B}} - \theta$ denote the estimation error. Recalling that $\mathscr{E}[Y - \Phi\mu_\theta] = \mathscr{E}[E] = 0$, from (4.10) it follows that $\mathscr{E}X = 0$. Note also that X and Y are jointly Gaussian and

$$
\text{Cov}(X, Y) = \mathscr{E}[X(Y - \mathscr{E}Y)^T] = \mathscr{E}[XY^T] - \mathscr{E}X\mathscr{E}Y^T = \mathscr{E}[XY^T].
$$

Now, using (4.7), we have

$$
\begin{aligned}
\mathscr{E}[XY^T] &= \mathscr{E}\left[(\theta^{\text{B}} - \theta)Y^T\right] \\
&= \Sigma_{\theta Y} \Sigma_Y^{-1} \mathscr{E}\left[(Y - \mu_Y)Y^T\right] - \mathscr{E}\left[(\theta - \mu_\theta)Y^T\right] \\
&= \Sigma_{\theta Y} \Sigma_Y^{-1} (\mathscr{E}\left[YY^T\right] - \mu_Y \mu_Y^T) - \mathscr{E}\left[\theta Y^T\right] - \mu_\theta \mu_Y^T \\
&= \Sigma_{\theta Y} \Sigma_Y^{-1} \Sigma_Y - \Sigma_{\theta Y} = 0.
\end{aligned}
$$

4.13.3 Proof of Lemma 4.1

By applying the matrix inversion lemma (3.145) and proceeding with simple matrix manipulations,

$$\Sigma_\theta \Phi^T (\Sigma_E + \Phi \Sigma_\theta \Phi^T)^{-1} = \Sigma_\theta \Phi^T (\Sigma_E^{-1} - \Sigma_E^{-1} \Phi (\Phi^T \Sigma_E^{-1} \Phi + \Sigma_\theta^{-1})^{-1} \Phi^T \Sigma_E^{-1})$$
$$= \Sigma_\theta \Phi^T \Sigma_E^{-1} - \Sigma_\theta \Phi^T \Sigma_E^{-1} \Phi (\Phi^T \Sigma_E^{-1} \Phi + \Sigma_\theta^{-1})^{-1} \Phi^T \Sigma_E^{-1}$$
$$= \Sigma_\theta (I - \Phi^T \Sigma_E^{-1} \Phi (\Phi^T \Sigma_E^{-1} \Phi + \Sigma_\theta^{-1})^{-1}) \Phi^T \Sigma_E^{-1}$$
$$= \Sigma_\theta (\Phi^T \Sigma_E^{-1} \Phi + \Sigma_\theta^{-1} - \Phi^T \Sigma_E^{-1} \Phi)(\Phi^T \Sigma_E^{-1} \Phi + \Sigma_\theta^{-1})^{-1} \Phi^T \Sigma_E^{-1}$$
$$= (\Phi^T \Sigma_E^{-1} \Phi + \Sigma_\theta^{-1})^{-1} \Phi^T \Sigma_E^{-1}.$$

4.13.4 Proof of Theorem 4.3

In view of (4.13), the conditional variance is

$$\mathrm{Var}(\theta|Y) = \left(\frac{\Phi^T \Phi}{\sigma^2} + \Sigma_\theta^{-1} \right)^{-1} = \sigma^2 \left(\Phi^T \Phi + \sigma^2 \begin{bmatrix} a^{-1} I_{n-p} & 0 \\ 0 & \Sigma^{-1}. \end{bmatrix} \right)^{-1}.$$

In view of (4.7)

$$\mathscr{E}(\theta|Y) = \Sigma_\theta \Phi^T (\Phi \Sigma_\theta \Phi^T + \sigma^2 I_n)^{-1} Y = \begin{bmatrix} \Sigma \Omega^T \\ a \Psi^T \end{bmatrix} (a \Psi \Psi^T + M)^{-1} Y.$$

By replicating the passages of Lemma 4.1

$$a \Psi (a \Psi \Psi^T + M)^{-1} = \left(\Psi^T M^{-1} \Psi + \frac{I_{n-p}}{a} \right)^{-1} \Psi^T M^{-1}.$$

Moreover, by applying the matrix inversion lemma, see (3.145),

$$(a \Psi \Psi^T + M)^{-1} = M^{-1} - M^{-1} \Psi \left(\Psi^T M^{-1} \Psi + \frac{I_{n-p}}{a} \right)^{-1} \Psi^T M^{-1}$$
$$= M^{-1} - M^{-1} \Psi (\Psi^T M^{-1} \Psi)^{-1} \left(I_{n-p} + \frac{1}{a} (\Psi^T M^{-1} \Psi)^{-1} \right)^{-1} \Psi^T M^{-1}.$$

Then, letting $a \to \infty$ complete the proof. Observe that all the inverse matrices appearing in the proof exist due to the full-rank assumptions made on Φ and Ψ.

4.13.5 Proof of Theorem 4.6

The expectation in (4.41) can be rewritten as

$$\mathcal{E}\left[\left\|\theta - \theta^{B} + \theta^{B} - g(\zeta)\right\|^{2}\Big|Y\right]$$

$$= \mathcal{E}\left[\left\|\theta - \theta^{B}\right\|^{2} + 2\left(\theta - \theta^{B}\right)^{T}\left(\theta^{B} - g(\zeta)\right) + \left\|\theta^{B} - g(\zeta)\right\|^{2}\Big|Y\right]$$

$$= \mathcal{E}\left[\left\|\theta - \theta^{B}\right\|^{2}\Big|Y\right] + \mathcal{E}\left\|\theta^{B} - g(\zeta)\right\|^{2}.$$

The proof follows by observing that in the last equation the first term does not depend on ζ. In the last passage, we have exploited the fact that $\theta^{B}|Y$ is deterministic and equal to $\mathcal{E}(\theta|Y)$.

4.13.6   Proof of Proposition 4.3

First observe that

$$\text{WRSS} = \|\bar{\varepsilon}\|^{2} = \sum_{i=1}^{N} \frac{\gamma^{2}\bar{y}_{i}^{2}}{(\gamma + \bar{d}_{i}^{2})^{2}}. \tag{4.63}$$

Hence, in view of (4.52)

$$\mathcal{E}\text{WRSS} = \sigma^{2}\left(N - \text{trace}(D(D^{T}D + \gamma I_{N})^{-1}D^{T})\right).$$

On the other hand, by simple matrix manipulations, it turns out that

$$U^{T}\Psi^{-1/2}H\Psi^{1/2}U = D(D^{T}D + \gamma I_{N})^{-1}D^{T}.$$

Finally, recalling that $\text{trace}(AB) = \text{trace}(BA)$,

$$\text{trace}(U^{T}\Psi^{-1/2}H\Psi^{1/2}U) = \text{trace}(\Psi^{1/2}UU^{T}\Psi^{-1/2}H) = \text{trace}(H)$$

thus proving the thesis.

4.13.7   Proof of Theorem 4.8

Without loss of generality, the proof refers to the diagonalized Bayesian estimation problem (4.48). The marginal loglikelihood function is

$$\sum_{i=1}^{N}\log(\bar{d}_{i}^{2}\lambda + \sigma^{2}) + \sum_{i=1}^{N}\frac{\bar{y}_{i}^{2}}{\bar{d}_{i}^{2}\lambda + \sigma^{2}} + \kappa,$$

where κ denotes a constant we are not concerned with. By equating to zero the partial derivatives with respect to σ^{2} and λ we obtain

$$\sum_{i=1}^{N} \frac{1}{\bar{d}_i^2 \lambda + \sigma^2} - \sum_{i=1}^{N} \frac{\bar{y}_i^2}{(\bar{d}_i^2 \lambda + \sigma^2)^2} = 0$$

$$\sum_{i=1}^{n} \frac{d_i^2}{d_i^2 \lambda + \sigma^2} - \sum_{i=1}^{n} \frac{d_i^2 \bar{y}_i^2}{(d_i^2 \lambda + \sigma^2)^2} = 0.$$

In view of (4.54) and (4.63),

$$\sigma^2 \left(N - \mathrm{dof}(\gamma) \right) - \mathrm{WRSS} = 0$$
$$\lambda \mathrm{dof}(\gamma) - \mathrm{WPSS} = 0,$$

which concludes the proof.

References

1. Aravkin A, Burke JV, Chiuso A, Pillonetto G (2014) Convex vs non-convex estimators for regression and sparse estimation: the mean squared error properties of ARD and GLASSO. J Mach Learn Res 15(1):217–252
2. Buja A, Hastie T, Tibshirani R (1989) Linear smoothers and additive models. Ann Stat 453–510
3. Craven P, Wahba G (1979) Smoothing noisy data with spline functions. Numer Math 31:377–403
4. De Finetti B (2017) Theory of probability: a critical introductory treatment, vol 6. Wiley
5. De Nicolao G, Sparacino G, Cobelli C (1997) Nonparametric input estimation in physiological systems: problems, methods, and case studies. Automatica 33(5):851–870
6. Efron B, Morris C (1973) Stein's estimation rule and its competitors-an empirical Bayes approach. J Am Stat Assoc 68(341):117–130
7. Geman S, Geman D (1984) Stochastic relaxation, Gibbs distributions, and the Bayesian restoration of images. IEEE Trans Pattern Anal Mach Intell 6:721–741
8. Girosi F, Jones M, Poggio T (1995) Regularization theory and neural networks architectures. Neural Comput 7(2):219–269
9. Golub GH, Heath M, Wahba G (1979) Generalized cross-validation as a method for choosing a good ridge parameter. Technometrics 21(2):215–223
10. Green PJ (1995) Reversible jump Markov chain Monte Carlo computation and Bayesian model determination. Biometrika 82(4):711–732
11. Hjalmarsson H (2005) From experiment design to closed loop control. Automatica 41(3):393–438
12. Jaynes ET (1982) On the rationale of maximum-entropy methods. Proc IEEE 70(9):939–952
13. Jeffreys H (1946) An invariant form for the prior probability in estimation problems. Proc Math Phys Eng Sci 186(1007):453–461
14. Kass RE, Raftery AE (1995) Bayes factors. J Amer Statist Assoc 90:773–795
15. Kimeldorf G, Wahba G (1970) A correspondence between Bayesian estimation on stochastic processes and smoothing by splines. Ann Math Stat 41(2):495–502
16. Lindley DV (2013) Understanding uncertainty. Wiley
17. MacKay DJC (1992) Bayesian interpolation. Neural Comput 4:415–447

18. Magni P, Bellazzi R, De Nicolao G (1998) Bayesian function learning using MCMC methods. IEEE Trans Pattern Anal Mach Intell 20(12):1319–1331
19. Papoulis A (1984) Probability, random variables and stochastic processes. Mc Graw-Hill
20. Pillonetto G, De Nicolao G (2010) A new kernel-based approach for linear system identification. Automatica 46(1):81–93
21. Pillonetto G, Chiuso A (2015) Tuning complexity in regularized kernel-based regression and linear system identification: the robustness of the marginal likelihood estimator. Automatica 58:106–117
22. Wahba G (1990) Spline models for observational data. SIAM, Philadelphia
23. Zhu H, Rohwer R (1996) Bayesian regression filters and the issue of priors. Neural Comput Appl 4:130–142

Chapter 5
Regularization for Linear System Identification

Abstract Regularization has been intensively used in statistics and numerical analysis to stabilize the solution of ill-posed inverse problems. Its use in System Identification, instead, has been less systematic until very recently. This chapter provides an overview of the main motivations for using regularization in system identification from a "classical" (Mean Square Error) statistical perspective, also discussing how structural properties of dynamical models such as stability can be controlled via regularization. A Bayesian perspective is also provided, and the language of maximum entropy priors is exploited to connect different form of regularization with time-domain and frequency-domain properties of dynamical systems. Some numerical examples illustrate the role of hyper parameters in controlling model complexity, for instance, quantified by the notion of Degrees of Freedom. A brief outlook on more advanced topics such as the connection with (orthogonal) basis expansion, McMillan degree, Hankel norms is also provided. The chapter is concluded with an historical overview on the early developments of the use of regularization in System Identification.

5.1 Preliminaries

As we have discussed in the preceding chapters, system identification can be framed as an inverse problem which aims at finding a dynamical model $\mathcal{M}$ from a set of measured input output "training" data $\mathcal{D}_T := \{u(t), y(t)\}_{t=1,\dots,N}$. The field of inverse problems [5] has motivated the development of, and is pervaded by, regularization techniques; as such it is evident that regularization could and should play a major role also in the system identification arena.

On the contrary, we believe it is fair to say that regularization has not had a pervasive impact in system identification until very recently. To introduce its use in this field, we will refer to linear models $\mathcal{M} = \{\mathcal{M}(\theta)|\theta \in D_{\mathcal{M}}\}$ introduced in Chap. 2, Eq. (2.1). Note that this notation not only includes classical parametric structures,

G. Pillonetto et al., *Regularized System Identification*, Communications and Control
Engineering, https://doi.org/10.1007/978-3-030-95860-2_5

such as ARX, ARMAX, Box–Jenkins models but also so-called nonparametric ones where the "parameter" θ may be infinite dimensional, e.g., containing all the impulse response coefficients of the filters $W_y(q)$ and $W_u(q)$ which characterize the predictor

$$\hat{y}(t|\theta) = W_y(q)y(t) + W_u(q)u(t). \tag{5.1}$$

The transfer functions $W_y(q)$ and $W_u(q)$ are related to the input–output model

$$y(t) = G(q, \theta)u(t) + H(q, \theta)e(t)$$

by the relation

$$W_y(q) := [1 - H^{-1}(q, \theta)] \quad W_u(q) := H^{-1}(q, \theta)G(q, \theta), \tag{5.2}$$

see also (2.4).

For simplicity here, we consider the single-output case $y(t) \in \mathbb{R}$. In the prediction error framework described in Chap. 2, the model fit is typically measured by the negative log likelihood

$$V_N(\theta) = -2\log \mathrm{p}(\mathscr{D}_T|\theta) = -2\sum_{t=1}^{N} \log(\mathrm{p}(y(t) - \hat{y}(t|\theta))),$$

which in the Gaussian case is, up to constants, proportional to the sum of squared prediction errors

$$V_N(\theta) \propto \sum_{t=1}^{N}(y(t) - \hat{y}(t|\theta))^2.$$

As discussed in Chap. 3, regularization can be added to make the inverse problem of estimating the model $\mathcal{M}(\theta)$ from data well-posed, and therefore regularized estimators $\hat{\theta}_R$ of the form

$$\hat{\theta}_R := \arg\min_{\theta} W_N(\theta) = \arg\min_{\theta} V_N(\theta) + J_\gamma(\theta) \tag{5.3}$$

are considered. This framework has been extensively discussed in the previous chapter in the context of linear regression under the squared loss $V_N(\theta) = \|Y - \Phi\theta\|_2^2$, see e.g., Eq. (3.57).

The function $J_\gamma(\theta)$ is usually referred to as the *penalty function*, and possibly depends on some (hyper-)parameter γ. In the simplest case $J_\gamma(\theta)$ takes the multiplicative form

$$J_\gamma(\theta) := \gamma J(\theta)$$

and γ acts a scaling factor which controls the "amount" of regularization. The most famous example is the so-called ridge regression problem, in which a quadratic loss $V_N(\theta)$ is used and $J(\theta) := \|\theta\|^2$ so that (see also (3.61a)):

$$\hat{\theta}^R := \arg\min_{\theta} \ \|Y - \Phi\theta\|_2^2 + \gamma \|\theta\|^2 = \left(\Phi^T \Phi + \gamma I\right)^{-1} \Phi^T Y.$$

However, ridge regression has not had a significant impact in the context of System Identification, i.e., when the vector θ contains the impulse response coefficients of a (linear) dynamical system. To understand why, it is important to discuss the choice of $J_\gamma(\theta)$. We will see that it plays a fundamental role and strongly influences the properties of the estimator $\hat{\theta}_R$. In particular, we will see how $J_\gamma(\theta)$ should be designed to encode properties of dynamical systems such as BIBO stability, smoothness in time domain and frequency domain, oscillatory behaviour and so on; this is a form of "inductive bias" well known and studied in the machine learning community, see e.g., [61].

As argued in Chap. 4, regularization can be given a Bayesian interpretation. In fact, introducing a probabilistic prior on model parameters θ of the form

$$p_\gamma(\theta) \propto e^{-\frac{J_\gamma(\theta)}{2}} \tag{5.4}$$

and the Likelihood function:

$$p(\mathscr{D}_T|\theta) \propto e^{-\frac{V_N(\theta)}{2}} \tag{5.5}$$

the maximum a posteriori (MAP) estimator of θ (see (4.2)), becomes

$$\hat{\theta}^{MAP} := \arg\max_\theta \ p(\theta|\mathscr{D}_T) \tag{5.6}$$

$$- \arg\max_\theta \ p(\mathscr{D}_T|\theta)p_\gamma(\theta) \tag{5.7}$$

$$= \arg\max_\theta \ \log\left[p(\mathscr{D}_T|\theta)p_\gamma(\theta)\right] \tag{5.8}$$

$$= \arg\min_\theta \ -\log\left[p(\mathscr{D}_T|\theta)\right] - \log\left[p_\gamma(\theta)\right] \tag{5.9}$$

$$= \arg\min_\theta \ V_N(\theta) + J_\gamma(\theta) \tag{5.10}$$

$$= \hat{\theta}_R. \tag{5.11}$$

In what follows, we will therefore use interchangeably the "regularization" framework, and thus think of $J_\gamma(\theta)$ as a penalty function, or the "Bayesian" framework, and thus think of $p_\gamma(\theta)$ as a prior (with some caution in the infinite-dimensional case).

5.2 MSE and Regularization

The final goal of modelling is to perform some task, e.g., prediction or control, on future unseen data. As such the estimated model quality should be measured having the objective in mind. For simplicity, we will consider a prediction task, referring the reader to the literature discussed in Sect. 5.9 for extensions. To this purpose, in addition to the training data $\mathscr{D}_T$, let us introduce testing data:

$$\mathscr{D}_{test} := \{u_{test}(t), y_{test}(t)\}_{t=1,\ldots,N_{test}}.$$

A model $\hat{\mathscr{M}} := \mathscr{M}(\hat{\theta})$ estimated using the training data $\mathscr{D}_T$ should then predict well testing data $\mathscr{D}_{test}$. In particular, let $\hat{y}(t|\hat{\theta})$ be the output prediction at instant t constructed using the estimated model. Then, we can measure the performance of $\hat{\mathscr{M}}$ using the Mean Squared Error (MSE) on output (Y) prediction and assuming that data are generated by some "true", yet unknown parameter vector θ_0. This is defined as

$$MSE_Y(\hat{\mathscr{M}}, \theta^o) = \mathscr{E}\left(\frac{1}{N_{test}} \sum_{t=1}^{N_{test}} (y_{test}(t) - \hat{y}_{test}(t|\hat{\theta}))^2\right) = \mathscr{E}\left(y_{test}(t) - \hat{y}_{test}(t|\hat{\theta})\right)^2,$$
(5.12)

where, for simplicity, we have assumed stationary statistics for the couples $u_{test}(t)$, $y_{test}(t)$ in the last passage. In this section, we will argue that using regularization in estimating $\hat{\theta}$ can indeed help in obtaining a small $MSE_Y(\hat{\mathscr{M}}, \theta_0)$. Let us first assume that data are generated by an unknown "true" linear time-invariant (LTI) causal model:

$$y(t) = \sum_{k=1}^{\infty} g_k u(t-k) + e(t),$$
(5.13)

where the "true" "parameter" $\theta_0 = [g_1, g_2, g_3, \ldots, g_n, \ldots]$ is an infinite sequence in ℓ^1, i.e.,

$$\sum_{k=1}^{\infty} |g_k| < \infty.$$

We now consider the model class $\mathscr{M}(\theta)$ of Finite Impulse Response (FIR) Output Error (OE) models

$$y(t) = \sum_{k=1}^{n} \theta_k u(t-k) + e(t),$$
(5.14)

where the parameter vector $\theta \in \mathbb{R}^n$ contains the coefficients of an nth-order finite impulse response model. Under the assumption that the input process is unit variance white noise, independent of the measurement noise, and defining

$$\hat{g}_k := \begin{cases} \hat{\theta}_k & k = 1, \ldots, n \\ 0 & \text{otherwise} \end{cases}$$

the MSE (5.12) has the expression

$$
\begin{aligned}
MSE_Y(\hat{\mathscr{M}}, \theta_0) &= \mathscr{E}(y_{test}(t) - \hat{y}_{test}(t|\hat{\theta}))^2 \\
&= \mathscr{E}\left(\textstyle\sum_{k=1}^{\infty}(g_k - \hat{g}_k)u_{test}(t-k) + e(t)\right)^2 \\
&= \underbrace{\sum_{k=1}^{\infty} \mathscr{E}(g_k - \hat{g}_k)^2}_{\mathscr{E}\|g-\hat{g}\|^2} + \sigma^2 \\
&= \underbrace{\sum_{k=1}^{\infty} \mathscr{E}(\hat{g}_k - \mathscr{E}[\hat{g}_k])^2}_{Variance} + \underbrace{\sum_{k=1}^{\infty}(g_k - \mathscr{E}[\hat{g}_k])^2}_{Bias^2} + \sigma^2 \\
&= \underbrace{\sum_{k=1}^{n} \mathscr{E}(\hat{\theta}_k - \mathscr{E}[\hat{\theta}_k])^2}_{Variance} + \underbrace{\sum_{k=1}^{n}(g_k - \mathscr{E}[\hat{\theta}_k])^2 + \sum_{k=n+1}^{\infty} g_k^2}_{Bias^2} + \sigma^2.
\end{aligned}
$$

(5.15)

This is nothing but the usual *bias-variance trade-off* discussed in Chap. 1: the model (θ in this case) has to be rich enough (i.e., n large) to capture the "true" data generating mechanism (low bias) but also simple enough (i.e., n small) to be estimated using the available data with low variability (low variance). The squared loss

$$
\mathscr{E}\|g - \hat{g}\|^2 = \sum_{k=1}^{\infty} \mathscr{E}(g_k - \hat{g}_k)^2
$$

present on the right-hand side of (5.15), after the third equality, is called a *compound loss* on the (possibly infinite) vector θ [60, 63] and defines the MSE.

Considering compound losses of this type allows us to connect with the discussion made in Chap. 1 on Stein's effect. To simplify exposition, let us assume that the identification input is a discrete impulse $u(t) = \delta(t)$ so that we can think of $y(t)$ as direct noisy measurements of all the (nonzero) impulse response coefficients

$$
y(t) = g_t + e(t) \qquad t = 1, \dots, n. \tag{5.16}
$$

Defining $Y := [y(1), \dots, y(n)]^T$ and $E := [e(1), \dots, e(n)]^T$ the measurement model (5.16) can be written in vector form

$$
Y = \theta + E, \qquad E \sim \mathscr{N}(0, \sigma^2 I_n). \tag{5.17}
$$

As we have seen in Chap. 1, the least squares estimator $\hat{\theta}_{LS}$ for model (5.17) is dominated (for $n > 2$) by the James–Stein estimator discussed in Sect. 1.1.1. As argued in Chap. 1, the James–Stein estimator (1.3) is a special case of a regularized estimator (5.3) where $J_\gamma(\theta) = \gamma \|\theta\|^2$ and γ takes the data-dependent form (1.4)

$$
\gamma = \frac{(n-2)\sigma^2}{\|y\|^2 - (n-2)\sigma^2}.
$$

Following this route, the James–Stein estimator favours "small" parameters values (the regularization term $J_\gamma(\theta) = \gamma \|\theta\|^2$ penalises large $\|\theta\|$) and therefore it is to be expected that the gap w.r.t. the least square estimator is larger under these circumstances; this has been illustrated in Fig. 1.1.

As pointed out in Sect. 1.1.2, there is actually nothing special in having chosen the origin as a reference. In fact, the penalty term can be replaced with $J_\gamma(\theta) = \gamma \|\theta - a\|^2$ for any $a \in \mathbb{R}^n$ yielding to estimators which always dominate least squares provided γ is chosen as

$$\frac{(n-2)\sigma^2}{\|y - a\|^2 - (n-2)\sigma^2}.$$

This teaches us that under certain circumstances it is possible to steer the estimators, using a suitable penalty functional, towards certain regions of the parameter space (or more generally model space); most importantly, this can be done without any loss (actually with a gain) for any possible occurrence of the "true" yet unknown system. However the reader should remind that this only holds for the compound loss (5.15) and should not be seen as *panacea*. For instance, James–Stein estimators may provide only marginal improvements over Least Squares in situations where the signal-to-noise ratio is highly non-uniform over parameter space, a situation often encountered in system identification when input signals are not white and poor excitation may be present, e.g., in certain frequency bands. This has been illustrated in Example 1.2.

Therefore, as a take home message from Chap. 1 and the discussion above, we should remind that regularization has potential to offer, yet its use in system identification is not straightforward. The main reasons are as follows:

1. Often one cannot restrict to Output Error models (i.e., also noise models should be included) and the input process is neither impulsive nor white. Thus, the MSE (5.12) takes a different form than (5.15). This calls for extensions of James–Stein estimators to weighted losses and non-orthogonal design; to some extent this has been pursued in the statistics literature, the reader is referred to [4, 9, 43, 64] and references therein. See also [13, Sect. 6].
2. While James–Stein estimators have been built with the purpose of showing that the least squares estimator is not admissible (see Sect. 1.1.1, for a formal definition), it may not necessarily be our primary goal to dominate least squares (or another estimator) uniformly over parameter space. In order to cure the ill-conditioning phenomenon widely discussed in Chap. 3, it could be advantageous to tailor regularization to certain "dynamical-system" oriented properties, thus gaining a lot in certain regions of the model space, while possibly incurring in minor losses in other regions which are very unlikely.

The latter is one of the main goals of this book, i.e., to provide the reader with a thorough understanding of the role of regularization in estimating dynamical systems so as to optimally design regularization methods depending on the intended use of the model. In the remaining part of the chapter, we will first introduce the concept of "optimal" prior and derive its expression. We will then connect the structure of the optimal prior to the notion of BIBO stability for linear dynamical systems and also its link with smoothness in time and frequency domains. Connection with the Bayesian setting will also be provided. The chapter will be concluded with an historical overview of how the use of regularization in the context of estimation of dynamical systems has evolved, illustrating also the role played by time- and frequency-domain smoothness.

5.3 Optimal Regularization for FIR Models

Let us consider the problem of estimating the impulse response $\{\theta_k\}_{k=1,\dots,n}$ of the FIR model (5.14) using data $\{y(t)\}_{t=1,\dots,N}$. The FIR model can be compactly written as

$$Y = \Phi\theta + E, \tag{5.18}$$

where $Y := [y(1), \dots, y(N)]^T$, $E := [e(1), \dots, e(N)]^T$ and Φ contains the input samples, which are assumed to be available for all times needed to avoid issues related to the initial condition. Then, we will still use θ_0 to denote the "true" value that has generated the data.

We now consider the class of regularized estimators

$$\hat{\theta}^R := \arg\min_{\theta \in \mathbb{R}^n} \ \|Y - \Phi\theta\|^2 + \sigma^2\theta^T P^{-1}\theta$$

parametrized by the regularization matrix $P = P^T > 0$. As shown in Chap. 3, see Eq. (3.60), the generalized ridge regression estimator $\hat{\theta}^R$ can be extended also to the case P is singular so that we can assume $P = P^T \succeq 0$. As a matter of fact, in the Bayesian framework introduced in Chap. 4, θ^R can be also interpreted as the MAP estimator

$$\hat{\theta}^{\text{MAP}} := \arg\max \ p(\theta|Y)$$

obtained under the assumption that the noise E is Gaussian, zero mean and variance $\sigma^2 I$ and that θ is independent of E, zero-mean Gaussian with (possibly singular) variance $P = P^T \succeq 0$ (the singular case was described in (4.19)).

In this section, to emphasize the dependence of the estimator $\hat{\theta}^R$ on $P = P^T \succeq 0$, we will use the notation

$$\hat{\theta}^P := \hat{\theta}^R = \hat{\theta}^{\text{MAP}}.$$

Our objective now is to study the performance of the estimator $\hat{\theta}^P$, in terms of MSE, as a function of $P = P^T \succeq 0$, under the assumption that Y has been generated by a "true model" of the form (5.18) with a deterministic and unknown parameter θ_0. Thus, the only source of "randomness" is the noise vector E and the system input which is seen as a stochastic process (independent of E) in this section.

We consider a test experiment with a new input $u_{test}(t)$, independent of the input $u(t)$ used for identification; for convenience of notation, we define the lagged test input vector

$$\phi_{test}(t) := [u_{test}(t), \ldots, u_{test}(t - n + 1)]^T$$

so that under (5.14) the test output is given by

$$y_{test}(t) = \phi_{test}^T(t)\theta_0 + e_{test}(t).$$

Let us also define the covariance matrix

$$W_u = Var\{\phi_{test}(t)\} = \mathscr{E}\phi_{test}(t)\phi_{test}^T(t)$$

(note that stationary assumptions are present here, in fact W_u does not depend on time t) and the MSE matrix

$$M_{\theta_0}(P) := \mathscr{E}(\theta_0 - \hat{\theta}^P)(\theta_0 - \hat{\theta}^P)^T.$$

If we now consider the output mean squared error $MSE_Y(\hat{\mathscr{M}}, \theta_0)$ in (5.12) computed for the model $\hat{\mathscr{M}}$, we obtain

$$
\begin{aligned}
MSE_Y(\hat{\mathscr{M}}, \theta^\circ) &= \mathscr{E}\left(y_{test}(t) - \hat{y}_{test}(t|\hat{\theta}^P)\right)^2 \\
&= \mathscr{E}\left[\phi_{test}^T(t)\theta_0 + e_{test} - \phi_{test}^T(t)\hat{\theta}^P\right]^2 \\
&= \mathscr{E}\left[(\theta_0 - \hat{\theta}^P)^T \phi_{test}(t)\phi_{test}^T(t)(\theta_0 - \hat{\theta}^P)^T\right] + \sigma^2 \\
&= Tr\{\mathscr{E}(\theta_0 - \hat{\theta}^P)(\theta_0 - \hat{\theta}^P)^T \mathscr{E}\phi_{test}(t)\phi_{test}^T(t)\} + \sigma^2 \\
&= Tr\{M_{\theta_0}(P)W_u\} + \sigma^2,
\end{aligned}
\tag{5.19}
$$

where in the second to last equation, we have used that the test inputs and noises are assumed to be independent of the training inputs and noise in the identification data used for estimating $\hat{\theta}^P$.

A direct consequence of this fact is that, given two prior covariance matrices P and P^*, if $M_{\theta_0}(P) \succeq M_{\theta_0}(P^*)$, then

$$MSE_Y(\hat{\theta}^P, \theta_0) \geq MSE_Y(\hat{\theta}^{P^*}, \theta_0) \qquad \forall W_u,$$

i.e., estimator θ^{P^*} outperforms θ^P in terms of output prediction for any possible choice of the test input covariance W_u. Thus, if the modelling purpose is output prediction, it is of interest to minimize, w.r.t. all possible $P = P^T \succeq 0$, the matrix $M_{\theta_0}(P)$, i.e., to find

$$P^* := \underset{P=P^T \succeq 0}{\arg\min} \ M_{\theta_0}(P), \tag{5.20}$$

so that $\hat{\theta}^{P^*}$ outperforms any other $\hat{\theta}_P$ in terms of output error (5.15) for any choice of the (test) input covariance W_u. Under the assumption that the true model generating the data is an FIR model of length n with impulse response

$$g_k = \begin{cases} \theta_{0,k} & k \leq n \\ 0 & k > n, \end{cases}$$

the solution P^* of the minimization problem in (5.20) has been derived in Proposition 3.1, and takes the form

$$P^* = \theta_0 \theta_0^T, \tag{5.21}$$

where θ_0 is the "true" impulse response of the data-generating mechanism (5.14). An alternative proof of the optimal solution (5.21) to problem (5.20) can be found in Sect. 5.10.1. Since P^* depends on the unknown true system, this result is not of practical interest; however, if we think of the FIR model (5.14) as the approximation of a BIBO stable infinite impulse response model

$$y(t) = \sum_{k=1}^{\infty} \theta_{0,k} u(t - k) + e(t), \tag{5.22}$$

the impulse response θ_0 should have finite ℓ_1 norm $\|\theta_0\|_1$, i.e.,

$$\|\theta_0\|_1 := \sum_{k=1}^{\infty} |\theta_{0,k}| < \infty, \tag{5.23}$$

and therefore $\theta_{0,k}$ should decay as a function of the index k. As a result, the entries $[P^*]_{ij} = \theta_{0,i} \theta_{0,j}$ of optimal kernel decay as functions of the row and column indexes i and j. In Bayesian terms, it is thus expected that also the elements $[P]_{ij}$ of any "good" candidate prior variance should do the same. As we will see later in this chapter, recent forms of regularization for system identification include a decay rate condition on the elements $[P]_{ij}$, so as to guarantee that the estimated system is BIBO stable. Therefore, we will often refer to conditions on the decay rate of P as "stability conditions". While condition (5.23) is obviously satisfied when θ is a finite dimensional vector, this loose connection between decay rate of the kernel and stability needs to be tightened. We will see in the next section that this can be properly formulated in a Bayesian framework.

5.4 Bayesian Formulation and BIBO Stability

In the previous section, we have considered only FIR models which are reasonable approximations of any BIBO LTI system in most practical scenarios. However, it is of interest to formulate the estimation of LTI BIBO stable systems in full generality, without assuming the impulse response to be of finite support. This entails working with infinite dimensional impulse responses $\{\theta_k\}_{k\in\mathbb{N}}$. In this chapter, we first consider the Bayesian framework, while regularization in infinite-dimensional Hilbert spaces will be addressed in Chap. 6. To start with, we model the unknown impulse response $\{\theta_k\}_{k\in\mathbb{N}}$ as a stochastic process indexed over time k; this is the straightforward extension to the infinite-dimensional case of (5.18) where θ was a finite-dimensional random vector. In this context, it is of interest to introduce the concept of "stable" priors:

Definition 5.1 (*Stable priors*) A prior on $\{\theta_k\}_{k\in\mathbb{N}}$ is said to be stable if realizations are sequences almost surely in ℓ_1, i.e.,

$$\sum_{k=1}^{\infty}|\theta_k| < \infty \quad a.s.$$

In most of this book, mostly for computational reasons, we will also assume that $\{\theta_k\}_{k\in\mathbb{N}}$ be Gaussian (i.e., that any finite collection of random variables $\{\theta_k\}_{k\in I}$, $I = \{i_1, \ldots, i_\ell\}$, $i_k \in \mathbb{N}$, $\ell \in \mathbb{N}$ are jointly Gaussian). This is formalized in the following assumption.

Assumption 5.1 Under the Bayesian framework, we assume $\{\theta_k\}_{k\in\mathbb{N}}$ to be a Gaussian stochastic process with mean $\{m_k\}_{k\in\mathbb{N}}$ and covariance function $K(t, s), t, s \in \mathbb{N}$. □

It is an interesting fact that, under additional assumptions on the mean and covariance functions, the prior is stable according to Definition 5.1, as formalized in the following lemma whose proof is in Sect. 5.10.2.

Lemma 5.1 *Under Assumption 5.1 and if the following additional conditions hold*

$$\sum_{k=1}^{\infty}|m_k| = M_{\ell_1} < \infty \qquad \sum_{k=1}^{\infty}K(k,k)^{1/2} = K_{\ell_1} < \infty, \qquad (5.24)$$

then the prior is stable as per Definition 5.1, i.e.,

$$\sum_{k=1}^{\infty}|\theta_k| < \infty \quad a.s.$$

In most of this book, we will also make the assumption that the a priori mean m_t is identically zero, and thus only the condition on the covariance $K(t, s)$ should be checked to ensure stability. We will now discuss different form of prior covariances K encountered in the literature.

5.5 Smoothness and Contractivity: Time- and Frequency-Domain Interpretations

As seen in Sect. 5.3, the optimal regularizer should mimic the "true" impulse response, which is clearly unfeasible since the impulse response is unknown. However, as already discussed in Sect. 5.4, we can use the prior to encode qualitative behaviour of impulse responses of BIBO stable linear systems. In particular we have seen in Lemma 5.1 that a certain decay condition on the prior mean and covariance guarantees the description of only (almost surely) BIBO stable linear systems. The simplest example of such a prior model is the following.

Example 5.2 (*Diagonal (DI) prior*) Assume the prior mean to be zero $m_t = 0$, $\forall t \in \mathbb{N}$ and the covariance function to be diagonal with exponentially decaying entries

$$K(t, s) = \lambda \alpha^t \delta(t - s) \qquad t, s \in \mathbb{N} \qquad \lambda > 0, \quad 0 \le \alpha < 1.$$

The parameters λ (scale factor) and α (decay rate) are treated as hyperparameters to be estimated from data, using e.g., marginal likelihood maximization, as described in Sect. 4.4. It is worth observing that the assumptions of Lemma 5.1 are satisfied, indeed

$$\sum_{t \in \mathbb{N}} |m_t| = 0 \qquad \sum_{t \in \mathbb{N}} K(t, t)^{1/2} = \sum_{t \in \mathbb{N}} \sqrt{\lambda} \alpha^{t/2} = \sqrt{\lambda} \frac{\sqrt{\alpha}}{1 - \sqrt{\alpha}} < \infty$$

and hence this is a *stable prior*.

$\square$

It is interesting to observe that a decay rate condition on the impulse response coefficients is equivalent to assuming a smoothness condition in the frequency domain. To see this, let us introduce the frequency response function

$$G(e^{j\omega}) := \sum_{k=1}^{\infty} \theta_k e^{-j\omega k}.$$

The L_2-norm of the first derivative $\frac{dG(e^{j\omega})}{d\omega}$ can be considered

$$\left\| \frac{dG(e^{j\omega})}{d\omega} \right\|^2 := \frac{1}{2\pi} \int_0^{2\pi} \left| \frac{dG(e^{j\omega})}{d\omega} \right|^2 d\omega$$

which using Parseval's theorem can be expressed in time domain

$$\left\| \frac{dG(e^{j\omega})}{d\omega} \right\|^2 = \sum_{k=1}^{\infty} k^2 |\theta_k|^2. \tag{5.25}$$

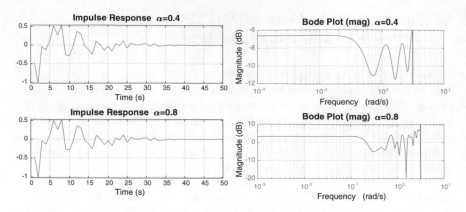

Fig. 5.1 Sample realizations from the diagonal kernel prior for $\alpha = 0.4$ (top) and $\alpha = 0.8$ (bottom). Impulse response is on the left, frequency response (magnitude only) on the right

Computing higher-order derivatives, and using again Parseval's theorem, the L_2-norm of the mth-order derivative is given by

$$\left\| \frac{d^{(m)}G(e^{j\omega})}{d\omega^{(m)}} \right\|^2 = \sum_{k=1}^{\infty} k^{2m}|\theta_k|^2. \tag{5.26}$$

Hence, the condition that the $\{\theta_k\}$ decay rapidly (and possibly exponentially as postulated by the Diagonal kernel) with k, implies a bound on the L_2 norm of the mth-order derivatives, i.e., smoothness in the frequency domain of the model.

As illustrated in Fig. 5.1, smoothness in the frequency domain decreases when α increases. However, under this prior, the impulse response coefficients are modelled as independent (yet not identically distributed) random variables. Thus no smoothness in the time domain is included, as for instance, is typically performed with priors based on random walk, which are the discrete-time counterpart of spline models as discussed in Sect. 4.9. A prior model that, in addition to stability, also includes a smoothness condition in the time domain, is the so-called TC-kernel:

Example 5.3 (*Tuned-Correlated (TC) prior*) Assume the prior mean is zero $m_t = 0$, $\forall t \in \mathbb{N}$ and the covariance function takes the form

$$K(t, s) = \lambda \alpha^{\max(t,s)} \qquad t, s \in \mathbb{N} \qquad \lambda > 0, \quad 0 \le \alpha < 1.$$

As in the previous example, the parameters λ (scale factor) and α (decay rate) are treated as hyperparameters to be estimated from data, using e.g., marginal likelihood maximization. It is worth observing that also in this case the assumptions of Lemma 5.1 are satisfied, indeed

$$\sum_{t \in \mathbb{N}} |m_t| = 0 < \infty \qquad \sum_{t \in \mathbb{N}} K(t, t)^{1/2} = \sum_{t \in \mathbb{N}} \sqrt{\lambda} \alpha^{t/2} = \sqrt{\lambda} \frac{\sqrt{\alpha}}{1 - \sqrt{\alpha}} < \infty$$

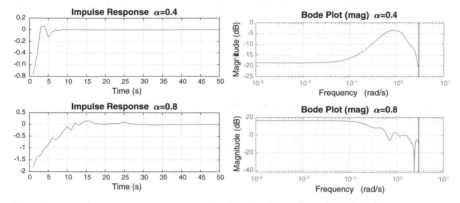

Fig. 5.2 Sample realizations from the Tuned-Correlated (TC) prior for $\alpha = 0.4$ (top) and $\alpha = 0.8$ (bottom). Impulse response is on the left, frequency response (magnitude only) on the right

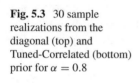

Fig. 5.3 30 sample realizations from the diagonal (top) and Tuned-Correlated (bottom) prior for $\alpha = 0.8$

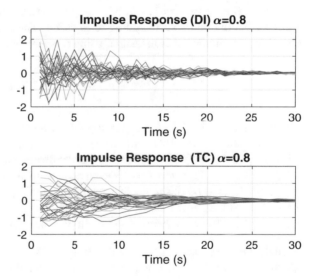

and hence this is a *stable prior*. In addition, the TC prior now introduces correlation between impulse response coefficients, e.g., one has

$$\mathscr{E}\theta_t\theta_s = \lambda\alpha^t \quad \forall t \geq s.$$

So, the correlation is different from zero and exponentially decays to zero.   □

Figure 5.2 shows two typical realizations from the TC prior, both in time domain and frequency domain, for $\alpha = 0.4$ (top) and $\alpha = 0.8$ (bottom), while Fig. 5.3 shows 30 sample realizations from the DI (top) and TC (bottom) priors, respectively.

Example 5.4 (*Importance of stable priors*) In order to illustrate the advantage of using stable priors, we now consider a simple example of identification of an output error model. In particular, we consider a system of the form

$$y(t) = \sum_{k=1}^{\infty} g_k u(t-k) + e(t),$$

where the measured input $u(t)$ and the noise $e(t)$ are realizations from white Gaussian noise with zero mean and unit variance. The impulse response is

$$g_k = \begin{cases} \left(\frac{k}{2}\right)^2 e^{-\frac{k}{4}} & k \geq 1 \\ 0 & k < 1 \end{cases}.$$

For the purpose of identification, we assume the input is available at all time instances needed. For illustration purposes, the impulse response has been truncated at $k = 50$, since it is in practice zero for $k > 50$. We also assume that output measurements $y(t)$ are available for $t = 1, \ldots, 35$. The hyperparameters are all estimated using marginal likelihood maximization, see Sect. 4.4. The results are shown in Fig. 5.4. The reconstruction error is measured using the percentage root mean square (RMS) error:

$$\sqrt{\frac{\sum_{k=1}^{\infty}(g_k - \hat{g}_k)^2}{\sum_{k=1}^{\infty} g_k^2}} \times 100\%. \tag{5.27}$$

As illustrated in Fig. 5.4, it is apparent that the results obtained by using the stable priors, see panels (b) and (c), outperform those returned by the spline (random walk) prior, see panel a, that does not include the stability constraint. The best relative error is obtained by the TC priors ($\simeq 10\%$) and goes up to as much as $\simeq 33\%$ for the spline priors. It can also be observed that while for stable priors (b) and (c) confidence intervals shrink as time index k grows, the same does not hold for the spline prior. The same behaviour had been observed in Sect. 4.9, see Fig. 4.1.   □

In the next section, a class of stable priors, which includes TC as a special case, will be derived following a first-principle maximum entropy framework.

5.5.1 Maximum Entropy Priors for Smoothness and Stability: From Splines to Dynamical Systems

The class of Stable Spline priors introduced in the paper [49] extends smoothness priors ideas used in splines models introduced in Sect. 4.9, embedding exponential decay conditions on the impulse response prior. They ultimately lead to estimated models which are BIBO stable with probability 1.

In this section, we will introduce a simple construction of these stable spline priors in discrete time. In particular, we will exploit a very natural axiomatic derivation in the maximum entropy framework introduced in Chap. 4. For the sake of illustration,

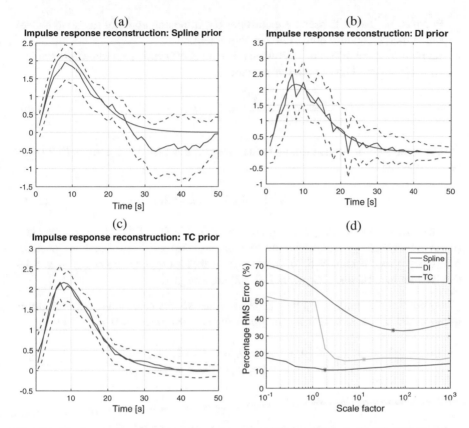

Fig. 5.4 Panels **a–c**: impulse response reconstruction (blue) and true (red) with 95% Bayesian confidence intervals (dashed). Panel **d** is the relative RMS error (5.27) on impulse response reconstruction as a function of the scale factor λ. For DI and TC priors for each scale factor, the optimal decay rate α is estimated using marginal likelihood. The star denotes the performance obtained using the scale factor selected using marginal likelihood optimization. It is remarkable that the relative error achieved by maximizing the marginal likelihood is close to the minimum achievable by an oracle who would have access to the true impulse response and thus could minimize the relative RMS error

we will only consider the so-called stable spline prior of order one (also known as the TC prior, see Example 5.3) and its extension known as DC prior. Possible extensions will be discussed, but not developed in full detail.

The most natural construction, inspired by smoothing spline ideas, is based on the following two observations:

1. Stability: the variance of θ_k should decay "sufficiently fast" (see Lemma 5.1), possibly exponentially, with the lag k. Assuming a zero-mean process, this can be expressed using a condition on second-order moments of the form:

$$\mathscr{E}\left[\theta_k^2\right] = \lambda_S \alpha^k \quad k = 1, \ldots, n \quad 0 < \alpha < 1. \tag{5.28}$$

For reasons that will become clear later on, imposing equality (as done above) rather than inequality constraints is convenient.

2. Smoothness: the difference between adjacent coefficients should be constrained, e.g., as measured by the relative variance,

$$\frac{\mathscr{E}\left[(\theta_{k-1} - \theta_k)^2\right]}{\mathscr{E}\left[\theta_{k-1}^2\right]} = \lambda_R \quad k = 2, \dots, n. \tag{5.29}$$

Using the stability constraint and redefining the constant λ_R, condition (5.29) can be rewritten as

$$\mathscr{E}\left[(\theta_{k-1} - \theta_k)^2\right] = \lambda_R \alpha^{k-1} \quad k = 2, \dots, n. \tag{5.30}$$

The following theorem (whose proof is reported in Sect. 5.10.3) derives the class of maximum entropy priors under the constraints (5.28) and (5.29). Next, in Corollary 5.1 (whose proof is in Sect. 5.10.4), we will see that for special choices of λ_S and λ_R the well-known TC and DC priors [10, 52] are obtained.

Theorem 5.5 *Let $\{\theta_k\}_{k=1,\dots,n}$ be a zero mean, absolutely continuous random vector with density $p_\theta(\theta)$, that satisfies the following constraints (with $0 < \alpha < 1$):*

$$\begin{aligned} \mathscr{E}\left[\theta_k^2\right] &= \lambda_S \alpha^k \quad k = 1, \dots, n \\ \mathscr{E}\left[(\theta_{k-1} - \theta_k)^2\right] &= \lambda_R \alpha^{k-1} \quad k = 2, \dots, n \end{aligned} \tag{5.31}$$

with $\lambda_S \in \mathbb{R}$ and $\lambda_R \in \mathbb{R}$ such that

$$\lambda_S(1 - \sqrt{\alpha})^2 < \lambda_R < \lambda_S(1 + \sqrt{\alpha})^2. \tag{5.32}$$

Then, the solution $p_{\theta,ME}(\theta)$ of the maximum entropy problem

$$p_{\theta,ME} := \underset{p(\cdot) \ s.t. \ (5.31)}{\arg \max} \quad -\mathscr{E} \log(p_\theta(\theta)) \tag{5.33}$$

has the following form:

$$p_{\theta,ME}(\theta) = C e^{-\frac{1}{2}\theta^T \Sigma^{-1} \theta}, \tag{5.34}$$

where the matrix Σ^{-1} has the band structure:

$$\Sigma^{-1} = \begin{bmatrix} * & * & 0 & \dots \dots & 0 \\ * & * & * & 0 & \dots & 0 \\ 0 & * & * & * & 0 & \dots \\ \vdots & \dots & \ddots & \ddots & \ddots & \dots \\ 0 & \dots & 0 & * & * & * \\ 0 & \dots \dots & 0 & * & * \end{bmatrix}.$$

The maximum entropy process admits the backward representation

$$\theta_{k-1} = a_B \theta_k + w_k \quad w_k \sim \mathcal{N}(0, \sigma_k^2) \quad k \in \{1, \dots, n\}$$

with

$$a_B = \frac{\lambda_S(1+\alpha) - \lambda_R}{2\lambda_S\alpha}, \tag{5.35}$$

$$\sigma_k^2 = \lambda_S \alpha^{k-1}(1 - a_B^2\alpha), \tag{5.36}$$

and terminal condition

$$\mathcal{E}\theta_n^2 = \lambda_S \alpha^n. \tag{5.37}$$

Last, the autocovariance of θ_k satisfies the relation:

$$\mathcal{E}\theta_k\theta_h = \lambda_S a_B^{|k-h|}\alpha^{\max\{k,h\}}. \tag{5.38}$$

Corollary 5.1   *Under the conditions of Theorem 5.5 and defining*

$$\rho := a_B\sqrt{\alpha} = \frac{\lambda_S(1+\alpha) - \lambda_R}{2\lambda_S\sqrt{\alpha}}, \tag{5.39}$$

the maximum entropy model in Theorem 5.5 corresponds to the so-called DC-kernel [10], i.e.,

$$\mathcal{E}\theta_k\theta_h = \lambda_S \rho^{|k-h|}\alpha^{\frac{k+h}{2}}. \tag{5.40}$$

In particular, for $\lambda_R = \lambda_S(1 - \alpha)$, this reduces to the so-called TC kernel [10] with

$$\mathcal{E}\theta_k\theta_h = \lambda_S \alpha^{\max\{k,h\}}, \tag{5.41}$$

while for $\lambda_R = \lambda_S(1 + \alpha)$, we obtain the covariance of the "diagonal" kernel

$$\mathcal{E}\theta_k\theta_h = \begin{cases} \lambda_S\alpha^P & k = h \\ 0 & k \neq h \end{cases}. \tag{5.42}$$

Remark 5.1   In the maximum entropy kernel derived in Theorem 5.5, which includes DC, TC and DI as special cases as stressed in Corollary 5.1, the constant λ_S plays only the role of a scale factor while α is a "decay rate". Therefore, by fixing $\lambda_S = 1$ and $\alpha = 0.8$ we can study the behaviour as the "regularity" constant λ_R varies in the interval $\lambda_S(1 - \sqrt{\alpha})^2 = \lambda_{R,min} \leq \lambda_R \leq \lambda_{R,max} = \lambda_S(1 + \sqrt{\alpha})^2$. This is entirely equivalent to studying the behaviour of the kernel as a function of the ratio λ_R/λ_S. We thus consider a grid of 9 possible values $\lambda_{R,min} = \lambda_{R,1} < \lambda_{R,2} < \cdots < \lambda_{R,9} = \lambda_{R,max}$. Then, Fig. 5.5 plots 5 sample realizations for each of these values with panel (i) corresponding to the value $\lambda_{R,i}$. In particular, $\lambda_{R,4} = \lambda_S(1 - \alpha)$ corresponds to the TC kernel and $\lambda_{R,6} = \lambda_S(1 + \alpha)$ induces the DI kernel. For each realization

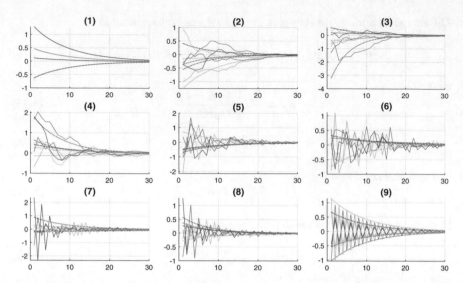

Fig. 5.5 Sample realizations (solid) and best (least squares) exponential fit as a function of the kernel parameters. In all figures $\alpha = 0.8$ and $\lambda_S = 1$. The regularity parameter λ_R varies, from its minimum value $\lambda_{R,min} = \lambda_S(1 - \sqrt{\alpha})^2 \simeq 0.011$ in panel (1) to the maximum value $\lambda_{R,max} = \lambda_S(1 + \sqrt{\alpha})^2 \simeq 3.589$ in panel (9). Panel (4), with $\lambda_R = 0.2$, corresponds to the TC kernel; panel (6) with $\lambda_R = 2.6$ to the DI kernel

from the prior (solid line) also its best single-exponential fit is shown in order to highlight the "overall" decay rate which can be thought of as an envelope of the curves. In panel (1), with λ_R taking the smallest possible value, hence imposing the "maximum" amount of regularity, all realizations are pure exponentials. In panel (9), with λ_R taking its maximum value, all realizations are pure damped oscillations. In fact, in both cases, it can be checked that the corresponding kernel is singular.

Degrees of Freedom of the DC Kernels

Theorem 5.5 provides a class of kernels K_η parametrized by the hyperparameter vector $\eta := [\lambda_S, \lambda_R, \alpha]$. In Fig. 5.5, we have illustrated how realizations from the prior change as a function of the regularity parameter λ_R having fixed $\lambda_S = 1$ (or equivalently as a function of the ratio λ_R/λ_S). As discussed in Chap. 4, choosing the prior is equivalent to describing the model class. In the linear system identification context, this then defines a penalty function on impulse responses. A way to measure the "size" of the model class is to use the concept of equivalent degrees of freedom, introduced in the Bayesian context in Sect. 4.8. Unfortunately, the degrees of freedom are defined in terms of the output predictor sensitivity and they thus require to specify not only the model class but also the experimental conditions under which the model is estimated. Only in limiting cases (such as improper prior on finitely and linearly parametrized model classes), degrees of freedom become independent of the experiment and coincide with the number of parameters. In this section, we thus consider the prototypical setup in Eq. (5.18):

$$Y = \Phi\theta_0 + E \quad Y \in \mathbb{R}^N, \quad N = 1000, \quad \theta_0 \in \mathbb{R}^n. \tag{5.43}$$

We recall that the matrix Φ is an Hankel matrix built with the input samples $\{u(t)\}$ so that $\Phi\theta_0$ implements the convolution of u with θ_0. The input $\{u(t)\}$ is now assumed to be a zero-mean unit variance white noise. We also assume the noise $\{e(t)\}$ is zero-mean unit variance white noise. We consider two scenarios in which the order of the system (length n of θ_0) is assumed to be either $n = 30$ or $n = 100$. Exploiting the derivation in Chap. 4 (see Definition 4.2 and Proposition 4.3), the degrees of freedom dof (η), as a function of the hyperparameter vector η, are given by

$$\text{dof}(\eta) = \text{trace}\left(\Phi(\Phi^T\Phi + K_\eta^{-1})^{-1}\Phi^T\right).$$

Assuming also here that $\lambda_S = 1$, we study how dof (η) varies as a function of λ_R for three different values of α (0.6, 0.8, and 0.95). The behaviour is illustrated in Fig. 5.6 where it is apparent that the maximum is achieved for the DI kernel, and the minimum (a bit smaller than 1) is attained at the extremum points, where the kernel has rank exactly equal to 1. It is interesting to observe the intertwining between the value of α (that controls the decay rate) and the length of the FIR model n. As the coefficient vector θ_0 changes from length $n = 30$ (left) to $n = 100$ (right) the effective "size" of the model doesn't change much for $\alpha = 0.6$ and $\alpha = 0.8$, while it does increase when $\alpha = 0.95$. This confirms the fact that the kernel, for α fixed, effectively controls the model complexity so that the estimator becomes insensitive to the chosen length, provided n is "big enough" w.r.t. α. In particular $n = 15$ would be sufficient for $\alpha = 0.6$, $n = 30$ for $\alpha = 0.8$ while for $\alpha = 0.95$ the effective size is about $n = 100$.

Extension to Smoothness Conditions on Filtered Versions ⋆

So far, we limited our attention to so-called "first-order" stable splines, which are derived imposing conditions on "first-order" differences, leading to first-order, i.e., AR(1), realizations. Of course these constructions can be generalized by replacing (5.31) with a higher-order constraint of the form

$$\begin{aligned} \mathcal{E}\|\theta_k\|^2 &\le \lambda_S\alpha^k \\ \mathcal{E}\|\theta_k - \textstyle\sum_{i=1}^p a_i\theta_{k+i}\|^2 &\le \lambda_R\alpha^k. \end{aligned} \tag{5.44}$$

While the first constraint is a "standard" stability condition, the second constraint can be interpreted as a filtered frequency domain smoothness condition. In fact, defining the filter $F(q) := 1 - \sum_{i=1}^p a_i q^i$, let us denote with θ_k^F the sequence obtained filtering θ_k with $F(q)$. The condition

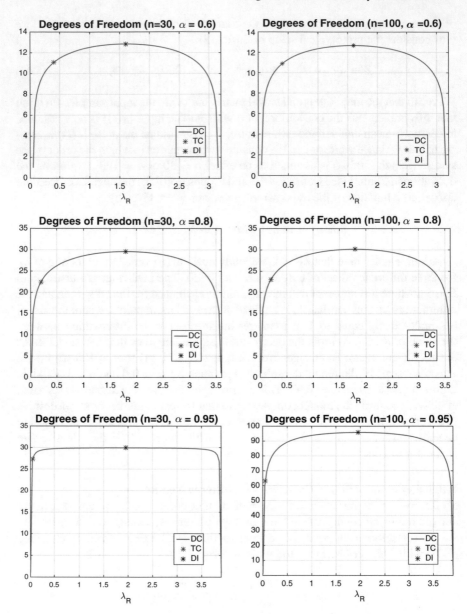

Fig. 5.6 Effective degrees of freedom of the DC kernel as a function of λ_R ($\lambda_S = 1$) for model (5.43); $n = 30$ (left), $n = 100$ (right). From top to bottom: $\alpha = 0.6$, $\alpha = 0.8$ and $\alpha = 0.95$

$$\mathscr{E}\|\theta_k^F\|^2 = \mathscr{E}\|\theta_k - \sum_{i=1}^{p} a_i \theta_{k+i}\|^2 \le \lambda_R \alpha^k$$

implies that θ_k^F should decay "fast" enough (in mean square) and thus

$$\mathscr{E}\sum_{k=0}^{\infty} k^{2m}\|\theta_k^F\|^2$$

should be small for any integer m. As a consequence, if

$$G(e^{j\omega}) := \sum_{k=1}^{\infty} \theta_k e^{-j\omega k},$$

using Parseval's theorem,

$$\mathscr{E}\int_{0}^{2\pi} \left\| F(e^{j\omega})\frac{d^{(m)}G(e^{j\omega})}{d\omega^{(m)}} \right\|^2 d\omega$$

should be small as well, implying that θ_k should concentrate most of his energy (variance) in frequency bands where the (absolute value of the) filter $F(e^{j\omega})$ is small.

We regard developments of this type, in principle, as a straightforward extension of the basic ideas discussed in this chapter to obtain DC kernels. In particular, the choice of the coefficients a in (5.44) is a design issue, which can be guided by prior knowledge on the candidate models, and its underlying principles and ideas are the same as those illustrated above. There are however additional complications due to the richer structure of the constraints, which might entail non-trivial issues to derive an analytic expression of the kernel.

5.6   Regularization and Basis Expansion ⋆

The ℓ_2 (ridge regression) regularized estimators that have been discussed in this chapter can also be framed in the context of basis expansion using the so-called Karhunen–Loève decomposition of the random process θ. For the sake of exposition, we will now consider the finite-dimensional case, i.e., we will study FIR models of length n of the form (5.14). Extension to the infinite-dimensional case will be discussed in the framework of Reproducing Kernel Hilbert Spaces illustrated in Chap. 6. Under this finite-dimensional assumption, we consider the covariance matrix $\mathbf{K} \in \mathbb{R}^{n \times n}$ whose entries satisfy $[\mathbf{K}]_{(t,s)} := K(t,s) = \mathrm{cov}(\theta_t, \theta_s)$. The matrix $\mathbf{K}$ can be written in terms of its spectral decomposition (Singular Value Decomposition) in the form:

$$\mathbf{K} = USU^T = \sum_{i=1}^{n} \xi_i u_i u_i^T \quad u_i \in \mathbb{R}^n \quad \|u_i\| = 1 \quad u_i \perp u_j \quad \forall i \neq j, \qquad (5.45)$$

where

$$U := [u_1, \ldots, u_n] \quad S := \text{diag}\{\xi_1, \ldots, \xi_n\}.$$

The set of vectors $u_i \in \mathbb{R}^n$ provides an orthonormal basis of $\mathbb{R}^n$ so that any impulse response $\theta \in \mathbb{R}^n$ can be written using the orthonormal basis expansion

$$\theta = \sum_{i=1}^{n} u_i \beta_i \qquad \beta_i := <\theta, u_i>, \qquad (5.46)$$

where the coefficients $\beta_i = <\theta, u_i> = u_i^T \theta$ are therefore zero-mean random vectors with covariances

$$\mathscr{E}\beta_i \beta_j = \mathscr{E} u_i^T \theta \theta^T u_j = u_i^T \mathbf{K} u_j = \xi_i \delta_{ij}.$$

Clearly, the argument above can be reversed. Namely, starting from (a possibly orthonormal) basis u_i, $i = 1, \ldots, n$ the random basis expansion

$$\theta = \sum_{i=1}^{n} u_i \beta_i, \qquad \beta_i \sim \mathcal{N}(0, \xi_i) \quad (\beta_1, \ldots, \beta_n) \quad \text{independent} \qquad (5.47)$$

induces a probability description of the candidate θ's which turns out to be zero mean and with covariance matrix as in (5.45). This interpretation provides a clear link between "standard" models described in terms of basis expansions, regularization and the Bayesian view.

Remark 5.2 (*Low-Rank Kernel Approximation*) The spectral decomposition of the kernel (5.45) suggests also that, when some singular values ξ_i are "very small", it can be easily approximated by a low-rank matrix

$$\mathbf{K} = \sum_{i=1}^{n} \xi_i u_i u_i^T \simeq \sum_{i=1}^{\hat{n}} \xi_i u_i u_i^T \qquad \hat{n} \leq n.$$

This is equivalent to approximating the ξ_i below a certain threshold with zero singular values. This threshold can be chosen by a standard SVD-truncation criterion, e.g., neglecting singular values below a certain fraction of the largest singular value ξ_1, i.e., that satisfy

$$\xi_i < \frac{\xi_1}{R}.$$

In Fig. 5.7, the value $R = 20$ has been chosen to plot the most relevant eigenfunctions. Low-rank kernel approximation can also be exploited to reduce the computational burden in computing the solutions.

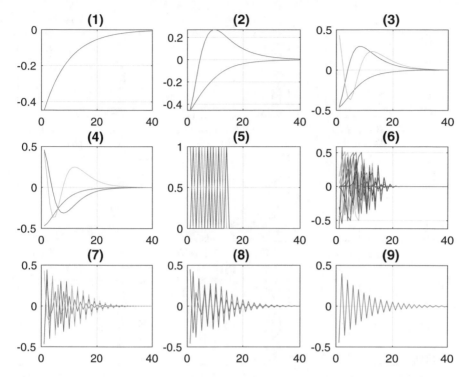

Fig. 5.7 First $\hat{n}$ eigenfunctions of the DC kernel. To enhance clarity, n is chosen for each combination of the parameters so that $\hat{n} = \arg \max_i \; i \;\; s.t. \;\; \sigma_i^2 > \sigma_1^2/20$ (see Remark 5.2). In all figures, $\alpha = 0.8$ and $\lambda_S = 1$. The regularity parameter λ_R varies, from its minimum value $\lambda_{R,min} = \lambda_S(1 - \sqrt{\alpha})^2 \simeq 0.011$ in panel (1) to the maximum value $\lambda_{R,max} = \lambda_S(1 + \sqrt{\alpha})^2 \simeq 3.589$ in panel (9). Panel (4), with $\lambda_R = 0.2$, corresponds to the TC kernel; panel (6) with $\lambda_R = 2.6$ to the DI kernel

Figure 5.7 shows the eigenfunctions of the DC kernel for different choices of the hyperparameters. As already studied in the previous section, the "complexity" of the kernel, measured e.g., by the degrees of freedom as illustrated in Fig. 5.6, varies as the hyperparameters change. In the context of basis expansions, this is clear from Fig. 5.8 where the singular values of the kernel, i.e., the variances of the basis expansion coefficients β_i, introduced in (5.47), vary as the hyperparameters change. For instance when $\lambda_R = \lambda_{R,min}$, see panel (1), and $\lambda_R = \lambda_{R,max}$, see panel (9), the kernel has rank 1. Instead the singular values decay slower for the DI kernel, see panel (5), that also has the largest number of degrees of freedom, see Fig. (5.6).

Even if this section is devoted to finite impulse response models (i.e., n finite, and therefore BIBO stable systems), it still makes sense to discuss what happens to the coefficients θ_n when n becomes "large" and its relation with BIBO stability. In Lemma 5.1, we have seen that a sufficient conditions for a.s. BIBO stability of realizations from the Gaussian prior, is that the diagonal elements of K satisfy the summability condition

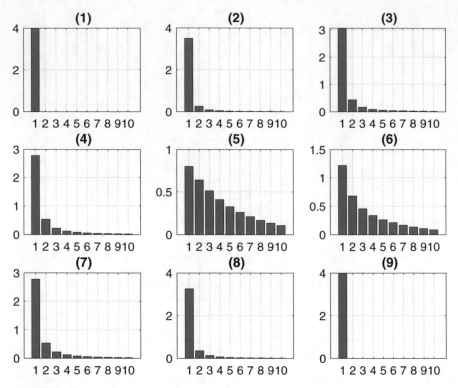

Fig. 5.8 First 10 singular values of the DC kernel. In all figures, $\alpha = 0.8$ and $\lambda_S = 1$. The regularity parameter λ_R varies, from its minimum value $\lambda_{R,min} = \lambda_S(1 - \sqrt{\alpha})^2 \simeq 0.011$ in panel (1) to the maximum value $\lambda_{R,max} = \lambda_S(1 + \sqrt{\alpha})^2 \simeq 3.589$ in panel (9). Panel (4), with $\lambda_R = 0.2$, corresponds to the TC kernel; panel (6) with $\lambda_R = 2.6$ to the DI kernel

$$\sum_{t=1}^{\infty} K(t, t)^{1/2} < \infty$$

which requires a "sufficiently fast" decay rate of the diagonal $K(t, t)$. A quite natural question concerns how the behaviour of $K(t, t)$ reflects on the basis vectors u_i. The following lemma, whose proof is in Sect. 5.10.5, gives the answer.

Lemma 5.2 *The basis vectors u_i introduced in (5.45), whose tth elements are denoted by u_{it}, satisfy the inequality*

$$|u_{it}| \leq \frac{1}{\xi_i} C[\mathbf{K}]_{t,t}^{1/2}, \quad C := \sum_{t=1}^{n} [\mathbf{K}]_{t,t}. \tag{5.48}$$

Condition (5.48) holds also in the infinite dimensional case, i.e., as $n \to \infty$, provided $K(t, s)$ admits the spectral decomposition

$$K(t, s) = \sum_{i=1}^{\infty} \xi_i u_{it} u_{is},$$

where the u_i are orthonormal sequences in ℓ_2 and the condition $\sum_{t=1}^{\infty} K(t, t) = C < \infty$ is satisfied.

While this result is essentially trivial for n finite, it becomes important when $n \to \infty$, since it provides a condition on the tail behaviour of the eigenvectors (eigenfunctions). For instance, if the diagonal entries (variances) of the kernel $K(t, t)$ decay exponentially fast as a function of t, also the u_{it} do so. The decay of the eigenfunctions can be visually inspected in Fig. 5.7.

5.7 Hankel Nuclear Norm Regularization

As discussed above, regularization can be used to enforce smoothness and stability of impulse responses. Yet this is just one way, and possibly not the most common in the field of dynamical systems, to control the "complexity" of model classes.

For instance, in the parametric approach to system identification, the complexity can be measured by the dimension of a minimal state-space realization of the unknown system. For ease of exposition, let us now only consider the single-input single-output output error case (i.e., $H(z) = 1$). In this case, the number of free parameters is $2n + 1$ where n is the degree of the denominator of the transfer function $G_\theta(z)$, that also equals the dimension n of a minimal state-space realization of $G_\theta(z)$ which is called the McMillan degree of $G(z, \theta)$, as seen in Sect. 2.2.1.1. To fix notation, let us introduce a minimal state-space realization of $G(z, \theta)$

$$\begin{aligned} x_{t+1} &= Ax_t + Bu_t \quad x_t \in \mathbb{R}^n, \\ y_t &= Cx_t \end{aligned} \tag{5.49}$$

which is such that $G(z, \theta) = C(zI - A)^{-1}B$. If $\{g(k, \theta)\}_{k \in \mathbb{N}}$ is the impulse response sequence, parametrized by θ, then one has $g(k, \theta) = CA^{k-1}B \; \forall k > 0$.

It is well known from realization theory that the McMillan degree has a close connection with the so-called *Hankel* matrix formed with the impulse response coefficients, i.e.,

$$\mathscr{H}_{r,c}(\theta) := \begin{bmatrix} g(1, \theta) & g(2, \theta) & g(3, \theta) & \cdots & g(c, \theta) \\ g(2, \theta) & g(3, \theta) & g(4, \theta) & \cdots & g(c + 1, \theta) \\ \vdots & \ddots & \ddots & \ddots & \vdots \\ g(r, \theta) & g(r + 1, \theta) & g(r + 2, \theta) & \cdots & g(r + c - 1, \theta) \end{bmatrix} \tag{5.50}$$

with r block rows and c block columns. The following lemma holds.

Lemma 5.3 (based on [65]) *The linear time-invariant system with impulse response* $\{g(k, \theta)\}_{k \in \mathbb{N}}$ *admits a minimal state-space realization of order n (i.e., has McMillan degree equal to n) if and only if, for some choice of r, c the following holds:*

$$n = \operatorname{rank}\{\mathscr{H}_{r,c}(\theta)\} = \operatorname{rank}\{\mathscr{H}_{r+j,c+i}(\theta)\} \quad \forall \ i, j \in \mathbb{N}. \tag{5.51}$$

In practice, only a finite number of impulse response (Markov) parameters $g(k, \theta)$, $k = 1, \ldots, p$ is available and the problem of finding a state-space model of the form (5.49) such that $g(k, \theta) = CA^{k-1}B \ \forall \ k = 1, \ldots, p$, is known as *partial realization problem.*

This shows that, indeed, a notion of "complexity" can be attached to the dimension n of a minimal state-space realization (5.49); therefore the rank of the Hankel matrix $\mathscr{H}_{c,r}(\theta)$ can be considered as a candidate for performing regularization. This leads to the choice of a penalty given by

$$J_{\mathscr{H},\gamma}(\theta) := \gamma \operatorname{rank}\{\mathscr{H}_{r,c}(\theta)\} \tag{5.52}$$

for suitable values of the integers c, r. Unfortunately, similarly to what happens for the 0 quasi-norm $\|x\|_0$ (defined as the number of non-zero entries in the vector x) discussed in Sect. 3.6.2.1, the rank functional is not convex; as a result solving optimization problems involving penalties of the form (5.52) is problematic. The very same issue arise in a variety of rank-constrained optimization problems.

As seen in Chap. 3, to overcome this limitations, inspired by work on ℓ_1 regularization, researchers have suggested to use the *nuclear norm* $\|A\|_*$ of a matrix $A \in \mathbb{R}^{m \times n}$ defined as

$$\|A\|_* := \operatorname{trace}\left(\sqrt{A^T A}\right) = \sum_i \sigma_i(A), \tag{5.53}$$

where $\sigma_i(A)$ denotes the ith singular value of the matrix A, as a surrogate for the rank of the matrix A. The nuclear norm is also known as Ky–Fan n-norm or trace norm. This choice is motivated by the following lemma.

Lemma 5.4 (based on [20]) *Given a matrix $A \in \mathbb{R}^{m \times n}$ the nuclear norm of A is the convex envelope of the rank function on the set $\mathscr{A} := \{A \in \mathbb{R}^{m \times n}, \|A\| \leq 1\}$.*

These considerations have led to a whole class of regularization methods which build upon the nuclear norm of the Hankel matrix

$$J_{\mathscr{H},\gamma}(\theta) := \gamma \|\mathscr{H}_{r,c}(\theta)\|_*$$

as a possible regularizer. Also several extensions have been considered, including weighted versions of the form

$$J_{\mathscr{H},\gamma}(\theta) := \gamma \|W_r \mathscr{H}_{r,c}(\theta) W_c\|_*$$

where W_c and W_r are, respectively, "column" and "row" weightings. These latter can be possibly adapted iteratively, in the framework of iteratively reweighted methods such as those commonly used in conjunction with ℓ_1 and/or ℓ_2 reweighted schemes, see e.g., [72].

The Hankel norm regularizer can also be studied from a Bayesian perspective, considering the prior

$$p_{\mathcal{H},\gamma}(\theta) \propto \exp\left(-\gamma \|\mathcal{H}_{r,c}(\theta)\|_*\right) \propto \exp\left(-\gamma \sum_i \sigma_i(\mathcal{H}_{r,c}(\theta))\right). \tag{5.54}$$

To gain some intuition on the structure of this prior, let $g(k, \theta) = \theta_k$ and consider the following modified prior which penalizes the nuclear norm of the squared Hankel matrix, i.e.,

$$\tilde{p}_{\mathcal{H},\gamma}(\theta) \propto \exp\left(-\gamma \|\mathcal{H}_{r,c}(\theta)\mathcal{H}_{r,c}(\theta)^T\|_*\right) \propto \exp\left(-\gamma \sum_i \sigma_i(\mathcal{H}_{r,c}(\theta)\mathcal{H}_{r,c}(\theta)^T)\right). \tag{5.55}$$

The reason for introducing $\tilde{p}$ is twofold. The first is related to the fact that the prior (5.55) is equivalent to assuming that the entries θ_k of the impulse response are independent zero mean Gaussians, as formalized in the following proposition.

Proposition 5.1 (based on [53]) *Let $\tilde{p}_{\mathcal{H},\gamma}(\theta)$ be as in (5.55) and let $\theta \in \mathbb{R}^m \sim \tilde{p}_{\mathcal{H},\gamma}(\theta)$, where $\mathcal{H}_{p,p}(\theta)$ is its $p \times p$ Hankel matrix (with $m = 2p - 1$). Then the θ_k's are zero mean, independent and Gaussian. In particular:*

$$\theta_k \sim \begin{cases} \mathcal{N}\left(0, \frac{1}{2\gamma k}\right) & \text{if } 1 \leqslant k \leqslant \frac{m+1}{2} \\ \mathcal{N}\left(0, \frac{1}{2\gamma(m-k+1)}\right) & \text{if } \frac{m+1}{2} < k \leqslant m \end{cases}. \tag{5.56}$$

As illustrated in Fig. 5.9, from (5.56) one sees that the variance of θ_k *is not* decaying with the lag k, and hence the prior $\tilde{p}_{\mathcal{H},\gamma}(\theta)$ does not induce a BIBO stable hypothesis space.

Second, the prior $\tilde{p}_{\mathcal{H},\gamma}(\theta)$ can be used as a proposal distribution for an MCMC scheme, as introduced in Sect. 4.10, to sample from the Hankel prior $p_{\mathcal{H},\gamma}(\theta)$ in (5.54) with $g(k, \theta) = \theta_k$. Samples from $p_{\mathcal{H},\gamma}(\theta)$ can then be used to approximate the variances $\text{Var}\{\theta_k\}$ and the correlations $\text{Corr}\{\theta_P, \theta_h\}$. These are shown in Fig. 5.9. In particular, the solid line in the left panel shows $\text{Var}\{\theta_k\}$ as a function of k, while the right panel $\text{Corr}\{\theta_P, \theta_{k+h}\}$ as a function of h for k fixed to 50. It is clear that, even though under $p_{\mathcal{H},\gamma}(\theta)$ the θ_k's are not Gaussian, the variances resemble those of $\tilde{p}_{\mathcal{H},\gamma}(\theta)$ (left panel, dashed line) as and also their correlations resemble those of independent variables. For the sake of comparison, the left panel plots also the profiles of the impulse response coefficients' variances using the TC prior for two different decay rates (dashdot lines).

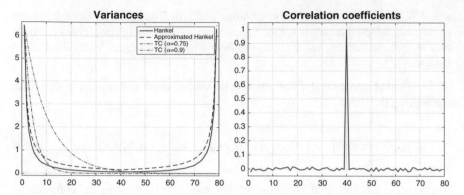

Fig. 5.9 Prior induced by the Hankel Nuclear Norm: the impulse response coefficients are contained in the vector $\theta \in \mathbb{R}^{79}$, modelled as a random vector with probability density function $p_{\mathscr{H},\gamma}(\theta) \propto \exp(-\|\mathscr{H}_{40,40}(\theta)\|_*)$. *Left*: variances of the impulse response coefficients θ_k reconstructed by MCMC (solid line) and approximated using the prior (5.56) (dashed line). The figure also displays the variances of θ_k when θ is a Gaussian random vector with stable spline (TC) covariance (5.41) for two different values of α (dashdot lines). All the profiles are rescaled so that they share the same initial value. *Right*: 40th row of the matrix containing the correlation coefficients returned by the MATLAB command `corrcoef(M)` where each column of the 79×10^6 matrix M contains one MCMC realization of θ under the Hankel prior $p_{\mathscr{H},\gamma}(\theta)$. The adopted MCMC scheme was a random walk Metropolis with increments proportional to the variances (5.56) divided by a factor equal to 4

These observations suggest that, while the nuclear norm regularization (prior) accounts for system-theoretic notions of model complexity as defined by the McMillan degree, it fails to include decay rate and smoothness constraints. One would expect, therefore, that Hankel regularization alone may not give satisfactory results as it is not able to properly bound the candidate set of models. It turns out that the maximum entropy framework discussed in Sect. 5.5.1 can be used to build prior distribution which account for stability, smoothness as well as "complexity". The following theorem (whose proof is given in Sect. 5.10.6) gives the structure of the MaxEnt prior under a simple "TC"-like condition on the stability-smoothness constraint.

Theorem 5.6 *Let* $\{\theta_P\}_{k=1,\ldots,m}$ *be a zero mean, absolutely continuous random vector with density* $p_\theta(\theta)$, *which satisfies the following constraints:*

$$
\begin{aligned}
\mathscr{E}\left[\theta_m^2\right] &\leq \sigma^2 \alpha^{m-1} \\
\mathscr{E}\left[(\theta_{k-1} - \theta_k)^2\right] &\leq \sigma^2 \alpha^{k-2}(1-\alpha) \quad k = 2, \ldots, m \\
\mathscr{E}\|\mathscr{H}_{r,c}(\theta)\|_* &\leq h.
\end{aligned}
\tag{5.57}
$$

Then, the solution $p_{\theta,MEH}(\theta)$ *of the maximum entropy problem*

$$p_{\theta,MEH} := \underset{p(\cdot) \ s.t. \ (5.31)}{\arg\max} \quad -\mathscr{E}\log(p_\theta(\theta)) \tag{5.58}$$

has the following form:

$$p_{\theta,MEH}(\theta) \propto e^{-\mu_H \|\mathscr{H}_{r,c}(\theta)\|_*} \left[\prod_{k=2}^{m} e^{-\frac{1}{2}\mu_{k-1}(\theta_{k-1}-\theta_P)^2} \right] e^{-\frac{1}{2}\mu_m \theta_m^2}, \tag{5.59}$$

where the Lagrange multipliers $\mu_H, \mu_1, \ldots, \mu_m$ *are determined so that the constraints (5.57) are satisfied.*[1]

Hankel nuclear norm discussed in this chapter is only one possible way to favour "simple" (in the sense of having small McMillan degree) models. Indeed, it is by no means trivial to use priors of the form (5.59), that involve nuclear norm terms, in conjunction with marginal likelihood optimization to estimate hyperparameters. Several variations are possible and, indeed, matricial reweighting schemes such as those used in [55] can be used in a Bayesian context, leading to iteratively reweighted schemes that remind of ℓ_1/ℓ_2 reweighting [72].

5.8 Historical Overview

The framework discussed in this chapter has indeed a long history that can be traced back, by and large outside the control community, until the early '70s of the last century. In this section, we will review these developments and point to similarities and differences with the theory developed in this chapter.

5.8.1 The Distributed Lag Estimator: Prior Means and Smoothing

To the best of our knowledge Bayesian methods for estimating dynamical systems have first been advocated in the early '70s in the econometrics literature for FIR models of the form (5.14), which were referred to as *distributed lag* models. The length n of the FIR model was actually left unspecified, and possibly let going to infinity.

In particular, [40, 62] were the first to talk about (and apply) Bayesian methods for system identification, arguing that "rigid parametric" structures may be inadequate,

---

[1] Using the complementary slackness conditions it follows that a multiplier may be nonzero only if the corresponding inequality in (5.57) holds with an equality sign.

extending arguments which can be found in [66] for "static" linear regression models to the "dynamical" systems scenario. In the paper [40], having in mind that modes of linear time-invariant systems have an exponentially decaying behaviour of the type α^t, it was suggested to describe the unknown impulse response θ with a process having an exponentially decaying prior mean

$$\{m_t\}_{t \in \mathbb{N}} \quad m_t := \lambda \alpha^t \quad |\alpha| < 1. \tag{5.60}$$

Other possible response patterns had been considered, such as the hump, composed of the response build-up, the maximum and its decay, see [40] for details and alternative patterns. The covariance function $K(t, s)$ in [40] was taken so that the ratio

$$\frac{std(\theta_t)}{m_t}$$

remains constant over time t. This was called the "proportionality principle". and can be achieved with the choice

$$K(t, s) = cov(\theta_t, \theta_s) := v w_{ts} \alpha^{t+s-2} \quad |w_{ts}| \leq 1 \tag{5.61}$$

so that the normalized standard deviation

$$\frac{std(\theta_t)}{m_t} = \frac{\sqrt{K(t, t)}}{m_t} = \frac{\sqrt{v w_{ts} \alpha^{2t-2}}}{\lambda \alpha^t} = \frac{\sqrt{v w_{ts} \alpha^{-2}}}{c}$$

is indeed constant if w_{ts} is so. This would imply that prior credible intervals have constant relative size w.r.t. their means, see p. 1065 of [40].

The choice (5.61) left the coefficients w_{ts} unspecified and, indeed in [40], it was emphasized that *"the selection of the values of the set of w_{ij} still remains a relatively difficult task"*; one suggestion, inspired by work on smoothing [34], has been to take

$$w_{ij} = w^{|i-j|} \quad 0 < w < 1 \tag{5.62}$$

leading to

$$K_{ij} = v \alpha^{i+j-2} w^{|i-j|}, \tag{5.63}$$

which is exactly the DC kernel introduced in Corollary 5.1. It is also interesting to observe that [40] already suggested the use of marginal likelihood to choose the most suitable prior distribution in the class.

Of course, postulating a prior mean m introduces in the estimation procedures a remarkable prejudice and requires quite accurate knowledge on the expected θ. The paper [62], inspired by "smoothing priors" arguments, suggested instead that the prior mean should be zero, and only smoothness conditions on the lags should be enforced; this leads to a zero mean prior, i.e., $c = 0$ in (5.60), with a dth degree smoothing covariance. For instance, for $d = 2$, the prior model can be expressed in

Fig. 5.10 50 realizations form Shiller's prior (with penalty on initial condition and first difference). It is clear from the picture that the realizations are smooth, as expected, but certainly do not resemble impulse responses of (stable) linear systems

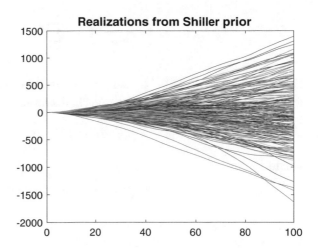

terms of the second-order differences:

$$\beta := \underbrace{\begin{bmatrix} 1 & -2 & 1 & 0 & \dots & 0 \\ 0 & 1 & -2 & \ddots & \vdots & 0 \\ \vdots & \vdots & \ddots & \ddots & \ddots & \vdots \\ 0 & \dots & \dots & 1 & -2 & 1 \end{bmatrix}}_{:=S} \theta = S\theta$$

postulating $\mathscr{E}\beta\beta^T = S\mathscr{E}\theta\theta^T S^T = I$.

It is clear from Fig. 5.10 that this prior guarantees smoothness in time domain (and therefore low-pass behaviour in frequency domain) but no guarantee on stability.

5.8.2 Frequency-Domain Smoothing and Stability

The "time-domain" smoothing discussed in the previous section has been criticized by Akaike [1] who posed the question whether time-domain smoothness conditions would "be the most natural ones". Akaike suggested that instead smoothness should be enforced in the frequency domain, i.e., considering the frequency response

$$G(e^{j\omega}) := \sum_{k=1}^{n} \theta_k e^{-j\omega k}.$$

To this purpose, the L_2-norm of the first derivative $\frac{dG(e^{j\omega})}{d\omega}$ can be considered and we have already seen in (5.25) that one obtains

$$\left\| \frac{dG(e^{j\omega})}{d\omega} \right\|^2 = \sum_{k=1}^{n} k^2 |\theta_k|^2. \tag{5.64}$$

Discouraging large $\left\| \frac{dG(e^{j\omega})}{d\omega} \right\|^2$ can thus be obtained using the right-hand side of (5.64) as a penalty, which can be written in the form:

$$\mathrm{p}(\gamma, \theta) := \theta^T K_\gamma^{-1} \theta,$$

where

$$K_\gamma := \frac{1}{\gamma} \mathrm{diag} \left\{ 1, \ \frac{1}{4}, \ \frac{1}{9}, \dots, \frac{1}{n^2} \right\}. \tag{5.65}$$

This is of course equivalent to assuming that the impulse response vector θ has a zero-mean normal prior with covariance K_γ.

Unfortunately, in the limit $n \to \infty$, the covariance function (5.65) does not meet the (more stringent) sufficient conditions of Lemma 5.1; of course rather straightforward extensions include setting penalties on higher-order derivatives, which would result in a faster decay rate of the diagonal elements of (5.65). This is a manifestation of the well-known link between regularity in the frequency domain and decay rate of the impulse response already discussed in Sect. 5.5.

5.8.3 Exponential Stability and Stochastic Embedding

More recently, Gaussian priors for dynamical systems have been considered in the control literature; in particular, a zero-mean Gaussian prior with diagonal and exponentially decaying covariance

$$\mathscr{E}\theta\theta^T = K_{\rho,\alpha} := \alpha \, \mathrm{diag} \left\{ 1, \ \rho, \ \rho^2, \dots, \rho^{n-1} \right\} \tag{5.66}$$

has been proposed in the so-called "stochastic embedding" framework [25, 26]. Let us now briefly introduce the problem: consider an Output Error model of the form

$$y(t) = \sum_{k=1}^{\infty} g_k(\theta) u(t - k) + e(t),$$

where $g_k(\theta)$, $\theta \in \mathbb{R}^n$ is a parametric description of the unknown impulse response $\{g_k\}_{k=1,\dots,\infty}$ in the model class $\mathscr{M}_n(\theta)$. Let $\hat{\theta}$ be some parametric estimator of θ, e.g., the PEM estimator

$$\hat{\theta} = \arg\min_{\theta} \sum_{k=1}^{N} \|y(t) - G(z, \theta)u(t)\|^2. \tag{5.67}$$

Let now

$$\hat{G}(z) := G(z, \hat{\theta}) = \sum_{k=1}^{\infty} g_k(\hat{\theta}) z^{-k}$$

be the corresponding estimator of the transfer function $G(z, \theta) = \sum_{k=1}^{\infty} g_k(\theta) z^{-k}$.

In the Model Error Modelling framework, it is assumed that the "true" transfer function $G(z)$ is only partially captured by the chosen model class $\mathcal{M}_n(\theta)$ so that

$$G(z) = G(z, \theta_0) + \tilde{G}(z) \quad G(z, \theta_0) \in \mathcal{M}_n(\theta) \tag{5.68}$$

and $\tilde{G}(z)$ represents a model error. The purpose of Model Error Modelling is to obtain a statistical description of the model error, say

$$\tilde{G}(z) := G(z) - \hat{G}(z)$$

which may be used, for instance, to estimate the model order, e.g., the dimension n of the parameter vector θ. This can be achieved by minimizing an estimate of the MSE

$$\mathcal{E}\|G(z) - G(z, \hat{\eta})\|^2$$

while accounting for the model error model $\tilde{G}(z)$, see e.g., Eqs. (89)–(92) in [26].

The model error $\tilde{G}(z)$ is estimated in [26] starting from the least squares residuals $v_{\hat{\theta}}(t) := y(t) - G(z, \hat{\theta})u(t)$ which, under assumption (5.68), is expected to be described by the model

$$v(t) = \tilde{G}(z)u(t) + e(t).$$

It is remarkable that [26] propose to estimate the parameters α and ρ that characterize the covariance (5.66) resorting to marginal likelihood maximization

$$(\hat{\alpha}, \hat{\rho}) := \arg \max_{\alpha, \rho} \int p(V_{\hat{\eta}}|\tilde{g}) p(\tilde{g}|\alpha, \rho) \, d\tilde{g}, \tag{5.69}$$

where $V_{\hat{\eta}} := [v_{\hat{\eta}}(1), \dots, v_{\hat{\eta}}(N)]$. It is also interesting to observe that the exponential decay of the covariance sequence (5.66) implies a smoothness condition in the frequency response function similar in spirit to that advocated in [1]. This is formalized in the following result whose proof is in Sect. 5.10.7.

Lemma 5.5 *Let $\{g_{k,\alpha}\}_{k=0,\dots,\infty}$ be a zero-mean Gaussian process with covariance* (5.66) *and let*

$$G_\alpha(e^{j\omega}) := \sum_{k=0}^{\infty} g_{k,\alpha} e^{-jk\omega} \quad \omega \in [0, 2\pi)$$

be its Fourier transform. Then the Lipschitz-like condition

$$\mathscr{E}[\|G_\alpha(e^{j\omega_1}) - G_\alpha(e^{j\omega_2})\|^2] \leq \frac{c}{1-\alpha}(\omega_1 - \omega_2)^2 \qquad |\alpha| < 1 \qquad (5.70)$$

holds.

5.9 Further Topics and Advanced Reading

Section 1.3 already reported a list of topics and readings on inverse problems, Stein estimators and their link with the Empirical Bayes framework.

The use of regularization and Bayesian priors can be probably dated back to the paper [71] were smoothing ideas have been advocated for a denoising problem in the field of Actuarial Science. See also the much later reference [34]. The later developments are essentially impossible to survey in this short section and we refer the reader to [66] for an early overview on the use of Bayes priors in the context of linear regression; the interested reader may also consult [22, 31, 32, 42, 59] where generalized ridge regression has been proposed to stabilize ill-conditioned inverse problems.

To the best of our knowledge, [40, 62] have been the first to use these ideas in the context of dynamical systems, named "distributed-lag" models in these early references. This work has been subsequently taken up by Akaike [1] and later on by Kitagawa and Gersh in a series of papers, see e.g., [35, 36], which culminated in the well-known book [37]. The seminal papers by Leamer and Shiller have also been continued by the econometrics community, starting with the work by Doan, Litterman and Sims, see e.g., [18] for an overview and further references. This has lead to the so-called "Minnesota prior", which has been discussed quite extensively in the econometrics literature; several variations and extensions are found, see for instance [23, 41].

The econometrics literature has since then studied Bayesian procedures for system identification rather intensively, mostly under the acronym *Bayesian VARs*; the main driving motivation was that of handling high-dimensional time series (i.e., *p* large, called *cross-sectional dimension* in the econometrics literature) with possibly many explicative variables (*m* large), see for instance [2, 17, 23, 38].

The problem of tuning the regularization parameters (or equivalently the hyper-parameters describing the prior in a Bayesian setting) has received relatively little attention in the econometrics literature: [40] already suggested the use of Empirical Bayes procedure, while [2, 18] propose tuning the hyperparameters using out-of-sample and in-sample errors, respectively. The paper [38] and the most recent work [23] adopt again an Empirical Bayes approach using the marginal likelihood; [23] claims the superiority of this approach w.r.t. previous "ad-hoc" techniques [2, 18].

Despite this long history, the use of Bayesian priors for system identification has only gained popularity in relatively recent times, e.g., see the survey [52]. We believe it is fair to say that reason for this is to be attributed to the fact that much more efforts have been recently devoted to developing prior models tailored to estimating dynamical system. In the remaining part of the book, these issues will be dealt with in some details. The reader is referred to [10, 11, 49, 50, 55] for various classes of prior models and to [6, 7, 12, 55] for more details on Maximum Entropy derivations. Extensions include prior models to estimate sparse models for high-dimensional time series [14, 74] as well as classes of priors for nonlinear dynamical models [51], that will be thoroughly discussed in Chap. 8. In particular, the techniques described in this chapter can be also used to identify the so-called *dynamic networks* that consist of a large set of interconnected dynamic systems. Modelling such complex physical systems is important in several fields of science and engineering, including also biomedicine and neuroscience [27, 30, 46, 56]. Estimation is difficult since they are often large scale and the network topology is typically unknown [14, 44, 67]. One typically postulates the existence of many connections and then has to understand from data which are really active. Since in real physical systems often only a small fraction of links is really working, the estimation process needs to exploit sparsity regularizers as those introduced in Chap. 3 and their stochastic interpretation like the Bayesian Lasso [47]. In the context of *linear dynamic networks*, where modules are defined by impulse responses, many approaches have been recently designed e.g., relying on local multi-input single-output (MISO) models [16, 19, 45]. Contributions based on variational Bayesian inference and/or nonparametric regularization, deeply connected with the techniques discussed in this book, are in [14, 33, 58, 73]. Methods to infer the full network dynamics using (structured) multiple-input multiple-output (MIMO) models can be found instead in [21, 69], with estimates consistency analyzed in [57]. A contribution based on the combination of the stable spline kernel and the so-called horseshoe sparsity prior [8, 54, 68] has been developed in [48]. See also [3, 24, 29, 70] for insights on identifiability issues and [28] where compressed sensing is exploited.

5.10 Appendix

5.10.1 Optimal Kernel

Theorem 5.7 *The solution P^* of problem* (5.20) *is given by*

$$P^* = \theta_0 \theta_0^T, \tag{5.71}$$

where θ is the "true" impulse response of the data-generating mechanism (5.14).

Proof The proof will proceed as follows: let us denote with $\hat{\theta}^{P^*}$ the estimator obtained with P as in (5.71). Consider the error

$$\tilde{\theta}^P := \hat{\theta}^P - \theta_0$$

which can be written as

$$\tilde{\theta}^P = \hat{\theta}^P - \theta_0$$
$$= \hat{\theta}^{P^*}\theta_0 + \hat{\theta}^P - \hat{\theta}^{P^*}$$
$$= \tilde{\theta}^{P^*} + \left(\hat{\theta}_P - \hat{\theta}^{P^*}\right).$$

We shall show that the following orthogonality property holds:

$$\mathscr{E}\tilde{\theta}^{P^*}\left(\hat{\theta}_P - \hat{\theta}^{P^*}\right)^T = 0 \tag{5.72}$$

so that

$$\mathscr{E}\tilde{\theta}^P(\theta^P)^T = \mathscr{E}\tilde{\theta}^{P^*}(\tilde{\theta}^{P^*})^T + \mathscr{E}\left(\hat{\theta}_P - \hat{\theta}^{P^*}\right)\left(\hat{\theta}_P - \hat{\theta}^{P^*}\right)^T \tag{5.73}$$

and therefore:

$$M_\theta(P) - M_\theta(P^*) = \mathscr{E}\tilde{\theta}^P(\tilde{\theta}^P)^T - \mathscr{E}\tilde{\theta}^{P^*}((\tilde{\theta}^{P^*})^T = \mathscr{E}\left(\hat{\theta}_P - \hat{\theta}^{P^*}\right)\left(\hat{\theta}_P - \hat{\theta}^{P^*}\right)^T \geq 0$$

which will prove the claim that $P^* = \theta_0\theta_0^T$ is the optimal solution to (5.20).

It now just remains to show that (5.72) holds. To do so, let us rewrite (4.7) assuming null μ_θ and using the matrix inversion lemma as (3.145):

$$\hat{\theta}^P = \left(\sigma^2 I + P\Phi^T\Phi\right)^{-1}P\Phi^T Y$$
$$= \left(\sigma^2 I + P\Phi^T\Phi\right)^{-1}P\Phi^T(\Phi\theta_0 + E)$$
$$= \left(\sigma^2 I + P\Phi^T\Phi\right)^{-1}\left[\left(P\Phi^T\Phi + \sigma^2 I - \sigma^2 I\right)\theta_0 + P\Phi^T E\right]$$
$$= \theta_0 - \left(\sigma^2 I + P\Phi^T\Phi\right)^{-1}\left[\sigma^2\theta_0 - P\Phi^T E\right].$$

Therefore, the error $\tilde{\theta}^P := \theta_0 - \hat{\theta}^P$ can be written in the form:

$$\tilde{\theta}^P = \underbrace{\left(\sigma^2 I + P\Phi^T\Phi\right)^{-1}}_{:=W_P}\left[\sigma^2\theta_0 - P\Phi^T E\right] = W_P\left[\sigma^2\theta_0 - P\Phi^T E\right]. \tag{5.74}$$

Now, using (5.74), we have:

$$\hat{\theta}^{P^*} - \hat{\theta}_P = \tilde{\theta}_P - \tilde{\theta}^{P^*} = \sigma^2\left(W_P - W_{P^*}\right)\theta_0 + \left(W_{P^*}P - W_P P\right)\Phi^T E.$$

Now, let us compute

$$\mathscr{E}\left(\hat{\theta}^{P^*} - \hat{\theta}_P\right)(\tilde{\theta}^{P^*})^T = \sigma^4\left(W_P - W_{P^*}\right)\theta_0\theta^T W_{P^*}^T - \sigma^2\left(W_{P^*}P - W_P K\right)\Phi^T\Phi P^* W_{P^*}^T.$$
$$= \sigma^2\left[\sigma^2\left(W_P - W_{P^*}\right) - \left(W_{P^*}P^* - W_P P\right)\Phi^T\Phi\right]P^* W_{P^*}^T \tag{5.75}$$

If we now use the identity

$$W_P\left(\sigma^2 I + P\Phi^T\Phi\right) = I \quad \Rightarrow \quad \sigma^2 W_P = I - W_P P\Phi^T\Phi$$

we obtain

$$\sigma^2 \left(W_P - W_{P^*} \right) = \left(W_{P^*} P^* - W_P P \right) \Phi^T \Phi$$

so that, using (5.75),

$$\mathscr{E} \left(\hat{\theta}^{P^*} - \hat{\theta}_P \right) (\tilde{\theta}^{P^*})^T = 0$$

which proves (5.72) and thus the theorem. □

5.10.2 Proof of Lemma 5.1

Consider the following upper bound on the probability that the ℓ_1 norm of θ be larger than a given threshold T_{ℓ_1}:

$$\mathbb{P} \left[\sum_{t=1}^{\infty} |\theta_t| \geq T_{\ell_1} \right] \leq \frac{1}{T_{\ell_1}} \mathscr{E} \sum_{t=1}^{\infty} |\theta_t| = \frac{1}{T_{\ell_1}} \sum_{t=1}^{\infty} \mathscr{E} |\theta_t| \leq \frac{1}{T_{\ell_1}} \sum_{t=1}^{\infty} \left(m_t + \sqrt{2/\pi} K(t,t)^{1/2} \right)$$

where we have used the equality $\mathscr{E}|X| = \sigma \sqrt{2/\pi}$ for $X \sim \mathscr{N}(0, \sigma^2)$. Using the hypothesis (5.24) we have that

$$\mathbb{P} \left[\sum_{t=1}^{\infty} |\theta_t| \geq T_{\ell_1} \right] \leq \frac{M_{\ell_1} + K_{\ell_1} \sqrt{2/\pi}}{T_{\ell_1}}$$

and therefore

$$\mathbb{P} \left[\sum_{t=1}^{\infty} |\theta_t| < T_{\ell_1} \right] \geq 1 - \frac{M_{\ell_1} + K_{\ell_1} \sqrt{2/\pi}}{T_{\ell_1}}.$$

Taking the limit as $T_{\ell_1} \to +\infty$ we have

$$\mathbb{P} \left[\sum_{t=1}^{\infty} |\theta_t| < +\infty \right] = 1$$

which concludes the proof.

5.10.3 Proof of Theorem 5.5

The proof is based on the fact that the Maximum Entropy distribution p(θ) under constrains $\mathscr{E} f_k(\theta) = F_k$ and $\mathscr{E} g_k(\theta) = G_k$ has the "Gibbs" structure, i.e., it is the exponential of a weighted sum of the constraint functionals (see e.g., [15]):

$$p(\theta) \propto e^{-\sum_i \mu_i f_i(\theta) + \gamma_i g_i(\theta)}.$$

In our case, we have $f_k(\theta) = \theta_k^2$ and $g_k(\theta) = (\theta_{k-1} - \theta_k)^2$, and therefore the max-ent solution has the form

$$p_{\theta,ME}(\theta) = Ce^{-\frac{1}{2}\left(\mu_1\theta_1^2 + \sum_{k=2}^n \mu_k\theta_k^2 + \gamma_k(\theta_{k-1}-\theta_k)^2\right)}. \tag{5.76}$$

Using a well-known result in graphical models (see e.g., Lauritzen [39]), the variables θ_k and $\{\theta_{k+2}, \ldots, \theta_n\}$ are conditionally independent given θ_{k+1} (because θ_{k+1} is the only neighbour of θ_P in the graph representing $p(\theta_k, \theta_{k+1}, \ldots, \theta_n)$ (or equivalently θ_{k+1} separates θ_k from $\theta_{k+2}, \theta_{k+3}, \ldots, \theta_n$).

   In our case, this conditional independence implies that the best linear estimator $\hat{\theta}_{k-1}$ of θ_{k-1} given $\theta_k, \theta_{k+1}, \ldots, \theta_n$ depends only θ_P (i.e., $\hat{\theta}_{k-1} = a_{B,k}\theta_k$) so that the vector θ admits the f^2 representation:

$$\theta_{k-1} = a_{B,k}\theta_k + w_k \tag{5.77}$$

with $w_k := \theta_{k-1} - \hat{\theta}_{k-1} = \theta_{k-1} - a_{B,k}\theta_k$ zero mean and uncorrelated of θ_k, $\theta_{k+1}, \ldots, \theta_n$. Let us define $\sigma_k^2 := \mathscr{E}w_k^2$. In order to express $a_{B,k}$ and σ_k^2 as a function of $\lambda_R, \lambda_S, \alpha$, we exploit the constraints (5.31) and the dynamical model (5.77). In particular we have

$$\begin{aligned}\lambda_S\alpha^{k-1} &= \mathscr{E}\theta_{k-1}^2 \\ &= a_{B,k}^2\mathscr{E}\theta_P^2 + \sigma_k^2 \\ &= a_{B,k}^2\lambda_S\alpha^P + \sigma_k^2\end{aligned} \tag{5.78}$$

$$\begin{aligned}\lambda_R\alpha^{k-1} &= \mathscr{E}(\theta_{k-1} - \theta_k)^2 \\ &= \mathscr{E}((a_{B,k}-1)\theta_k^2 + w_k)^2 \\ &= \mathscr{E}(a_{B,k}-1)^2\theta_k^2 + \mathscr{E}w_k^2 \\ &= (a_{B,k}-1)^2\lambda_S\alpha^k + \sigma_k^2\end{aligned} \tag{5.79}$$

Substracting (5.78) from (5.79) we obtain

$$(\lambda_S - \lambda_R)\alpha^{k-1} = a_{B,k}^2\lambda_S\alpha^P - (a_{B,k}-1)^2\lambda_S\alpha^k = (2a_{B,k}-1)\lambda_S\alpha^k$$

which implies that

$$a_{B,k} = \frac{\lambda_S(1+\alpha) - \lambda_R}{2\lambda_S\alpha} =: a_B$$

that is independent of k, thus denoted with a_B as in (5.35). From (5.79) we also have that

---

[2] We prefer here to work with backward representations since, as we will see, with this choice we will have $a_{B,k} = a_B$, independent of k. Forward representations are discussed in Sect. 5.10.8.

$$\sigma_k^2 = \lambda_R \alpha^{k-1} - (a_B - 1)^2 \lambda_S \alpha^k = (\lambda_R - (a_B - 1)^2 \lambda_S \alpha)\alpha^{k-1} = (1 - a_B^2 \alpha)\lambda_S \alpha^{k-1}$$

where the last equality follows after a few manipulations and proves (5.36). Replacing

$$a_B - 1 = \frac{\lambda_S(1 - \alpha) - \lambda_R}{2\lambda_S \alpha}$$

in the previous equation we have:

$$\sigma_k^2 = \left[\lambda_R - \left(\frac{\lambda_S(1 - \alpha) - \lambda_R}{2\lambda_S \alpha} \right)^2 \lambda_S \alpha \right] \alpha^{k-1}.$$

Of course σ_k^2, and thus the right hand side, should be positive (for simplicity we exclude the singular case $\sigma_k^2 = 0$):

$$\lambda_R - \left(\frac{\lambda_S(1 - \alpha) - \lambda_R}{2\lambda_S \alpha} \right)^2 \lambda_S \alpha = \frac{4\lambda_R \lambda_S \alpha - (\lambda_S(1 - \alpha) - \lambda_R)^2}{4\lambda_S \alpha} > 0$$

which in turn is equivalent to

$$4\lambda_R \lambda_S \alpha - (\lambda_S(1 - \alpha) - \lambda_R)^2 > 0.$$

This happens if and only if

$$\lambda_R^2 - 2\lambda_R \lambda_S(1 + \alpha) + (1 - \alpha)^2 \lambda_S^2 < 0.$$

This is a degree two polynomial in λ_R with two positive roots

$$\lambda_{R,i} = \lambda_S(1 + \alpha) \pm \sqrt{\lambda_S^2(1 + \alpha)^2 + \lambda_S^2(1 - \alpha)^2} = \lambda_S(1 + \alpha \pm 2\sqrt{\alpha}) \quad i = 1, 2$$

and therefore our problem is feasible if and only if

$$\lambda_{R,min} = \lambda_{R,1} = \lambda_S(1 + \alpha - 2\sqrt{\alpha}) < \lambda_R < \lambda_S(1 + \alpha + 2\sqrt{\alpha}) = \lambda_{R,2} = \lambda_{R,max}$$

thus proving (5.32). Now it remains to prove that (5.76) takes the form (5.34). First let us observe that the exponent of (5.76) is a quadratic form in θ, and therefore (5.76) can be written in the form

$$p_{\theta,ME}(\theta) = Ce^{-\frac{1}{2}\theta^T \Phi \theta}.$$

Last, since in (5.76) only products of the form $\theta_k \theta_h$ for $h \in [k - 1, k, k + 1]$ appear, the matrix $\Phi = \Phi^T$ has the following band structure:

$$\Phi = \begin{bmatrix} * & * & 0 & \ldots & \ldots & 0 \\ * & * & * & 0 & \ldots & 0 \\ 0 & * & * & * & 0 & \ldots \\ \vdots & \ddots & \ddots & \ddots & \ddots & \ddots \\ 0 & \ldots & 0 & * & * & * \\ 0 & \ldots & \ldots & 0 & * & * \end{bmatrix}.$$

In addition, for $p_{\theta,ME}(\theta)$ to be a density, Φ needs to be positive semidefinite (otherwise there would be directions in which the density grows indefinitely). Since θ admits the backward AR representation (5.77) with $\mathscr{E}w_k^2 = \sigma_k^2 > 0$, the covariance matrix $\Sigma = \mathscr{E}\theta\theta^T$ is positive definite and thus $\Phi = \Sigma^{-1}$. To compute the autocovariance function $\mathscr{E}\theta_h\theta_k$ we consider the following cases: if $k = h$ we have

$$\mathscr{E}\theta_h\theta_h = \lambda_S\alpha^h.$$

If $k > h$ we have

$$\mathscr{E}\theta_h\theta_k = a_B\mathscr{E}\theta_{h+1}\theta_k$$

and iterating the relation we find

$$\mathscr{E}\theta_h\theta_k = a_B^{k-h}\mathscr{E}\theta_k\theta_k = \lambda_S a_B^{k-h}\alpha^k.$$

Analogously, if $h > k$ we have

$$\mathscr{E}\theta_h\theta_k = a_B^{h-k}\mathscr{E}\theta_h\theta_h = \lambda_S a_B^{h-k}\alpha^h.$$

Combining the three cases we obtain

$$\mathscr{E}\theta_k\theta_h = \lambda_S a_B^{|k-h|}\alpha^{\max\{k,h\}}$$

proving (5.38).

5.10.4   Proof of Corollary 5.1

Using the definition (5.39) in Eq. (5.38) we obtain:

$$\mathscr{E}\theta_k\theta_h = \lambda_S a_B^{|k-h|}\alpha^{\max\{k,h\}} = \lambda_S\frac{\rho^{|k-h|}}{\alpha^{\frac{|k-h|}{2}}}\alpha^{\max\{k,h\}} = \lambda_S\rho^{|k-h|}\alpha^{\frac{k+h}{2}}.$$

In addition, if the matching condition $\lambda_R = \lambda_S(1 - \alpha)$ is satisfied, then from (5.35) $a_B = 1$ and from (5.39) $\rho = \sqrt{\alpha}$; substituting in (5.40) we obtain

$$\mathscr{E}\theta_k\theta_h = \lambda_S \rho^{|k-h|}\alpha^{\frac{k+h}{2}} = \lambda_S \alpha^{\max\{k,h\}}$$

i.e., the covariance sequence of the well known TC kernel.

5.10.5 Proof of Lemma 5.2

The proof of this lemma is a simple application of Schwartz inequality. In particular we have:

$$|u_{it}| = \frac{1}{\xi_i}\left|\sum_{s=1}^{n}[\mathbf{K}]_{t,s}u_{is}\right| \leq \sum_{s=1}^{n}\sqrt{[\mathbf{K}]_{t,t}}\sqrt{[\mathbf{K}]_{s,s}}|u_{is}|$$

$$\leq \frac{1}{\xi_i}\sqrt{[\mathbf{K}]_{t,t}}\sum_{s=1}^{n}\sqrt{[\mathbf{K}]_{s,s}}|u_{is}| \leq \frac{1}{\xi_i}K(t,t)^{1/2}\underbrace{\sum_{s=1}^{n}\sqrt{[\mathbf{K}]_{s,s}}}_{=C<\infty}\underbrace{\sum_{s=1}^{n}|u_{is}|^2}_{=1},$$

where the last inequality follows from the fact that u_{it} has 2-norm equal to 1 for all i. The same condition clearly holds also in the infinite dimensional case, i.e., as $n \to \infty$ if $K(t,s)$ admits the spectral decomposition

$$K(t,s) = \sum_{i=1}^{\infty}\xi_i u_{it}u_{is}$$

and the condition $\sum_t K(t,t) = C < \infty$ holds. In particular this latter condition holds true if the more stringent condition $\sum_t K^{1/2}(t,t) < \infty$ in Lemma 5.1 is satisfied.

5.10.6 Proof of Theorem 5.6

The proof follows from fact that the Maximum Entropy distribution $p(x)$ under constrains $\mathscr{E} f_i(x) \leq \gamma_i$ has the "Gibbs" structure, i.e., it is the exponential of a weighted sum of the constraint functionals (see e.g., [15]):

$$p(x) \propto e^{-\sum_i \mu_i f_i(x)}.$$

5.10.7 Proof of Lemma 5.5

Since $\{g_{k,\alpha}\}_{k=0,\dots,\infty}$ is zero mean, then clearly also $G_\alpha(e^{j\omega})$ is so, i.e., $\mathscr{E} G_\alpha(e^{j\omega}) = 0$. If we now consider the difference

$$G_\alpha(e^{j\omega_1}) - G_\alpha(e^{j\omega_2}) = \sum_{k=0}^{\infty} g_{k,\alpha} \left[e^{-jk\omega_1} - e^{-jk\omega_2} \right],$$

taking the expected value of the squared norm, and using the fact the $\mathscr{E} g_{k,\alpha} g_{k,\alpha} = c\alpha^k \delta_{k-h}$, we have

$$\mathscr{E} \| G_\alpha(e^{j\omega_1}) - G_\alpha(e^{j\omega_2}) \|^2 = \sum_{k=0}^{\infty} c\alpha^k \| e^{-jk\omega_1} - e^{-jk\omega_2} \|^2.$$

Now, using

$$\| e^{-jk\omega_1} - e^{-jk\omega_2} \|^2 = 2 \left(1 - \cos(\omega_1 - \omega_2) \right) \le (\omega_1 - \omega_2)^2$$

and the expression for the sum of the geometric series α^k the thesis follows.

5.10.8 Forward Representations of Stable-Splines Kernels ⋆

A major drawback of the backward construction is that it is not straightforward to extend it to an infinite interval, i.e., to let $n \to \infty$ in order to consider infinitely long impulse response models $\{\theta_k\}_{k \in \mathbb{N}}$. However this difficulty can be circumvented exploiting the "forward" representation of (5.77), which turns out to be again a time varying AR(1) model.[3] Theorem 5.8 derives the forward AR(1) representation of the maximum entropy process found in Theorem 5.5.

Theorem 5.8 *The maximum entropy solution to (5.33) found in Theorem 5.5 admits the forward AR(1) representation*

$$\theta_{k+1} = a_F \theta_k + w_k \qquad k \ge 0 \tag{5.80}$$

with zero-mean initial condition such that $\mathscr{E} \theta_0^2 = \lambda_S$, and where

$$a_F = \rho \alpha^{1/2} = a_B \alpha \tag{5.81}$$

and w_k is a sequence of zero mean variables, uncorrelated with the initial condition θ_0 and such that

$$\mathscr{E} w_k w_h = \begin{cases} \sigma_{F,k}^2 & k = h \\ 0 & k \ne h \end{cases} \tag{5.82}$$

with $\sigma_{F,k}^2 = \lambda_S \alpha^{k+1} (1 - \rho^2)$.

---

[3] There are several ways to see this: perhaps the simplest is to recall that the inverse covariance matrix of an AR(1) process has a band (tridiagonal) structure, which implies that forward and backward models share the same conditional dependence structure.

Proof First of all let us observe that, if θ_k admits an AR(1) forward representation of the form (5.80) (with w_k that satisfies (5.82)), a_F should satisfy the relation

$$a_F = \mathscr{E}\theta_{k+1}\theta_k \left(\mathscr{E}\theta_k^2\right)^{-1}.$$

Using the expression (5.38), we obtain:

$$\mathscr{E}\theta_{k+1}\theta_k \left(\mathscr{E}\theta_k^2\right)^{-1} = \lambda_S a_B \alpha^{k+1} \left(\lambda_S \alpha^k\right)^{-1} = a_B \alpha$$

and recalling that $\rho = a_B \alpha^{1/2}$ we also obtain

$$a_F = a_B \alpha = \rho \alpha^{1/2}.$$

In addition, denoting $\sigma_{F,k}^2 := \mathscr{E}w_k^2$,

$$\mathscr{E}\theta_{k+1}^2 = a_F^2 \mathscr{E}\theta_k^2 + \sigma_{F,k}^2$$

must hold. Therefore,

$$\sigma_{F,k}^2 = \mathscr{E}\theta_{k+1}^2 - a_F^2 \mathscr{E}\theta_k^2 = \lambda_S \alpha^{k+1} - \rho^2 \alpha \alpha^k = \lambda_S \alpha^{k+1}(1 - \rho^2).$$

It also straightforward to verify that, if θ_k is generated by (5.80), then

$$\mathscr{E}\theta_{k+\tau}\theta_k = a_F^\tau \mathscr{E}\theta_k^2 = a_F^\tau \lambda_S \alpha^k = \lambda_S a_B^\tau \alpha^{k+\tau} \quad \tau > 0$$

which is exactly of the form

$$\mathscr{E}\theta_h \theta_k = \lambda_S a_B^{|h-k|} \alpha^{\max(k,h)}$$

provided $h = k + \tau, \tau > 0$. This concludes the proof.   □

References

1. Akaike H (1979) Smoothness priors and the distributed lag estimator. Technical report, Department of Statistics, Stanford University
2. Banbura M, Giannone D, Reichlin L (2010) Large Bayesian VARs. J Appl Econ 25(1):71–92
3. Bazanella AS, Gevers M, Hendrickx JM, Parraga A (2017) Identifiability of dynamical networks: which nodes need be measured? In: 2017 IEEE 56th annual conference on decision and control (CDC), pp 5870–5875
4. Berger JO (1982) Selecting a minimax estimator of a multivariate normal mean. Ann Stat 10:81–92
5. Bertero M (1989) Linear inverse and ill-posed problems. Adv Electron Electron Phys 75:1–120

6. Carli F (2014) On the maximum entropy property of the first-order stable spline kernel and its implications. In: Proceedings of the 2014 IEEE multi-conference on systems and control, pp 409–414
7. Carli FP, Chen T, Ljung L (2017) Maximum entropy kernels for system identification. IEEE Trans Autom Control 62(3):1471–1477
8. Carvalho C, Polson N, Scott J (2010) The horseshoe estimator for sparse signals. Biometrika 97(2):465–480
9. Casella G (1980) Minimax ridge regression estimation. Ann Stat 8:1036–1056
10. Chen T, Ohlsson H, Ljung L (2012) On the estimation of transfer functions, regularizations and Gaussian processes - revisited. Automatica 48:1525–1535
11. Chen T, Andersen MS, Ljung L, Chiuso A, Pillonetto G (2014) System identification via sparse multiple kernel-based regularization using sequential convex optimization techniques. IEEE Trans Autom Control 59(11):2933–2945
12. Chen T, Ardeshiri T, Carli FP, Chiuso A, Ljung L, Pillonetto G (2016) Maximum entropy properties of discrete-time first-order stable spline kernel. Automatica 66:34–38
13. Chiuso A (2016) Regularization and Bayesian learning in dynamical systems: past, present and future. Annu Rev Control 41:24–38
14. Chiuso A, Pillonetto G (2012) A Bayesian approach to sparse dynamic network identification. Automatica 48(8):1553–1565
15. Cover TM, Thomas JA (2006) Elements of information theory (Wiley series in telecommunications and signal processing). Wiley-Interscience, New York
16. Dankers AG, Van den Hof PMJ, Heuberger PSC, Bombois X (2016) Identification of dynamic models in complex networks with prediction error methods: predictor input selection. IEEE Trans Autom Control 61(4):937–952
17. De Mol C, Giannone D, Reichlin L (2008) Forecasting using a large number of predictors: is Bayesian shrinkage a valid alternative to principal components? J Econ 146(2):318–328
18. Doan T, Litterman R, Sims CA (1984) Forecasting and conditional projection using realistic prior distributions. Econ Rev 3:1–100
19. Everitt N, Galrinho M, Hjalmarsson H (2018) Open-loop asymptotically efficient model reduction with the Steiglitz-Mcbride method. Automatica 89:221–234
20. Fazel M, Hindi H, Boyd SP (2001) A rank minimization heuristic with application to minimum order system approximation. In: Proceedings of the American control conference, pp 4734–4739
21. Fonken SJM, Ferizbegovic M, Hjalmarsson H (2020) Consistent identification of dynamic networks subject to white noise using weighted null-space fitting. In: Proceedings of the 21st IFAC world congress, Berlin, Germany
22. Foster M (1961) An application of the Wiener-Kolmogorov smoothing theory to matrix inversion. J Soc Ind Appl Math 9(3):387–392
23. Giannone D, Lenza M, Primiceri GE (2015) Prior selection for vector auto regressions. Rev Econ Stat 97(2):436–451
24. Goncalves J, Warnick S (2008) Necessary and sufficient conditions for dynamical structure reconstruction of LTI networks. IEEE Trans Autom Control 53(7):1670–1674
25. Goodwin GC, Salgado M (1989) A stochastic embedding approach for quantifying uncertainty in estimation of restricted complexity models. Int J Adapt Control Signal Process 3:333–356
26. Goodwin GC, Gevers M, Ninness B (1992) Quantifying the error in estimated transfer functions with application to model order selection. IEEE Trans Autom Control 37(7):913–929
27. Hagmann P, Cammoun L, Gigandet X, Meuli R, Honey CJ, Wedeen VJ, Sporns O (2008) Mapping the structural core of human cerebral cortex. PLOS Biol 6(7):1–15
28. Hayden D, Hwan Chang Y, Goncalves J, Tomlin CJ (2016) Sparse network identifiability via compressed sensing. Automatica 68:9–17
29. Hendrickx JM, Gevers M, Bazanella AS (2019) Identifiability of dynamical networks with partial node measurements. IEEE Trans Autom Control 64(6):2240–2253

30. Hickman R, Van Verk MC, Van Dijken AJH, Mendes MP, Vroegop-Vos IA, Caarls L, Steenbergen M, Van der Nagel I, Wesselink GJ, Jironkin A, Talbot A, Rhodes J, De Vries M, Schuurink RC, Denby K, Pieterse CMJ, Van Wees SCM (2017) Architecture and dynamics of the jasmonic acid gene regulatory network. Plant Cell 29(9):2086–2105

31. Hoerl AE (1962) Application of ridge analysis to regression problems. Chem Eng Prog 58:54–59

32. Hoerl AE, Kennard RW (1970) Ridge regression: biased estimation for nonorthogonal problems. Technometrics 12:55–67

33. Jin J, Yuan Y, Goncalves J (2020) High precision variational Bayesian inference of sparse linear networks. Automatica 118:109017

34. Kimeldorf GS (1965) Applications of Bayesian statistics to actuarial graduation. PhD dissertation, University of Michigan

35. Kitagawa G, Gersh H (1984) A smoothness priors-state space modeling of time series with trends and seasonalities. J Am Stat Assoc 79(386):378–389

36. Kitagawa G, Gersh H (1985) A smoothness priors long AR model methods for spectral estimation. IEEE Trans Autom Control 30(1):57–65

37. Kitagawa G, Gersch W (1996) Smoothness priors analysis of time series. IMA volumes in mathematics and its applications. Springer, New York

38. Knox T, Stock JH, Watson MW (2001) Empirical Bayes forecast of one time series using many predictors. Technical report, National Bureau of Economic Research

39. Lauritzen SL (1996) Graphical models. Oxford University Press, Oxford

40. Leamer E (1972) A class of informative priors and distributed lag analysis. Econometrica 40(6):1059–1081

41. Lütkepohl H (2007) New introduction to multiple time series analysis. Springer Publishing Company, Incorporated, New York

42. Marquardt DW, Snee RD (1975) Ridge regression in practice. Am Stat 29(1):3–20

43. Maruyama Y, Strawderman WE (2005) A new class of generalized Bayes minimax ridge regression estimators. Ann Stat 1753–1770

44. Materassi D, Innocenti G (2010) Topological identification in networks of dynamical systems. IEEE Trans Autom Control 55(8):1860–1871

45. Materassi D, Salapaka MV (2020) Signal selection for estimation and identification in networks of dynamic systems: a graphical model approach. IEEE Trans Autom Control 65(10):4138–4153

46. Pagani GA, Aiello M (2013) The power grid as a complex network: a survey. Phys A Stat Mech Appl 392(11):2688–2700

47. Park T, Casella G (2008) The Bayesian Lasso. J Am Stat Assoc 103(482):681–686

48. Pillonetto G (2021) Estimation of sparse linear dynamic networks using the stable spline horseshoe prior. arXiv:2107.11155

49. Pillonetto G, De Nicolao G (2010) A new kernel-based approach for linear system identification. Automatica 46(1):81–93

50. Pillonetto G, Chiuso A, De Nicolao G (2011) Prediction error identification of linear systems: a nonparametric Gaussian regression approach. Automatica 47(2):291–305

51. Pillonetto G, Quang MH, Chiuso A (2011) A new kernel-based approach for nonlinear system identification. IEEE Trans Autom Control 56(12):2825–2840

52. Pillonetto G, Dinuzzo F, Chen T, De Nicolao G, Ljung L (2014) Kernel methods in system identification, machine learning and function estimation: a survey. Automatica 50

53. Pillonetto G, Chen T, Chiuso A, De Nicolao G, Ljung L (2016) Regularized linear system identification using atomic, nuclear and kernel-based norms: the role of the stability constraint. Automatica 69:137–149

54. Polson NG, Scott JG (2012) On the half-Cauchy prior for a global scale parameter. Bayesian Anal 7(4):887–902

55. Prando G, Chiuso A, Pillonetto G (2017) Maximum entropy vector kernels for MIMO system identification. Automatica 79:326–339

56. Prando G, Zorzi M, Bertoldo A, Corbetta M, Zorzi M, Chiuso A (2020) Sparse DCM for whole-brain effective connectivity from resting-state FMRI data. NeuroImage 208:116367
57. Ramaswamy KR, Van den Hof PMJ (2021) A local direct method for module identification in dynamic networks with correlated noise. IEEE Trans Autom Control
58. Ramaswamy KR, Bottegal G, Van den Hof PMJ (2021) Learning linear models in a dynamic network using regularized kernel-based methods. Automatica 129(109591)
59. Riley JD (1955) Solving systems of linear equations with a positive definite, symmetric, but possibly ill-conditioned matrix. Math Tables Other Aids Comput 9(51):96–101
60. Robbins H (1951) Asymptotically subminimax solutions of compound statistical decision problems. In: Berkeley symposium on mathematical statistics and probability, pp 131–149
61. Shalev-Shwartz S, Ben-David S (2014) Understanding machine learning: from theory to algorithms. Cambridge University Press, Cambridge
62. Shiller RJ (1973) A distributed lag estimator derived from smoothness priors. Econometrica 41(4):775–788
63. Stein C (1956) Inadmissibility of the usual estimator for the mean of a multivariate distribution. In: Proceedings of the 3rd Berkeley symposium on mathematical statistics and probability, vol I. University of California Press, pp 197–206
64. Strawderman WE (1978) Minimax adaptive generalized ridge regression estimators. J Am Stat Assoc 73:623–627
65. Tether A (1970) Construction of minimal linear state-variable models from finite input-output data. IEEE Trans Autom Control 15(4):427–436
66. Tiao GC, Zellner A (1964) Bayes's theorem and the use of prior knowledge in regression analysis. Biometrika 51(1/2):219–230
67. Van den Hof PMJ, Dankers AG, Heuberger PSC, Bombois X (2013) Identification of dynamic models in complex networks with prediction error methods: basic methods for consistent module estimates. Automatica 49(10):2994–3006
68. Van der Pas SL, Kleijn BJK, van der Vaart AW (2014) The horseshoe estimator: posterior concentration around nearly black vectors. Electron J Stat 8(2):2585–2618
69. Weerts HHM, Van den Hof PMJ, Dankers AG (2018) Prediction error identification of linear dynamic networks with rank-reduced noise. Automatica 98:256–268
70. Weerts HM, Van den Hof PMJ, Dankers AG (2018) Identifiability of linear dynamic networks. Automatica 89:247–258
71. Whittaker ET (1922) On a new method of graduation. Proc Edinb Math Soc 41:63–75
72. Wipf DP, Nagarajan SS (2010) Iterative reweighted ℓ_1 and ℓ_2 methods for finding sparse solutions. IEEE J Sel Top Signal Process 4(2):317–329
73. Yue Z, Thunberg J, Pan W, Ljung L, Goncalves J (2021) Dynamic network reconstruction from heterogeneous datasets. Automatica 123:109339
74. Zorzi M, Chiuso A (2017) Sparse plus low rank network identification: a nonparametric approach. Automatica 76:355–366

Chapter 6
Regularization in Reproducing Kernel Hilbert Spaces

Abstract Methods for obtaining a function g in a relationship $y = g(x)$ from observed samples of y and x are the building blocks for black-box estimation. The classical parametric approach discussed in the previous chapters uses a function model that depends on a finite-dimensional vector, like, e.g., a polynomial model. We have seen that an important issue is the model order choice. This chapter describes some regularization approaches which permit to reconcile flexibility of the model class with well-posedness of the solution exploiting an alternative paradigm to traditional parametric estimation. Instead of constraining the unknown function to a specific parametric structure, the function will be searched over a possibly infinite-dimensional functional space. Overfitting and ill-posedness are circumvented by using reproducing kernel Hilbert spaces as hypothesis spaces and related norms as regularizers. Such kernel-based approaches thus permit to cast all the regularized estimators based on quadratic penalties encountered in the previous chapters as special cases of a more general theory.

6.1 Preliminaries

Techniques for reconstructing a function g in a functional relationship $y = g(x)$ from observed samples of y and x are the fundamental building blocks for black-box estimation. As already seen in Chap. 3 when treating linear regression, given a finite set of pairs (x_i, y_i) the aim is to determine a function g having a good prediction capability, i.e., for a new pair (x, y) we would like the prediction $g(x)$ close to y (e.g., in the MSE sense).

The classical parametric approach discussed in Chap. 3 uses a model g_θ that depends on a finite-dimensional vector θ. A very simple example is a polynomial model, treated in Example 3.1, given, e.g., by $g_\theta(x) = \theta_1 + \theta_2 x + \theta_3 x^2$ whose coefficients θ_i can be estimated by fitting the data via least squares. In this parametric scenario, we have seen that an important issue is the model order choice. In fact, the least squares objective improves as the dimension of θ increases, eventually leading

G. Pillonetto et al., *Regularized System Identification*, Communications and Control Engineering, https://doi.org/10.1007/978-3-030-95860-2_6

181

to data interpolation. But overparametrized models, as a rule, perform poorly when used to predict future output data, even if benign overfitting may sometimes happen, as e.g., described in the context of deep networks [17, 55, 75]. Another drawback related to overparameterization is that the problem may become ill-posed in the sense of Hadamard, i.e., the solution may be non-unique, or ill-conditioned. This means that the estimate may be highly sensitive even to small perturbations of the outputs y_i as, e.g., illustrated in Fig. 1.3 of Sect. 1.2.

This chapter describes some regularization approaches which permit to reconcile flexibility of the model class with well-posedness of the solution exploiting an alternative paradigm to traditional parametric estimation. Instead of constraining the unknown function to a specific parametric structure, g will be searched over a possibly infinite-dimensional functional space. Overfitting and ill-posedness is circumvented by using reproducing kernel Hilbert spaces (RKHSs) as hypothesis spaces and related norms as regularizers. Such norms generalize the quadratic penalties seen in Chap. 3. In this scenario, the estimator is completely defined by a positive definite kernel which has to encode the expected function properties, e.g., the smoothness level. Furthermore we will see that, even when the model class is infinite dimensional, the function estimate turns out a finite linear combination of basis functions computable from the kernel. The estimator also enjoys strong asymptotic properties, permitting (under reasonable assumptions on data generation) to achieve the optimal predictor as the data set size grows to infinity.

The kernel-based approaches described in the following sections thus permit to cast all the regularized estimators based on quadratic penalties encountered in the previous chapters as special cases of a more general theory. In addition, RKHS theory paves the way to the development of other powerful techniques, e.g., for estimation of an infinite number of impulse response coefficients (IIR models estimation), for continuous-time linear system identification and also for nonlinear system identification.

The reader not familiar with functional analysis finds in the first part of the appendix of this chapter a brief overview on the basic results used in the next sections, like, e.g., the concept of linear and bounded functional which is key to define a RKHS.

6.2   Reproducing Kernel Hilbert Spaces

In what follows, we use $\mathscr{X}$ to indicate domains of functions. In machine learning, this set is often referred to as the *input space* with its generic element $x \in \mathscr{X}$ called *input location*. Sometimes, $\mathscr{X}$ is assumed to be a compact metric space, e.g., one can think of $\mathscr{X}$ as a closed and bounded set in the familiar space $\mathbb{R}^m$ equipped with the Euclidean norm. In what follows, all the functions are real valued, so that $f : \mathscr{X} \to \mathbb{R}$.

Reproducing kernel Hilbert spaces We now introduce a class of Hilbert spaces $\mathscr{H}$ which play a fundamental role as hypothesis spaces for function estimation problems. Our goal is to estimate maps which permit to make predictions over the whole $\mathscr{X}$. Thus, a basic requirement is to search for the predictor in a space containing functions which are well-defined pointwise for any $x \in \mathscr{X}$. In particular, we assume that all the pointwise evaluators $g \to g(x)$ are linear and bounded over $\mathscr{H}$. This means that $\forall x \in \mathscr{X}$ there exists $C_x < \infty$ such that

$$|g(x)| \leq C_x \|g\|_{\mathscr{H}}, \quad \forall g \in \mathscr{H}. \tag{6.1}$$

The above condition is stronger than requiring $g(x) < \infty \ \forall x$ since C_x can depend on x but not on g. This property already leads to the function spaces of interest. The following definitions are taken from [13].

Definition 6.1 *(RKHS, based on [13])* A reproducing kernel Hilbert space (RKHS) over a non-empty set $\mathscr{X}$ is a Hilbert space of functions $g : \mathscr{X} \to \mathbb{R}$ such that (6.1) holds.

As suggested by the name itself, RKHSs are related to the concept of positive definite kernel [13, 20], a particular function defined over $\mathscr{X} \times \mathscr{X}$. In the literature it is also called positive semidefinite kernel, hence in what follows positive definite kernel and positive semidefinite kernel will define the same mathematical object. This is also specified in the next definition.

Definition 6.2 *(Positive definite kernel, Mercer kernel and kernel section, based on [13])* Let $\mathscr{X}$ denote a non-empty set. A symmetric function $K : \mathscr{X} \times \mathscr{X} \to \mathbb{R}$ is called *positive definite kernel* or *positive semidefinite kernel* if, for any finite natural number p, it holds

$$\sum_{i=1}^{p} \sum_{j=1}^{p} a_i a_j K(x_i, x_j) \geq 0, \quad \forall (x_k, a_k) \in (\mathscr{X}, \mathbb{R}), \quad k = 1, \ldots, p.$$

If strict inequality holds for any set of p distinct input locations x_k, i.e.,

$$\sum_{i=1}^{p} \sum_{j=1}^{p} a_i a_j K(x_i, x_j) > 0,$$

then the kernel is *strictly positive definite*.

If $\mathscr{X}$ is a metric space and the positive definite kernel is also continuous, then K is said to be a *Mercer kernel*.

Finally, given a kernel K, the *kernel section* K_x centred at x is the function $\mathscr{X} \to \mathbb{R}$ defined by

$$K_x(y) = K(x, y) \quad \forall y \in \mathscr{X}.$$

Hence, in the sense given above, a positive definite kernel "contains" matrices which are all at least positive semidefinite.

We are now in a position to state a fundamental theorem from [13] here specialized to Mercer kernels which lead to RKHSs containing continuous functions (the proof is reported in Sect. 6.9.2).

Theorem 6.1 (RKHSs induced by Mercer kernels, based on [13]) *Let $\mathscr{X}$ be a compact metric space and let $K : \mathscr{X} \times \mathscr{X} \to \mathbb{R}$ be a Mercer kernel. Then, there exists a unique (up to isometries) Hilbert space $\mathscr{H}$ of functions, called RKHS associated to K, such that*

1. *all the kernel sections belong to $\mathscr{H}$, i.e.,*

$$K_x \in \mathscr{H} \quad \forall x \in \mathscr{X} ; \tag{6.2}$$

2. *the so-called* reproducing property *holds, i.e.,*

$$\langle K_x, g \rangle_{\mathscr{H}} = g(x) \quad \forall (x, g) \in (\mathscr{X}, \mathscr{H}) . \tag{6.3}$$

In addition, $\mathscr{H}$ is contained in the space $\mathscr{C}$ of continuous functions.

Remark 6.1 Note that the space $\mathscr{H}$ characterized in Theorem 6.1 is indeed a RKHS according to Definition 6.1. In fact, for any input location x the kernel section K_x belongs to the space and, according to the reproducing property, represents the evaluation functional at x. Then, Theorem 6.27 (Riesz representation theorem), reported in the appendix to this chapter, permits the conclusion that all the pointwise evaluators over $\mathscr{H}$ are linear and bounded.

While Theorem 6.1 establishes a link between Mercer kernels (which enjoy continuity properties) and RKHSs, it is possible also to state a one-to-one correspondence with the entire class of positive definite kernels (not necessarily continuous). In particular, the following result holds.

Theorem 6.2 (Moore–Aronszajn, based on [13]) *Let $\mathscr{X}$ be any non-empty set. Then, to every RKHS $\mathscr{H}$ there corresponds a unique positive definite kernel K such that the reproducing property (6.3) holds. Conversely, given a positive definite kernel K, there exists a unique RKHS of real-valued functions defined over $\mathscr{X}$ where (6.2) and (6.3) hold.*

The proof can be quite easily obtained using Theorem 6.27 (Riesz representation theorem) and arguments similar to those contained in the proof of Theorem 6.1.

Further notes and RKHSs examples Thus, a RKHS $\mathscr{H}$ can be defined just by specifying a kernel K, also called the *reproducing kernel* of $\mathscr{H}$. In particular, any RKHS is generated by the kernel sections. More specifically, let $S = \text{span}(\{K_x\}_{x \in \mathscr{X}})$ and define the following norm in S

$$\|f\|_{\mathscr{H}}^2 = \sum_{i=1}^{p} \sum_{j=1}^{p} c_i c_j K(x_i, x_j), \tag{6.4}$$

where

$$f(\cdot) = \sum_{i=1}^{p} c_i K_{x_i}(\cdot).$$

Then, one has

$$\mathscr{H} = S \cup \{\text{all the limits w.r.t. } \|\cdot\|_{\mathscr{H}} \text{ of Cauchy sequences contained in } S\}.$$

Summarizing, one has

- all the kernel sections $K_x(\cdot)$ belong to the RKHS $\mathscr{H}$ induced by K;
- $\mathscr{H}$ contains also all the finite linear combinations of kernel sections along with some particular infinite sums, convergent w.r.t. the norm (6.4);
- every $f \in \mathscr{H}$ is thus a linear combination of a possibly infinite number of kernel sections.

Assume for instance $K(x_1, x_2) = \exp(-\|x_1 - x_2\|^2)$, which is the so-called Gaussian kernel. Then, all the functions in the corresponding RKHS are sums, or limits of sums, of functions proportional to Gaussians. As further elucidated later on, this means that every function of $\mathscr{H}$ inherits properties such as smoothness and integrability of the kernel, e.g., we have seen in Theorem 6.1 that kernel continuity implies $\mathscr{H} \subset \mathscr{C}$. This fact has an important consequence on modelling: instead of specifying a whole set of basis functions, it suffices to choose a single positive definite kernel that encodes the desired properties of the function to be synthesized.

Example 6.3 *(Norm in a two-dimensional RKHS)* We introduce a very simple RKHS to illustrate how the kernel K can be seen as a similarity function that establishes the norm (complexity) of a function by comparing function values at different input locations.

When $\mathscr{X}$ has finite cardinality m, the functions are evaluated just on a finite number of input locations. Hence, each function f is in one-to-one correspondence with the m-dimensional vector

$$\mathbf{f} = \begin{pmatrix} f(1) \\ f(2) \\ \vdots \\ f(m) \end{pmatrix}.$$

In addition, any kernel is in one-to-one correspondence with one symmetric positive semidefinite matrix $\mathbf{K} \in \mathbb{R}^{m \times m}$ with (i, j)-entry $\mathbf{K}_{ij} = K(i, j)$. Finally, the kernel sections can be seen as the columns of $\mathbf{K}$.

Assume, e.g., $m = 2$ with $\mathscr{X} = \{1, 2\}$. Then, the functions can be seen as two-dimensional vectors and any kernel K is in one-to-one correspondence with one symmetric positive semidefinite matrix $\mathbf{K} \in \mathbb{R}^{2 \times 2}$. The RKHS $\mathscr{H}$ associated to K is finite-dimensional being spanned just by the two kernel sections $K_1(\cdot)$ and $K_2(\cdot)$ which can be seen as the two columns of $\mathbf{K}$. Hence, the functions f in $\mathscr{H}$ are in one-to-one correspondence with the vectors

$$\mathbf{f} = \begin{pmatrix} f(1) \\ f(2) \end{pmatrix} = \mathbf{K}c, \quad c \in \mathbb{R}^2.$$

If $\mathbf{K}$ is full rank, $\mathscr{H}$ covers the whole $\mathbb{R}^2$ and from (6.4) we have

$$\|f\|_{\mathscr{H}}^2 = c^T \mathbf{K} c = \mathbf{f}^T \mathbf{K}^{-1} \mathbf{f}.$$

For the sake of simplicity, assume also that $\mathbf{K}_{11} = \mathbf{K}_{22} = 1$ so that it must hold $-1 < \mathbf{K}_{12} < 1$. Then, considering, e.g., the function $f(i) = i$, one has

$$\begin{aligned} \|f\|_{\mathscr{H}}^2 &= [1 \ 2] \, \mathbf{K}^{-1} \, [1 \ 2]^T \\ &= \frac{5 - 4\mathbf{K}_{12}}{1 - \mathbf{K}_{12}^2}, \quad -1 < \mathbf{K}_{12} < 1. \end{aligned}$$

Figure 6.1 displays $\|f\|_{\mathscr{H}}^2$ as a function of $\mathbf{K}_{12}$. One can see that the norm diverges as $|\mathbf{K}_{12}|$ approaches 1.

If, e.g., $\mathbf{K}_{12} = 1$ the kernel function becomes constant over $\mathscr{X} \times \mathscr{X}$. Hence, the two kernel sections $K_1(\cdot)$ and $K_2(\cdot)$ coincide, being constant with $K_1(i) = K_2(i) = 1$ for $i = 1, 2$. This means that $\mathbf{K}_{12} = 1$ induces a space $\mathscr{H}$ containing only constant functions.[1] This explains why the norm (complexity) of f becomes large if $\mathbf{K}_{12}$ is close to 1: the space becomes less and less "tolerant" of functions with $f(1) \neq f(2)$.

Letting now $f(1) = 1$ and $f(2) = a$, the joint effect of $\mathbf{K}_{12}$ and a is explained by the formula

$$\begin{aligned} \|f\|_{\mathscr{H}}^2 &= [1 \ a] \, \mathbf{K}^{-1} \, [1 \ a]^T \\ &= \frac{(a - \mathbf{K}_{12})^2}{1 - \mathbf{K}_{12}^2} + 1, \quad -1 < \mathbf{K}_{12} < 1. \end{aligned}$$

Note that, thinking now of $\mathbf{K}_{12}$ as fixed, the function with minimal RKHS norm (complexity) is obtained with $a = \mathbf{K}_{12}$ and has a norm equal to one. $\square$

Example 6.4 ($\mathscr{L}_2^\mu$ *and* ℓ_2) Let $\mathscr{X} = \mathbb{R}$ and consider the classical Lebesgue space of square summable functions with μ equal to the Lebesgue measure. Recall that this is a Hilbert space whose elements are equivalence classes of functions measurable

---

[1] One can then also easily check that the case $\mathbf{K}_{12} = -1$ instead induces a RKHS containing only functions satisfying $f(1) = -f(2)$.

Fig. 6.1 The figure plots $\|f\|_{\mathcal{H}}^2$, with $f(i) = i$ and $i \in \{1, 2\}$, as a function of the kernel value $K(1, 2)$, having set $K(1, 1) = K(2, 2) = 1$

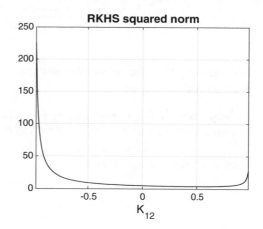

w.r.t. Lebesgue: any group of functions which differ only on a set of null measure (e.g., containing only a countable number of input locations) identifies the same vector. Hence, $\mathcal{L}_2^\mu$ cannot be a RKHS since pointwise evaluation is not even well defined.

Let instead $\mathcal{X} = \mathbb{N}$ (the set of natural numbers) and define the identity kernel

$$K(i, j) = \delta_{ij}, \quad (i, j) \in \mathbb{N} \times \mathbb{N}, \tag{6.5}$$

where δ_{ij} is the Kronecker delta. Clearly, K is symmetric and positive definite according to Definition 6.2 (it can be associated with an identity matrix of infinite size). Hence, it induces unique RKHS $\mathcal{H}$ that contains all the finite combinations of the kernel sections. In particular, any finite sum can be written as $f(\cdot) = \sum_{i=1}^m f_i K_i(\cdot)$, where some of the f_i may be null, and corresponds to a sequence with a finite number of non null components. To obtain the entire $\mathcal{H}$, we need also to add all the Cauchy sequences limits w.r.t. the norm (6.4) given by

$$\|f\|_{\mathcal{H}}^2 = \left\| \sum_{i=1}^m f_i K_i(\cdot) \right\|_{\mathcal{H}}^2$$
$$= \sum_{i=1}^m \sum_{j=1}^m f_i f_j K(i, j) = \sum_{i=1}^m f_i^2,$$

which coincides with the classical Euclidean norm of $[f_1 \ldots f_m]$. This allows us to conclude that the associated RKHS is the classical space ℓ_2 of square summable sequences.

As a finale note, Definition 6.1 easily confirms that ℓ_2 is a RKHS. In fact, for every $f = [f_1 \ f_2 \ \ldots] \in \ell_2$ one has

$$|f_i| \leq \sqrt{\sum_i f_i^2} = \|f\|_2 \quad \forall i,$$

and, recalling (6.1), this shows that all the evaluation functionals $f \to f_i$ with $i \in \mathbb{N}$ are bounded.   □

Example 6.5 *(Sobolev space and the first-order spline kernel)* While in the previous example we have seen that $\mathscr{L}_2^{\mu}$ is not a RKHS, consider now the space obtained by integrating the functions in this space. In particular, let $\mathscr{X} = [0, 1]$, set μ to the Lebesgue measure and consider

$$\mathscr{H} = \left\{ f \mid f(x) = \int_0^x h(y)dy \text{ with } h \in \mathscr{L}_2^{\mu} \right\}.$$

One thus has that any f in $\mathscr{H}$ satisfies $f(0) = 0$ and is absolutely continuous: its derivative $h = \dot{f}$ is defined almost everywhere and is Lebesgue integrable.

With the inner product given by

$$\langle f, g \rangle_{\mathscr{H}} = \langle \dot{f}, \dot{g} \rangle_{\mathscr{L}_2^{\mu}},$$

it is easy to see that $\mathscr{H}$ is a Hilbert space. In fact, $\mathscr{L}_2^{\mu}$ is Hilbert and we have established a one-to-one correspondence between functions in $\mathscr{H}$ and $\mathscr{L}_2^{\mu}$ which preserves inner product. Such $\mathscr{H}$ is an example of Sobolev space [2] since the complexity of a function is measured by the energy of its derivative:

$$\|f\|_{\mathscr{H}}^2 = \int_0^1 \dot{f}^2(x)dx.$$

Now, given $x \in [0, 1]$, let $\chi_x(\cdot)$ be the indicator function of the set $[0, x]$. Then, one has

$$|f(x)| = \left| \int_0^x \dot{f}(a)da \right| = |\langle \chi_x, \dot{f} \rangle_{\mathscr{L}_2^{\mu}}|$$
$$\leq \|\dot{f}\|_{\mathscr{L}_2^{\mu}} = \|f\|_{\mathscr{H}},$$

where we have used the Cauchy–Schwarz inequality. Hence, $\mathscr{H}$ is also a RKHS since all the evaluations functionals are bounded. We now prove that its reproducing kernel is the so-called first-order (linear) *spline kernel* given by

$$K(x, y) = \min(x, y). \tag{6.6}$$

In fact, every kernel section belongs to $\mathscr{H}$, being piecewise linear with $\dot{K}_x = \chi_x$. Furthermore, (6.6) satisfies the reproducing property since

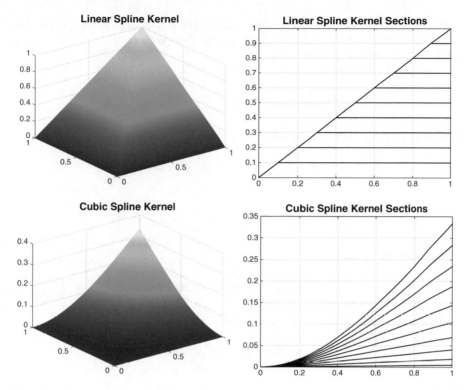

Fig. 6.2 Linear and cubic spline kernel with kernel sections $K_{x_i}(x)$ for $x_i - 0.1, 0.2, \ldots, 1$ (bottom)

$$\langle f, K_x \rangle_{\mathscr{H}} = \langle \dot{f}, \chi_x \rangle_{\mathscr{L}_2^{\mu}}$$
$$= \int_0^x \dot{f}(y)dy = f(x).$$

The linear spline kernel and some of its sections are displayed in the top panels of Fig. 6.2. □

6.2.1 Reproducing Kernel Hilbert Spaces Induced by Operations on Kernels ★

We report some classical results about RKHSs induced by operations on kernels which can be derived from [13]. The first theorem characterizes the RKHS induced by the sum or product of two kernels.

Theorem 6.6 (RKHS induced by sum or product of two kernels, based on [13]) *Let K and G be two positive definite kernels over the same domain $\mathcal{X} \times \mathcal{X}$, associated to the RKHSs $\mathcal{H}$ and $\mathcal{G}$, respectively.*
  The sum $K + G$, where

$$[K + G](x, y) = K(x, y) + G(x, y),$$

is the reproducing kernel of the RKHS $\mathcal{R}$ containing functions

$$f = h + g, \quad (h, g) \in \mathcal{H} \times \mathcal{G}$$

with

$$\|f\|_{\mathcal{R}}^2 = \min_{h \in \mathcal{H}, g \in \mathcal{G}} \|h\|_{\mathcal{H}}^2 + \|g\|_{\mathcal{G}}^2 \ \text{s.t.} \ f = h + g.$$

The product $K G$, where

$$[KG](x, y) = K(x, y)G(x, y)$$

is instead the reproducing kernel of the RKHS $\mathcal{R}$ containing functions

$$f = hg, \quad (h, g) \in \mathcal{H} \times \mathcal{G}$$

with

$$\|f\|_{\mathcal{R}}^2 = \min_{h \in \mathcal{H}, g \in \mathcal{G}} \|h\|_{\mathcal{H}}^2 \|g\|_{\mathcal{G}}^2 \ \text{s.t.} \ f = hg.$$

The second theorem instead provides the connection between two RKHSs, with the second one obtained from the first one by sampling its kernel.

Theorem 6.7 (RKHS induced by kernel sampling, based on [13]) *Let $\mathcal{H}$ be the RKHS induced by the kernel $K : \mathcal{X} \times \mathcal{X} \to \mathbb{R}$. Let $\mathcal{Y} \subset \mathcal{X}$ and denote with $\mathcal{R}$ the RKHS of functions over $\mathcal{Y}$ induced by the restriction of the kernel K on $\mathcal{Y} \times \mathcal{Y}$. Then, the functions in $\mathcal{R}$ correspond to the functions in $\mathcal{H}$ sampled on $\mathcal{Y}$. One also has*

$$\|f\|_{\mathcal{R}}^2 = \min_{g \in \mathcal{H}} \|g\|_{\mathcal{H}}^2 \ \text{s.t.} \ g_{\mathcal{Y}} = f, \tag{6.7}$$

where $g_{\mathcal{Y}}$ is g sampled on $\mathcal{Y}$.

The following theorem lists some operations which permit to build kernels (and hence RKHSs) from simple building blocks.

Theorem 6.8 (Building kernels from kernels, based on [13]) *Let K_1 and K_2 two positive definite kernels over $\mathcal{X} \times \mathcal{X}$ and K_3 a positive definite kernel over $\mathbb{R}^m \times \mathbb{R}^m$. Let also P an $m \times m$ symmetric positive semidefinite matrix and $\mathcal{P}(x)$ a polynomial with positive coefficients. Then, the following functions are positive definite kernels over $\mathcal{X} \times \mathcal{X}$:*

- $K(x, y) = K_1(x, y) + K_2(x, y)$ *(see also Theorem 6.6)*.
- $K(x, y) = aK_1(x, y), \quad a \geq 0$.
- $K(x, y) = K_1(x, y)K_2(x, y)$ *(see also Theorem 6.6)*.
- $K(x, y) = f(x)f(y), \quad f : \mathscr{X} \to \mathbb{R}$.
- $K(x, y) = K_3(f(x), f(y)), \quad f : \mathscr{X} \to \mathbb{R}^m$.
- $K(x, y) = x^T P y, \quad \mathscr{X} = \mathbb{R}^m$.
- $K(x, y) = \mathscr{P}(K_1(x, y))$.
- $K(x, y) = \exp(K_1(x, y))$.

6.3 Spectral Representations of Reproducing Kernel Hilbert Spaces

In the previous section we have seen that any RKHS is generated by its kernel sections. We now discuss another representation obtainable when the kernel can be diagonalized as follows

$$K(x, y) = \sum_{i \in \mathscr{I}} \zeta_i \rho_i(x) \rho_i(y), \quad \zeta_i > 0 \ \forall i, \tag{6.8}$$

where the set $\mathscr{I}$ is countable. This will lead to new insights on the nature of the RKHSs, generalizing to the infinite-dimensional case the connection between regularization and basis expansion reported in Sect. 5.6.

A simple situation holds when the input space has finite cardinality, e.g., $\mathscr{X} = \{x_1 \ldots x_m\}$. Under this assumption, any positive definite kernel is in one-to-one correspondence with the $m \times m$ matrix $\mathbf{K}$ whose (i, j)-entry is $K(x_i, x_j)$. The representation (6.8) then follows from the spectral theorem applied to $\mathbf{K}$. In fact, if ζ_i and v_i are, respectively, the eigenvalues and the orthonormal (column) eigenvectors of $\mathbf{K}$, (6.8) can be written as

$$\mathbf{K} = \sum_{i=1}^{m} \zeta_i v_i v_i^T,$$

where the functions $\rho_i(\cdot)$ have become the vectors v_i. One generalization of this result is described below.

Let L_K be the linear operator defined by the positive definite kernel K as follows:

$$L_K[f](\cdot) = \int_X K(\cdot, x) f(x) d\mu(x). \tag{6.9}$$

We also assume that μ is a σ-finite and nondegenerate Borel measure on $\mathscr{X}$. Essentially this means that $\mathscr{X}$ is the countable union of measurable sets with finite measure and that μ "covers" entirely $\mathscr{X}$. The reader can, e.g., consider $\mathscr{X} \subset \mathbb{R}^m$ and think of μ as the Lebesque measure or any probability measure with $\mu(A) > 0$ for any non-empty open set $A \subset \mathcal{X}$. The next classical result goes under the name of Mercer theorem whose formulations trace back to [60].

Theorem 6.9 (Mercer theorem, based on [60]) *Let $\mathcal{X}$ be a compact metric space equipped with a nondegenerate and σ-finite Borel measure μ and let K be a Mercer kernel on $\mathcal{X} \times \mathcal{X}$. Then, there exists a complete orthonormal basis of $\mathcal{L}_2^{\mu}$ given by a countable number of continuous functions $\{\rho_i\}_{i \in \mathcal{I}}$ satisfying*

$$L_K[\rho_i] = \zeta_i \rho_i, \quad i \in \mathcal{I}, \quad \zeta_1 \geq \zeta_2 \geq \cdots \geq 0, \tag{6.10}$$

with $\zeta_i > 0 \ \forall i$ if K is strictly positive and $\lim_{i \to \infty} \zeta_i = 0$ if the number of eigenvalues is infinite.

One also has

$$K(x, y) = \sum_{i \in \mathcal{I}} \zeta_i \rho_i(x) \rho_i(y), \tag{6.11}$$

where the convergence of the series is absolute and uniform on $\mathcal{X} \times \mathcal{X}$.

The following result characterizes a RKHS through the eigenfunctions of L_K. The proof is reported in Sect. 6.9.3.

Theorem 6.10 (RKHS defined by an orthonormal basis of $\mathcal{L}_2^{\mu}$) *Under the same assumption of Theorem 6.9, if the ρ_i and ζ_i satisfy (6.10), with also $\zeta_i > 0 \ \forall i$, one has*

$$\mathcal{H} = \left\{ f \ \middle| \ f(x) = \sum_{i \in \mathcal{I}} c_i \rho_i(x) \ s.t. \ \sum_{i \in \mathcal{I}} \frac{c_i^2}{\zeta_i} < \infty, \right\}. \tag{6.12}$$

In addition, if

$$f = \sum_{i \in \mathcal{I}} a_i \rho_i, \quad g = \sum_{i \in \mathcal{I}} b_i \rho_i,$$

one has

$$\langle f, g \rangle_{\mathcal{H}} = \sum_{i \in \mathcal{I}} \frac{a_i b_i}{\zeta_i}, \tag{6.13}$$

so that

$$\|f\|_{\mathcal{H}}^2 = \sum_{i \in \mathcal{I}} \frac{a_i^2}{\zeta_i}. \tag{6.14}$$

Hence, it also comes that $\{\sqrt{\zeta_i} \rho_i\}_{i \in \mathcal{I}}$ is an orthonormal basis of $\mathcal{H}$.

The representation (6.12) is not unique since the spectral maps, i.e., the functions that associate a kernel with a decomposition of the type (6.8), are not unique. They depend on the chosen measure μ even if they lead to the same RKHS.

Theorem 6.10 thus shows that any kernel admitting an expansion (6.11) coming from the Mercer theorem induces a separable RKHS, i.e., having a countable basis given by the ρ_i. Later on, Theorem 6.13 will show that such result holds under much milder assumptions. In fact, the representation (6.12) can be obtained starting from any diagonalized kernel (6.8) involving generic functions ρ_i, e.g., not necessarily independent of each other. One can also remove the compactness hypothesis on the input space, e.g., letting $\mathcal{X}$ be the entire $\mathbb{R}^m$.

Remark 6.2 *(Relationship between $\mathcal{H}$ and $\mathcal{L}_2^{\mu}$)* Theorem 6.10 points out an interesting connection between $\mathcal{H}$ and $\mathcal{L}_2^{\mu}$. Since the functions ρ_i form an orthonormal basis in $\mathcal{L}_2^{\mu}$, one has

$$f \in \mathcal{L}_2^{\mu} \iff f = \sum_{i \in \mathscr{I}} c_i \rho_i \text{ with } \sum_{i \in \mathscr{I}} c_i^2 < \infty \tag{6.15}$$

while (6.12) shows that

$$f \in \mathcal{H} \iff f = \sum_{i \in \mathscr{I}} c_i \rho_i \text{ with } \sum_{i \in \mathscr{I}} \frac{c_i^2}{\zeta_i} < \infty. \tag{6.16}$$

If $\zeta_i > 0 \ \forall i$, one has the set inclusion $\mathcal{H} \subset \mathcal{L}_2^{\mu}$ since the functions in the RKHS, must satisfy a more stringent condition on the expansion coefficients decay (the ζ_i decay to zero).
In addition, let $L_K^{1/2}$ denote the operator defined as the square root of L_K, i.e., for any $f \in \mathcal{L}_2^{\mu}$ with $f = \sum_{i \in \mathscr{I}} c_i \rho_i$, one has

$$L_K^{1/2}[f] = \sum_{i \in \mathscr{I}} \sqrt{\zeta_i} c_i \rho_i. \tag{6.17}$$

This is a smoothing operator: the function $L_K^{1/2}[f]$ is more regular than f since the expansion coefficients $\sqrt{\zeta_i} c_i$ decrease to zero faster than the c_i. In view of (6.15) and (6.16), we obtain

$$\mathcal{H} = \left\{ L_K^{1/2}[f] \mid f \in \mathcal{L}_2^{\mu} \right\}, \tag{6.18}$$

which shows that the RKHS can be thought of as the output of the linear system $L_K^{1/2}$ fed with the space $\mathcal{L}_2^{\mu}$, i.e., $\mathcal{H} = L_K^{1/2} \mathcal{L}_2^{\mu}$.

Example 6.11 *(Spline kernel expansion)* In Example 6.5, we have seen that the space of functions on the unit interval satisfying $f(0) = 0$ and $\int_0^1 \dot{f}^2(x)dx < \infty$ is the RKHS associated to the first-order spline kernel $\min(x, y)$. We now derive a representation of the type (6.12) for this space setting μ to the Lebesgue measure. For this purpose, consider the system

$$\int_0^1 \min(x, y)\rho(y)dy = \zeta\rho(x).$$

The above equation is equivalent to

$$\int_0^x y\rho(y)dy + x \int_x^1 \rho(y)dy = \zeta\rho(x),$$

which implies $\rho(0) = 0$. Taking the derivative w.r.t. x we also obtain

$$\int_x^1 \rho(y)dy = \zeta\dot{\rho}(x)$$

that implies $\dot{\rho}(1) = 0$. Differentiating again w.r.t. x gives

$$-\rho(x) = \zeta\ddot{\rho}(x),$$

whose general solution is

$$\rho(x) = a\sin(x/\sqrt{\zeta}) + b\cos(x/\sqrt{\zeta}), \quad a, b \in \mathbb{R}.$$

The boundary conditions $\rho(0) = \dot{\rho}(1) = 0$ imply $b = 0$ and lead to the following possible eigenvalues:

$$\zeta_i = \frac{1}{(i\pi - \pi/2)^2}, \quad i = 1, 2, \dots.$$

The orthonormality condition also implies $a = \sqrt{2}$ so that we obtain

$$\rho_i(x) = \sqrt{2}\sin\left(i\pi x - \frac{\pi x}{2}\right), \quad i = 1, 2, \dots.$$

This provides the formulation (6.12) of the Sobolev space $\mathscr{H}$. Figure 6.3 plots three eigenfunctions (left panel) and the first 100 eigenvalues ζ_i (right panel). It is evident that the larger i the larger is the high-frequency content of ρ_i and the RKHS norm of such basis function. In fact, a large value of i corresponds to a small eigenvalue ζ_i and one has $\|\rho_i\|_{\mathscr{H}}^2 = 1/\zeta_i$.   □

Example 6.12 *(Translation invariant kernels and Fourier expansion)* A translation invariant kernel depends only on the difference of its two arguments. Hence, there exists $h : \mathscr{X} \to \mathbb{R}$ such that $K(x, y) = h(x - y)$. Assume that $\mathscr{X} = [0, 2\pi]$ and that h can be extended to a continuous, symmetric and periodic function over $\mathbb{R}$. Then, it can be expanded in terms of the following uniformly convergent Fourier series

$$h(x) = \sum_{i=0}^{\infty} \zeta_i \cos(ix),$$

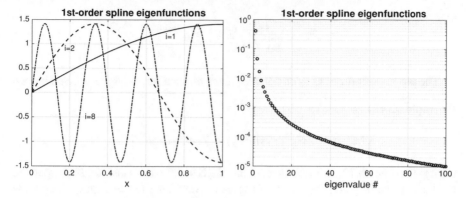

Fig. 6.3 Expansion of the first-order spline kernel $\min(x, y)$: eigenfunctions ρ_i for $i = 1, 2, 8$ (left panel) and eigenvalues ζ_i (right)

where ζ_0 accounts for the constant component and we assume $\zeta_i > 0 \ \forall i$. We thus obtain the kernel expansion

$$K(x, y) = \zeta_0 + \sum_{i=1}^{\infty} \zeta_i \cos(ix) \cos(iy) + \sum_{i=1}^{\infty} \zeta_i \sin(ix) \sin(iy),$$

in terms of functions which are all orthogonal in $\mathscr{L}_2^{\mu}$. Hence, these kernels induce RKHSs generated by the Fourier basis, with different inner products determined by ζ_i. □

6.3.1 More General Spectral Representation ⋆

Now, assume that the kernel K is available in the form $K(x, y) = \sum_{i \in \mathscr{I}} \zeta_i \rho_i(x) \rho_i(y)$ with $\zeta_i > 0 \ \forall i$, but with functions ρ_i not necessarily orthonormal. More generally, we do not even require that they are independent, e.g., ρ_1 could be a linear combination of ρ_2 and ρ_3. The following result shows that the RKHS associated to K is still generated by the ρ_i, but the relationship of the expansion coefficients with $\| \cdot \|_{\mathscr{H}}$ is more involved than in the previous case.

Theorem 6.13 (RKHS induced by a diagonalized kernel) *Let $\mathscr{H}$ be the RKHS induced by $K(x, y) = \sum_{i \in \mathscr{I}} \zeta_i \rho_i(x) \rho_i(y)$ with $\zeta_i > 0 \ \forall i$ and the set $\mathscr{I}$ countable. Then, $\mathscr{H}$ is separable and admits the representation*

$$\mathscr{H} = \left\{ f \ \middle| \ f(x) = \sum_{i \in \mathscr{I}} c_i \rho_i(x) \ s.t. \ \sum_{i \in \mathscr{I}} \frac{c_i^2}{\zeta_i} < \infty \right\} \tag{6.19}$$

and one has

$$\|f\|_{\mathscr{H}}^2 = \min_{\{c_i\}} \sum_{i \in \mathscr{I}} \frac{c_i^2}{\zeta_i} \quad s.t. \; f = \sum_{i \in \mathscr{I}} c_i \rho_i. \tag{6.20}$$

The proof is reported in Sect. 6.9.4 while an application example is given below.

Example 6.14 Let

$$K(x, y) = 2 \sin^2(x) \sin^2(y) + 2 \cos^2(x) \cos^2(y) + 1.$$

Using Theorem 6.13, we obtain that the RKHS $\mathscr{H}$ associated to K is spanned by $\sin^2(x)$, $\cos^2(x)$ and the constant function. Now, let $f(x) = 1$ and consider the problem of computing $\|f\|_{\mathscr{H}}^2$. To have a correspondence with (6.8) we can, e.g., fix the notation

$$\rho_1(x) = \sin^2(x), \quad \rho_2(x) = \cos^2(x), \quad \rho_3(x) = 1$$

and

$$\zeta_1 = 2, \quad \zeta_2 = 2, \quad \zeta_3 = 1.$$

Since the functions ρ_i are not independent, many different representation for $f(x) = 1$ can be found. In particular, one has

$$1 = c\rho_1(x) + c\rho_2(x) + (1 - c)\rho_3(x) \quad \forall c \in \mathbb{R},$$

so that

$$\|f\|_{\mathscr{H}}^2 = \min_c \frac{c^2}{2} + \frac{c^2}{2} + (1 - c)^2 = \min_c 2c^2 - 2c + 1 = \frac{1}{2}$$

with the minimum $1/2$ obtained at $c = 1/2$. Hence, according to the norm of $\mathscr{H}$, the "minimum energy" representation of $f(x) = 1$ is $1/2(\rho_1(x) + \rho_2(x) + \rho_3(x))$.
□

6.4   Kernel-Based Regularized Estimation

6.4.1   *Regularization in Reproducing Kernel Hilbert Spaces and the Representer Theorem*

A powerful approach to reconstruct a function $g : \mathscr{X} \to \mathbb{R}$ from sparse data $\{x_i, y_i\}_{i=1}^N$ consists of minimizing a suitable functional over a RKHS. An important generalization of the estimators based on quadratic penalties, denoted by ReLS-Q in Chap. 3, is defined by

$$\hat{g} = \arg\min_{f \in \mathscr{H}} \sum_{i=1}^{N} \mathscr{V}_i(y_i, f(x_i)) + \gamma \|f\|_{\mathscr{H}}^2. \qquad (6.21)$$

In (6.21), $\mathscr{V}_i$ are loss functions measuring the distance between y_i and $f(x_i)$. They can take only positive values and are assumed convex w.r.t. their second argument $f(x_i)$. As an example, when the quadratic loss is adopted for any i, one obtains

$$\mathscr{V}_i(y_i, f(x_i)) = (y_i - f(x_i))^2.$$

Then, the norm $\| \cdot \|_{\mathscr{H}}$ defines the regularizer, e.g., given by the energy of the first-order derivative

$$\|f\|_{\mathscr{H}}^2 = \int_0^1 \dot{f}^2(x)dx,$$

which corresponds to the spline norm introduced in Example 6.5. Finally, the positive scalar γ is the regularization parameter (already encountered in the previous chapters) which has to balance adherence to experimental data and function regularity. Indeed, the idea underlying (6.21) is that the predictor $\hat{g}$ should be able to describe the data without being too complex according to the RKHS norm. In particular, the scope of the regularizer is to restore the well-posedness of the problem, making the solution depend continuously on the data. It should also include our available information on the unknown function, e.g., the expected smoothness level.

The importance of the RKHSs in the context of regularization methods stems from the following central result, whose first formulation can be found in [52]. It shows that the solutions of the class of variational problems (6.21) admit a finite-dimensional representation, independently of the dimension of $\mathscr{H}$. The proof of an extended version of this result can be found in Sect. 6.9.5.

Theorem 6.15 (Representer theorem, adapted from [104]) *Let $\mathscr{H}$ be a RKHS. Then, all the solutions of (6.21) admit the following expression*

$$\hat{g} = \sum_{i=1}^{N} c_i K_{x_i}, \qquad (6.22)$$

where the c_i are suitable scalar expansion coefficients.

Thus, as in the traditional linear parametric approach, the optimal function is a linear combination of basis functions. However, a fundamental difference is that their number is now equal to the number of data pairs, and is thus not fixed a priori. In fact, the functions appearing in the expression of the minimizer $\hat{g}$ are just the kernel sections K_{x_i} centred on the input data. The representer theorem also conveys the message that, using estimators of the form (6.21), it is not possible to recover arbitrarily complex functions from a finite amount of data. The solution is always confined to a subspace with dimension equal to the data set size.

Now, let $\mathbf{K} \in \mathbb{R}^{N \times N}$ be the positive semidefinite matrix (called *kernel matrix*, or Gram matrix) such that $\mathbf{K}_{ij} = K(x_i, x_j)$. The ith row of $\mathbf{K}$ is denoted by $\mathbf{k}_i$. Using this notation, if $g = \sum_{i=1}^{N} c_i K_{x_i}$ then

$$g(x_i) = \mathbf{k}_i c \quad \text{and} \quad \|g\|_{\mathcal{H}}^2 = c^T \mathbf{K} c, \tag{6.23}$$

where $c = [c_1, \ldots, c_N]^T$ and the second equality derives from the reproducing property or, equivalently, from (6.4).

Using the representer theorem, we can plug the expression (6.22) of the optimal $\hat{g}$ into the objective (6.21). Then, exploiting (6.23), the variational problem (6.21) boils down to

$$\min_{c \in \mathbb{R}^N} \sum_{i=1}^{N} \mathcal{V}_i(y_i, \mathbf{k}_i c) + \gamma c^T \mathbf{K} c. \tag{6.24}$$

The regularization problem (6.21) has been thus reduced to a finite-dimensional optimization problem whose order N does not depend on the dimension of the original space $\mathcal{H}$. In addition, since each loss function $\mathcal{V}_i$ has been assumed convex, the objective (6.24) is convex overall. How to compute the expansion coefficients now depends on the specific choice of the $\mathcal{V}_i$, as discussed in the next section.

Remark 6.3 *(Kernel trick and implicit basis functions encoding)* Assume that the kernel admits the expansion $K(x, y) = \sum_{i=1}^{\infty} \zeta_i \rho_i(x) \rho_i(y)$, $\zeta_i > 0$. Then, as discussed in Sect. 6.3, any function in $\mathcal{H}$ has the representation

$$f = \sum_{i=1}^{\infty} a_i \rho_i \quad \text{with} \quad \|f\|_{\mathcal{H}}^2 = \sum_{j=1}^{\infty} \frac{a_j^2}{\zeta_j}.$$

Problem (6.21) can then be rewritten using the infinite-dimensional vector $a = [a_1 \ a_2 \ \ldots]$ as unknown:

$$\hat{a} = \arg\min_a \sum_{i=1}^{N} \mathcal{V}_i \left(y_i, \sum_{j=1}^{\infty} a_j \rho_j(x_i) \right) + \gamma \sum_{j=1}^{\infty} \frac{a_j^2}{\zeta_j},$$

and an equivalent representation of (6.22) becomes $\hat{g} = \sum_{i=1}^{\infty} \hat{a}_i \rho_i$. In comparison to this reformulation the use of the kernel and of the representer theorem subsumes modelling and computational advantages. In fact, through K one needs neither to choose the number of basis functions to be used (the kernel can already include in an implicit way an infinite number of basis functions) nor to store any basis function in memory (the representer theorem reduces inference to solving a finite-dimensional optimization problem based on the kernel matrix $\mathbf{K}$). These features are related to what is called the *kernel trick* in the machine learning literature.

6.4.2 Representer Theorem Using Linear and Bounded Functionals

A more general version of the representer theorem obtained in [52] can be obtained by replacing $f(x_i)$ with $L_i[f]$, where L_i is linear and bounded. In the first part of the following result $\mathcal{H}$ is just required to be Hilbert. In Sect. 6.9.5 we will see how Theorem 6.16 can be further generalized.

Theorem 6.16 (Representer theorem with functionals L_i, adapted from [104]) *Let $\mathcal{H}$ be a Hilbert space and consider the optimization problem*

$$\hat{g} = \underset{f \in \mathcal{H}}{arg\,min} \sum_{i=1}^{N} \mathcal{V}_i(y_i, L_i[f]) + \gamma \|f\|_{\mathcal{H}}^2, \qquad (6.25)$$

where each $L_i : \mathcal{H} \to \mathbb{R}$ is linear and bounded. Then, all the solutions of (6.25) admit the following expression

$$\hat{g} = \sum_{i=1}^{N} c_i \eta_i, \qquad (6.26)$$

where the c_i are suitable scalar expansion coefficients and each $\eta_i \in \mathcal{H}$ is the representer of L_i, i.e., for any i and $f \in \mathcal{H}$:

$$L_i[f] = \langle f, \eta_i \rangle_{\mathcal{H}}. \qquad (6.27)$$

In particular, if $\mathcal{H}$ is a RKHS with kernel K, each basis function is given by

$$\eta_i(x) = L_i[K(\cdot, x)]. \qquad (6.28)$$

The existence of η_i satisfying (6.27) is ensured by the Riesz representation theorem (Theorem 6.27). One can also prove that in a RKHS a linear functional L is linear and bounded if and only if the function f obtained by applying L to the kernel, i.e., $f(x) = L[K(x, \cdot)] \; \forall x$, belongs to the RKHS.

Note also that Theorem 6.15 is indeed a special case of the last result. In fact, let $\mathcal{H}$ be a RKHS and $L_i[f] = f(x_i) \; \forall i$. Then, each L_i is linear and bounded and each η_i becomes the kernel section K_{x_i} according to the reproducing property.

Example 6.17 *(Solution using the quadratic loss)* Let us adopt a quadratic loss in (6.25), i.e., $\mathcal{V}_i(y_i, L_i[f]) = (y_i - L_i[f])^2$. This makes the objective strictly convex so that a unique solution exists. To find it, plugging (6.26) in (6.25) and using also (6.28), the following quadratic problem is obtained

$$\|Y - Oc\|^2 + \gamma c^T Oc \qquad (6.29)$$

where $Y = [y_1, \ldots, y_N]^T$, $\| \cdot \|$ is the Euclidean norm, while the $N \times N$ matrix O has i, j entry given by

$$O_{ij} = \langle \eta_i, \eta_j \rangle_{\mathscr{H}} = L_i[L_j[K]]. \tag{6.30}$$

The minimizer $\hat{c}$ of (6.29) is unique if O is full rank. Otherwise, all the solutions lead to the same function estimate in view of the (already mentioned) strict convexity of (6.25). In particular, one can always use as optimal expansion coefficients the components of the vector

$$\hat{c} = (O + \gamma I_N)^{-1} Y. \tag{6.31}$$

In Sect. 6.5.1 this result will be further discussed in the context of the so-called regularization networks, where one comes back to assume $L_i[f] = f(x_i)$. $\square$

6.5   Regularization Networks and Support Vector Machines

The choice of the loss $\mathscr{V}_i$ in (6.21) yields regularization algorithms with different properties. We will illustrate four different cases below.

6.5.1   Regularization Networks

Let us consider the quadratic loss function $\mathscr{V}_i(y_i, f(x_i)) = r_i^2$, with the residual r_i defined by $r_i = y_i - f(x_i)$. Such a loss, also depicted in Fig. 6.4 (top left panel), leads to the problem

$$\hat{g} = \arg\min_{f \in \mathscr{H}} \sum_{i=1}^{N} (y_i - f(x_i))^2 + \gamma \| f \|_{\mathscr{H}}^2, \tag{6.32}$$

which is a generalization of the regularized least squares problem encountered in the previous chapters. In particular, it extends the estimator (3.58a) based on quadratic penalty called ReLS-Q in Chap. 3. The estimator (6.32) is known in the literature as *regularization network* [71] or also *kernel ridge regression*. The strict convexity of the objective (6.32) ensures that the minimizer $\hat{g}$ not only exists but is also unique (this issue is further discussed in the remark at the end of this subsection).

To find the solution, we can follow the same arguments developed in Example 6.17, just specializing the result to the case $L_i[f] = f(x_i)$. We will see that the matrix O has just to be replaced by the kernel matrix $\mathbf{K}$.

As previously done, let $Y = [y_1, \ldots, y_N]^T$ and use $\| \cdot \|$ to indicate the Euclidean norm. Then, the corresponding regularization problem (6.24) becomes

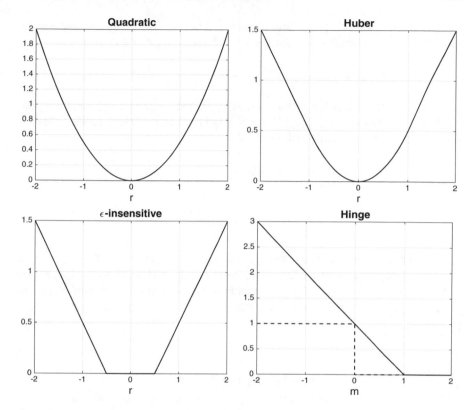

Fig. 6.4 Loss functions examples: quadratic (top left), Huber with $\delta = 1$ (top right), Vapnik with $\varepsilon = 0.5$ (bottom left) and Hinge (bottom right). The first three losses are all functions of the residual $r = y - f(x)$ while the hinge loss depends on the margin $m = yf(x)$

$$\min_{c \in \mathbb{R}^N} \|Y - \mathbf{K}c\|^2 + \gamma c^T \mathbf{K}c, \tag{6.33}$$

which is a finite-dimensional ReLS-Q. After simple calculations, one of the optimal solutions[2] is found to be

$$\hat{c} = (\mathbf{K} + \gamma I_N)^{-1} Y, \tag{6.34}$$

where I_N is the $N \times N$ identity matrix. The estimate from the regularization network is thus available in closed form, given by $\hat{g} = \sum_{i=1}^{N} \hat{c}_i K_{x_i}$ with the optimal coefficient vector $\hat{c}$ solving a linear system of equations.

Remark 6.4 *(Regularization network as projection)* An interpretation of the regularization network can be also given in terms of a projection. In particular, let $\mathscr{R}$

---

[2] Similarly to what discussed in Example 6.17, if $\mathbf{K}$ is not full rank, the solution of (6.33) is not unique. In fact, the minimizers are the sum of (6.34) and the null space of the kernel matrix. However, all of them lead to the same function estimate $\hat{g}$.

be the Hilbert space $\mathbb{R}^N \times \mathscr{H}$ (any element is a couple containing a vector v and a function f) with norm defined, for any $v \in \mathbb{R}^N$ and $f \in \mathscr{H}$, by

$$\|(v, f)\|_{\mathscr{R}}^2 = \|v\|^2 + \gamma \|f\|_{\mathscr{H}}^2, \quad \gamma > 0, \quad \| \cdot \| = \text{Euclidean norm}.$$

Let also S be the (closed) subspace given by all the couples (v, f) satisfying the constraint $v = [f(x_1) \dots f(x_N)]$. Then, if $g = (Y, 0)$ where 0 here denotes the null function in $\mathscr{H}$, the projection of g onto S is

$$g_S = \underset{h \in S}{\arg\min} \ \|g - h\|_{\mathscr{R}}^2$$

$$= \underset{(\{f(x_i)\}_{i=1}^N, f), \ f \in \mathscr{H}}{\arg\min} \ \sum_{i=1}^{N} (y_i - f(x_i))^2 + \gamma \|f\|_{\mathscr{H}}^2.$$

It is now immediate to conclude that g_S corresponds to $([\hat{g}(x_1) \dots \hat{g}(x_n)], \hat{g})$ where $\hat{g}$ is indeed the minimizer (6.32), which must thus be unique in view of Theorem 6.25 (Projection theorem). Note that this interpretation can be extended to all the variational problems (6.21) containing losses defined by a norm induced by an inner product in $\mathbb{R}^N$.

6.5.2   Robust Regression via Huber Loss ★

As described in Sect. 3.6.1, a shortcoming of the quadratic loss is its sensitivity to outliers because the influence of large residuals r_i grows quadratically. In presence of outliers, one would better use a loss function that grows linearly. These issues have been widely studied in the field of robust statistics [51], where loss functions such as the Huber's have been introduced. Recalling (3.115), one has

$$\mathscr{V}_i(y_i, f(x_i)) = \begin{cases} \frac{r_i^2}{2}, & |r_i| \leq \delta \\ \delta \left(|r_i| - \frac{\delta}{2} \right), & |r_i| > \delta \end{cases},$$

where we still have $r_i = y_i - f(x_i)$. The Huber loss function with $\delta = 1$ is shown in Fig. 6.4 (top right panel). Notice that it grows linearly and is thus robust to outliers. When $\delta \to +\infty$, one recovers the quadratic loss. On the other hand, we also have $\lim_{\delta \to 0^+} \mathscr{V}_i(r)/\delta = |r_i|$ that is the absolute value loss.

6.5.3   Support Vector Regression ★

Sometimes, it is desirable to neglect prediction errors, as long as they are below a certain threshold. This can be achieved, e.g., using the Vapnik's ε-insensitive loss given, for $r_i = y_i - f(x_i)$, by

$$\mathcal{V}_i(y_i, f(x_i)) = |r_i|_\varepsilon = \begin{cases} 0, & |r_i| \le \varepsilon \\ |r_i| - \varepsilon, & |r_i| > \varepsilon \end{cases}.$$

The Vapnik loss with $\varepsilon = 0.5$ is shown in Fig. 6.4 (bottom left panel). Notice that it has a null plateau in the interval $[-\varepsilon, \varepsilon]$ so that any predictor closer than ε to y_i is seen as a perfect interpolant. The loss then grows linearly, thus ensuring robustness. The regularization problem (6.21) associated with the ε-insensitive loss function turns out

$$\hat{g} = \arg\min_{f \in \mathcal{H}} \sum_{i=1}^{N} |y_i - f(x_i)|_\varepsilon + \gamma \|f\|_{\mathcal{H}}^2, \tag{6.35}$$

and is called *Support Vector Regression* (SVR), see, e.g., [37]. The SVR solution, given by $\hat{g} = \sum_{i=1}^{N} \hat{c}_i K_{x_i}$ according to the representer theorem, is characterized by sparsity in $\hat{c}$, i.e., some components $\hat{c}_i$ are set to zero. This feature is briefly discussed below.

In the SVR case, obtaining the optimal coefficient vector $\hat{c}$ by (6.24) is not trivial since the loss $|\cdot|_\varepsilon$ is not differentiable everywhere. This difficulty can be circumvented by replacing (6.24) with the following equivalent problem obtained considering two additional N-dimensional parameter vectors ξ and ξ^*:

$$\min_{c,\xi,\xi^*} \sum_{i=1}^{N} (\xi_i + \xi_i^*) + \gamma c^T \mathbf{K} c, \tag{6.36}$$

subject to the constraints

$$y_i - \mathbf{k}_i c \le \varepsilon + \xi_i, \quad i = 1, \ldots, N,$$
$$\mathbf{k}_i c - y_i \le \varepsilon + \xi_i^*, \quad i = 1, \ldots, N,$$
$$\xi_i, \xi_i^* \ge 0, \qquad i = 1, \ldots, N.$$

To see that its minimizer contains the optimal solution $\hat{c}$ of (6.24), it suffices noticing that (6.36) assigns a linear penalty only when $|y_i - \mathbf{k}_i c| > \varepsilon$.

Problem (6.36) is quadratic subject to linear inequality constraints, hence it is solvable by standard optimization approaches like interior point methods [64, 108]. Calculating the Karush–Kuhn–Tucker conditions, it is possible to show that the condition $|y_i - \mathbf{k}_i \hat{c}| < \varepsilon$ implies $\hat{c}_i = 0$. Indexes i for which $\hat{c}_i \ne 0$ instead identify the set of input locations x_i called *support vectors*.

6.5.4 Support Vector Classification ⋆

The three losses illustrated above were originally proposed for regression problems, with the output y real valued. When the outputs can assume only two values, e.g., 1 and -1, a classification problem arises. Here, the scope of the predictor is just to separate two classes. This problem can be seen as a special case of regression. In particular, even if the output space is binary, consider prediction functions $f : \mathscr{X} \to \mathbb{R}$ and assume that the input x_i is associated to the class 1 if $f(x_i) \geq 0$ and to the class -1 if $f(x_i) < 0$. Let the margin on an example (x_i, y_i) be $m_i = y_i f(x_i)$. Then, we will see that the value of m_i is a measure of how well we are classifying the available data. One can thus try to maximize the margin but still searching for a function not too complex according to the RKHS norm. In particular, we can exploit (6.21) with a loss that depends on the margin as described below.

The most natural classification loss is the $0 - 1$ loss defined for any i by

$$\mathscr{V}_i(y_i, f(x_i)) = \begin{cases} 0, & m_i > 0 \\ 1, & m_i \leq 0 \end{cases}, \quad m_i = y_i f(x_i),$$

and depicted in Fig. 6.4 (bottom right panel, dashed line). Adopting it, the first component of the objective in (6.21) returns the number of misclassifications. However, the $0 - 1$ loss is not convex and leads to an optimization problem of combinatorial nature.

An alternative is the so-called hinge loss [98] defined by

$$\mathscr{V}_i(y_i, f(x_i)) = |1 - y_i f(x_i)|_+ = \begin{cases} 0, & m > 1 \\ 1 - m, & m \leq 1 \end{cases}, \quad m = y_i f(x_i),$$

which thus provides a linear penalty when $m < 1$. Figure 6.4 (bottom right panel, solid line) illustrates that it is a convex upper bound on the $0 - 1$ loss. The problem associated with the hinge loss turns out

$$\hat{g} = \arg \min_{f \in \mathscr{H}} \sum_{i=1}^{N} |1 - y_i f(x_i)|_+ + \gamma \|f\|_{\mathscr{H}}^2, \tag{6.37}$$

and is called *support vector classification* (SVC).

Like in the SVR case, obtaining the optimal coefficient vector by (6.37) is not trivial since the hinge loss is not differentiable. But one can still resort to an equivalent problem, now obtained considering just an additional parameter vector ξ:

$$\min_{c, \xi} \sum_{i=1}^{N} \xi_i + \gamma c^T \mathbf{K} c, \tag{6.38}$$

subject to the constraints

$$y_i(\mathbf{k}_i c) \geq 1 - \xi_i, \quad i = 1, \ldots, N,$$
$$\xi_i \geq 0, \qquad\quad i = 1, \ldots, N.$$

As in the SVR case, the optimal solution $\hat{c}$ is sparse and indexes i for which $\hat{c}_i \neq 0$ define the *support vectors* x_i.

6.6 Kernels Examples

The reproducing kernel characterizes the hypothesis space $\mathscr{H}$. Together with the loss function, it also completely defines the key estimator (6.21) which exploits the RKHS norm as regularizer. The choice of K has thus a crucial impact on the ability of predicting future output data. Some important RKHSs are discussed below.

6.6.1 Linear Kernels, Regularized Linear Regression and System Identification

We now show that the regularization network (6.32) generalizes the ReLS-Q problem introduced in Chap. 3 which adopts quadratic penalties. The link is provided by the concept of *linear kernel*.

We start assuming that the input space is $\mathscr{X} = \mathbb{R}^m$. Hence, any input location x corresponds to an m-dimensional (column) vector. If $P \in \mathbb{R}^{m \times m}$ denotes a symmetric and positive semidefinite matrix, a linear kernel is defined as follows

$$K(y, x) = y^T P x, \quad (x, y) \in \mathbb{R}^m \times \mathbb{R}^m.$$

All the kernel sections are linear functions. Hence, their span defines a finite-dimensional (closed) subspace of linear functions that, in view of Theorem 6.1 (and subsequent discussion) coincides with the whole $\mathscr{H}$. Hence, the RKHS induced by the linear kernel is simply a space of linear functions and, for any $g \in \mathscr{H}$, there exists $a \in \mathbb{R}^m$ such that

$$g(x) = a^T P x = K_a(x).$$

If P is full rank, letting $\theta := Pa$, we also have

$$\begin{aligned}
\|g\|_{\mathscr{H}}^2 = \|K_a\|_{\mathscr{H}}^2 &= \langle K_a, K_a \rangle_{\mathscr{H}} \\
&= K(a, a) = a^T P a \\
&= \theta^T P^{-1} \theta.
\end{aligned}$$

Now, let us use such $\mathscr{H}$ in the regularization network (6.32). Without using the representer theorem, we can plug the representation $g(x) = \theta^T x$ in the regularization problem to obtain $\hat{g}(x) = \hat{\theta}^T x$ where

$$\hat{\theta} = \arg\min_{\theta \in \mathbb{R}^m} \|Y - \Phi\theta\|^2 + \gamma\theta^T P^{-1}\theta, \tag{6.39}$$

with the ith row of the regression matrix Φ equal to x_i^T. One can see that (6.39) coincides with ReLS-Q, with the regularization matrix P which defines the linear kernel K and, in turn, the penalty term $\theta^T P^{-1}\theta$.

We now derive the connection with linear system identification in discrete time. The data set consists of the output measurements $\{y_i\}_{i=1}^N$, collected on the time instants $\{t_i\}_{i=1}^N$, and of the system input u. We can form each input location using past input values as follows

$$x_i = [u_{t_i-1} \; u_{t_i-2} \; \ldots \; u_{t_i-m}]^T, \tag{6.40}$$

where m is the FIR order and an input delay of one unit has been assumed. Then, if Y collects the noisy outputs, $\hat{\theta}$ becomes the impulse response estimate. This establishes a correspondence between regularized FIR estimation and regularization in RKHS induced by linear kernels.

6.6.1.1   Infinite-Dimensional Extensions ★

In place of $\mathcal{X} = \mathbb{R}^m$, let now $\mathcal{X} \subset \mathbb{R}^\infty$, i.e., the input space contains sequences. We can interpret any input location as an infinite-dimensional column vector and use ordinary notation of algebra to handle infinite-dimensional objects. For instance, if $x, y \in \mathcal{X}$ then $x^T y = \langle x, y \rangle_2$ where $\langle \cdot, \cdot \rangle_2$ is the inner product in ℓ_2. Assume we are given a symmetric and infinite-dimensional matrix P such that the linear kernel

$$K(y, x) = y^T P x$$

is well defined over a subset of $\mathbb{R}^\infty \times \mathbb{R}^\infty$. For example, if P is absolutely summable, i.e., $\sum_{ij} |P_{ij}| < \infty$, the kernel is well defined for any input location $x \in \mathcal{X}$ with $\mathcal{X} = \ell_\infty$. The kernel section centred on x is the infinite-dimensional column vector Px. Following arguments similar to those seen in the finite-dimensional case, one can conclude that the RKHS associated to such K contains linear functions of the form $g(x) = a^T Px$ with $a \in \mathcal{X}$. Roughly speaking, the regularization network (6.32) relying on such hypothesis space is the limit of Problem (6.39) for $m \to \infty$. To compute the solution, in this case it is necessary to resort to the representer theorem (6.22). One obtains

$$\hat{g}(x) = \sum_{i=1}^N \hat{c}_i K_{x_i}(x) = \hat{\theta}^T x$$

where $\hat{c}$ is defined by (6.34) and

$$\hat{\theta} := \sum_{i=1}^{N} \hat{c}_i P x_i.$$

The link with linear system identification follows the same reasoning previously developed but x_i now contains an infinite number of past input values, i.e.,

$$x_i = [u_{t_i-1}\ u_{t_i-2}\ u_{t_i-3} \ldots]^T.$$

With this correspondence, the regularization network now implements regularized IIR estimation and $\hat{\theta}$ contains the impulse response coefficients estimates. In fact, note that the nature of x_i makes the value $\hat{g}(x_i)$ the convolution between the system input u and $\hat{\theta}$ evaluated at t_i (with one unit input delay).

In a more sophisticated scenario, in place of sequences, the input space $\mathcal{X}$ could contain functions. For instance, $\mathcal{X} \subset \mathcal{P}^c$ where $\mathcal{P}^c$ is the space of piecewise continuous functions on $\mathbb{R}^+$. Thus, each input location corresponds to a continuous function $x : \mathbb{R}^+ \to \mathbb{R}$. Given a suitable symmetric function $P : \mathbb{R}^+ \times \mathbb{R}^+ \to \mathbb{R}$, a linear kernel is now defined by

$$K(y, x) = \int_{\mathbb{R}^+ \times \mathbb{R}^+} y(t) P(t, \tau) x(\tau) dt d\tau.$$

The corresponding RKHS thus contains linear functionals: any $f \in \mathcal{H}$ maps x (which is a function) into $\mathbb{R}$. The solution of the regularization network (6.32) equipped with such hypothesis space is

$$\hat{g}(x) = \sum_{i=1}^{N} \hat{c}_i K_{x_i}(x) = \int_{\mathbb{R}^+} \hat{\theta}(\tau) x(\tau) d\tau,$$

where $\hat{c}$ is still defined by (6.34) and

$$\hat{\theta}(\tau) := \sum_{i=1}^{N} \hat{c}_i \int_{\mathbb{R}^+} P(\tau, t) x_i(t) dt.$$

The connection with linear system identification is obtained by defining

$$x_i(t) = u(t_i - t), \quad t \geq 0$$

(if the input $u(t)$ is continuous for $t \geq 0$ and causal, the functions $x_i(t)$ is piecewise continuous, making necessary the assumption $\mathcal{X} \subset \mathcal{P}^c$). In this way, each $g \in \mathcal{H}$ represents a different linear system. Furthermore, the regularization network (6.32) implements regularized system identification in continuous time and $\hat{\theta}$ is the continuous-time impulse response estimate. The class of kernels which include the BIBO stability constraint will be discussed in the next chapter.

6.6.2 Kernels Given by a Finite Number of Basis Functions

Assume we are given an input space $\mathscr{X}$ and m independent functions $\rho_i : \mathscr{X} \to \mathbb{R}$. Then, we define

$$K(x, y) = \sum_{i=1}^{m} \rho_i(x)\rho_i(y).$$

It is easy to verify that K is a positive definite kernel. Recalling Theorem 6.13, the associated RKHS coincides with the m-dimensional space spanned by the basis functions ρ_i. Each function in $\mathscr{H}$ has the representation $g(x) = \sum_{i=1}^{m} \theta_i \rho_i(x)$ and, in view of (6.20) and the independence of the basis functions, one has

$$\|g\|_{\mathscr{H}}^2 = \sum_{i=1}^{m} \theta_i^2.$$

Consider now the regularization network (6.32) equipped with such hypothesis space. The solution can be computed without using the representer theorem by plugging in (6.32) the expression of g as a function of θ. Letting $\Phi \in \mathbb{R}^{N \times m}$ with $\Phi_{ij} = \rho_j(x_i)$, we obtain $\hat{g} = \sum_{i=1}^{m} \hat{\theta}_i \rho_i$ with

$$\hat{\theta} = \arg\min_{\theta \in \mathbb{R}^m} \ \|Y - \Phi\theta\|^2 + \gamma\|\theta\|^2. \tag{6.41}$$

The solution (6.41) coincides with the ridge regression estimate introduced in Sect. 1.2.

6.6.3 Feature Map and Feature Space ⋆

Let $\mathscr{F}$ be a space endowed with an inner product, and assume that a representation of the form

$$K(x, y) = \langle \phi(x), \phi(y) \rangle_{\mathscr{F}}, \qquad \phi : \mathscr{X} \to \mathscr{F}, \tag{6.42}$$

is available. Then, it follows immediately that K is a positive definite kernel. In this context, ϕ is called a *feature map*, and $\mathscr{F}$ the *feature space*. For instance, to have the connection with the kernel discussed in the previous subsection, we can think of ϕ as a vector containing m functions. It is defined for any x by

$$\phi(x) = \begin{pmatrix} \rho_1(x) \\ \rho_2(x) \\ \vdots \\ \rho_m(x) \end{pmatrix}$$

so that $\mathscr{F} = \mathbb{R}^m$ with the Euclidean inner product. Then, we obtain

$$K(x, y) = \langle \phi(x), \phi(y) \rangle_2 = \phi^T(x)\phi(y) = \sum_{i=1}^{m} \rho_i(x)\rho_i(y).$$

Now, given any positive definite kernel K, Theorem 6.2 (Moore–Aronszajn theorem) implies the existence of at least one feature map, namely, the RKHS map $\phi_{\mathscr{H}} : \mathscr{X} \to \mathscr{H}$ such that

$$\phi_{\mathscr{H}}(x) = K_x,$$

where the representation (6.42) follows immediately from the reproducing property. These arguments show that K is a positive definite kernel iff there exists at least one Hilbert space $\mathscr{F}$ and a map $\phi : \mathscr{X} \to \mathscr{F}$ such that $K(x, y) = \langle \phi(x), \phi(y) \rangle_{\mathscr{F}}$.

Feature maps and feature spaces are not unique since, by introducing any linear isometry $I : \mathscr{H} \to \mathscr{F}$, one can obtain a representation in a different space:

$$K(x, y) = \langle \phi_{\mathscr{H}}(x), \phi_{\mathscr{H}}(y) \rangle_{\mathscr{H}} = \langle I \circ \phi_{\mathscr{H}}(x), I \circ \phi_{\mathscr{H}}(y) \rangle_{\mathscr{F}}.$$

Now, assume that the kernel admits the decomposition (6.8), i.e.,

$$K(x, y) = \sum_{i=1}^{\infty} \zeta_i \rho_i(x)\rho_i(y)$$

with $\zeta_i > 0 \ \forall i$. Then, a *spectral feature map* of K is

$$\phi_\mu : \mathscr{X} \to \ell_2$$

with

$$\phi_\mu(x) = \{\sqrt{\zeta_i}\rho_i(x)\}_{i=1}^{\infty}, \quad x \in \mathscr{X}.$$

In fact, we have

$$\langle \phi_\mu(x), \phi_\mu(y) \rangle_2 = \sum_{i=1}^{\infty} \zeta_i \rho_i(x)\rho_i(y) = K(x, y).$$

It is worth also pointing out the role of the feature map within the estimation scenario. In many applications, linear functions are not models powerful enough. Kernels define more expressive spaces by (implicitly) mapping the data into a high-dimensional feature space where linear machines can be applied. Then, the use of the estimator (6.21) does not require to know any feature map associated to K: the representer theorem shows that the only information needed to compute the estimate is the kernel matrix, as also discussed in Remark 6.3.

6.6.4 Polynomial Kernels

Another example of kernel is the (inhomogeneous) polynomial kernel [70]. For $x, y \in \mathbb{R}^m$, it is defined by

$$K(x, y) = (\langle x, y \rangle_2 + c)^p, \quad p \in \mathbb{N}, \quad c \geq 0,$$

with $\langle \cdot, \cdot \rangle_2$ to denote the classical Euclidean inner product. As an example, assume $c = 1$ and $m = p = 2$ with $x = [x_a \, x_b]$ and $y = [y_a \, y_b]$. Then, one obtains the kernel expansion

$$K(x, y) = 1 + x_a^2 y_a^2 + x_b^2 y_b^2 + 2x_a x_b y_a y_b + 2x_a y_a + 2x_b y_b,$$

of the type (6.8) with the $\rho_i(x_a, x_b)$ given by all the monomials of degree up to 2, i.e., the 6 functions

$$1, \ x_a^2, \ x_b^2, \ x_a x_b, \ x_a, \ x_b.$$

More in general, if $c > 0$, the polynomial kernel induces a $\binom{m+p}{p}$-dimensional RKHS spanned by all possible monomials up to the pth degree. The number of basis function is thus finite but exponential in p. This simple example is in some sense opposite to that described in Sect. 6.6.2. It shows how a kernel can be used to define implicitly a rich class of basis functions.

6.6.5 Translation Invariant and Radial Basis Kernels

A kernel is said translation invariant if there exists $h : \mathcal{X} \to \mathbb{R}$ such that $K(x, y) = h(x - y)$. This class has been already encountered in Example 6.12 where its relationship with the Fourier basis (in the case of one-dimensional input space) is illustrated. A general characterization is given below, see also [80].

Theorem 6.18 (Bochner, based on [23]) *A positive definite kernel K over $\mathcal{X} = \mathbb{R}^d$ is continuous and of the form $K(x, y) = h(x - y)$ if and only if there exists a probability measure μ and a positive scalar η such that:*

$$K(x, y) = \eta \int_{\mathcal{X}} \cos\left(\langle z, x - y \rangle_2\right) d\mu(z).$$

Translation invariant kernels include also the class of radial basis kernels (RBF) of the form $K(x, y) = h(\|x - y\|)$ where $\| \cdot \|$ is the Euclidean norm [85]. A notable example is the so-called *Gaussian kernel*:

$$K(x, y) = \exp\left(-\frac{\|x - y\|^2}{\rho}\right), \quad \rho > 0, \tag{6.43}$$

where ρ denotes the kernel width. This kernel is often used to model functions expected to be somewhat regular. Note however that ρ has an important role in tuning the smoothness level. A low value makes the kernel close to diagonal so that a low norm can be assigned also to rapidly changing functions. On the other hand, as ρ approaches zero, only functions close to be constant are given a low penalty. This is the same phenomenon illustrated in Fig. 6.1.

Another widely adopted kernel, which induces spaces of functions less regular than the Gaussian one, is the *Laplacian kernel* which uses the Euclidean norm in place of the squared Euclidean norm:

$$K(x, y) = \exp\left(-\frac{\|x - y\|}{\rho}\right), \quad \rho > 0. \tag{6.44}$$

Differently from the kernels described in the first part of Sect. 6.6.1, as well as in Sects. 6.6.2 and 6.6.4, the RKHS associated with any non-constant RBF kernel is infinite dimensional (it cannot be spanned by a finite number of basis functions). The associated RKHS can be shown to be dense in the space of all continuous functions defined on a compact subset $\mathscr{X} \subset \mathbb{R}^m$. This means that every continuous function can be represented in this space with the desired accuracy as measured by the sup-norm $\sup_{x \in \mathscr{X}} |f(x)|$. This property is called *universality*. This does not imply that the RKHS induced by a universal kernel includes any continuous function. For instance, the Gaussian kernel is universal but it has been proved that it does not contain any polynomial, including the constant function [69].

6.6.6 Spline Kernels

To simplify the exposition, let $\mathscr{X} = [0, 1]$ and let also $g^{(j)}$ denote the jth derivative of g, with $g^{(0)} := g$. Intuitively, in many circumstances an effective regularizer is obtained by penalizing the energy of the pth derivative of g, i.e., employing

$$\int_0^1 \left(g^{(p)}(x)\right)^2 dx.$$

An interesting question is whether this penalty term can be cast in the RKHS theory. For $p = 1$, a positive answer has been given by Example 6.5. Actually, the answer is positive for any integer p. In fact, consider the Sobolev space of functions g whose first $p - 1$ derivatives are absolutely continuous and satisfy $g^{(j)}(0) = 0$ for $j = 0, \ldots, p - 1$. The same arguments developed in Example 6.5 when $p = 1$ can be easily generalized to prove that this is a RKHS $\mathscr{H}$ with norm

$$\|g\|_{\mathscr{H}}^2 = \int_0^1 \left(g^{(p)}(x)\right)^2 dx.$$

The corresponding kernel is the pth-order spline kernel

$$K(x, y) = \int_0^1 G_p(x, u) G_p(y, u) du, \qquad (6.45)$$

where G_p is the so-called Green's function given by

$$G_p(x, u) = \frac{(x - u)_+^{p-1}}{(p - 1)!}, \qquad (u)_+ = \begin{cases} u & \text{if } u \geq 0 \\ 0 & \text{otherwise} \end{cases}. \qquad (6.46)$$

Note that the Laplace transform of $G_p(\cdot, 0)$ is $1/s^p$. Hence, the Green's function is connected with the impulse response of a p-fold integrator. When $p = 1$, we recover the linear spline kernel of Example 6.5:

$$K(x, y) = \min\{x, y\} \qquad (6.47)$$

whereas $p = 2$ leads to the popular cubic spline kernel [104]:

$$K(x, y) = \frac{xy \min\{x, y\}}{2} - \frac{(\min\{x, y\})^3}{6}. \qquad (6.48)$$

The linear and the cubic spline kernel are displayed in Fig. 6.2.

We can use the spline hypothesis space in the regularization problem (6.21). Then, from the representer theorem one obtains that the estimate $\hat{g}$ is a pth-order smoothing spline with derivatives continuous exactly up to order $2p - 2$ (the order's choice is thus related to the expected function smoothness). This can be seen also from the kernels sections plotted in Fig. 6.2 for p equal to 1 and 2. For $p = 2$ the (finite) sum of kernel sections provides the well-known cubic smoothing splines, i.e., piecewise third-order polynomials.

Spline functions enjoy many numerical properties originally studied in the interpolation scenario. In particular, piecewise polynomials circumvent Runge's phenomenon (large oscillations affecting the reconstructed function) which, e.g., arises when high-order polynomials are employed [81]. Fit convergence rates are discussed, e.g., in [3, 14].

6.6.7 The Bias Space and the Spline Estimator

Bias space As discussed in Sect. 4.5, in a Bayesian setting, in some cases it can be useful to enrich $\mathcal{H}$ with a low-dimensional parametric part, known in the literature as *bias space*. The bias space typically consists of linear combinations of functions $\{\phi_k\}_{k=1}^m$. For instance, if the unknown function exhibits a linear trend, one may let $m = 2$ and $\phi_1(x) = 1$, $\phi_2(x) = x$. Then, one can assume that g is sum of two functions, one in $\mathcal{H}$ and the other one in the bias space. In this way, the function space becomes $\mathcal{H} + \text{span}\{\phi_1, \ldots, \phi_m\}$. Using a quadratic loss, the regularization problem is given by

$$(\hat{f}, \hat{\theta}) = \underset{\substack{f \in \mathcal{H}, \\ \theta \in \mathbb{R}^m}}{\arg \min} \sum_{i=1}^{N} \left(y_i - f(x_i) - \sum_{k=1}^{m} \theta_k \phi_k(x_i) \right)^2 + \gamma \|f\|_{\mathcal{H}}^2, \qquad (6.49)$$

and the overall function estimate turns out $\hat{g} = \hat{f} + \sum_{k=1}^{m} \hat{\theta}_k \phi_k$. Note that the expansion coefficients in θ are not subject to any penalty term but a low value for m avoids overfitting. The solution can be computed exploiting an extended version of the representer theorem. In particular, it holds that

$$\hat{g} = \sum_{i=1}^{N} \hat{c}_i K_{x_i} + \sum_{k=1}^{m} \hat{\theta}_k \phi_k, \qquad (6.50)$$

where, assuming that $\Phi \in \mathbb{R}^{N \times m}$ is full column rank and $\Phi_{ij} = \phi_j(x_i)$,

$$\hat{\theta} = \left(\Phi^T A^{-1} \Phi \right)^{-1} \Phi^T A^{-1} Y \qquad (6.51a)$$

$$\hat{c} = A^{-1} \left(Y - \Phi \hat{\theta} \right) \qquad (6.51b)$$

$$A = \mathbf{K} + \gamma I_N. \qquad (6.51c)$$

Remark 6.5 (*Extended version of the representer theorem*) The correctness of formulas (6.51a–6.51c) can be easily verified as follows. Fix θ to the optimizer $\hat{\theta}$ in the objective present in the rhs of (6.49). Then, we can use the representer theorem with Y replaced by $Y - \Phi \hat{\theta}$ to obtain $\hat{f} = \sum_{i=1}^{N} \hat{c}_i K_{x_i}$ with

$$\hat{c} = A^{-1} \left(Y - \Phi \hat{\theta} \right)$$

with A indeed given by (6.51c). This proves (6.51b). Using the definition of A this also implies

$$Y - \mathbf{K} \hat{c} = \Phi \hat{\theta} + \gamma \hat{c}.$$

Now, if we fix f to $\hat{f}$, the optimizer $\hat{\theta}$ is just the least squares estimate of θ with Y replaced by $Y - \mathbf{K} \hat{c}$. Hence, we obtain

$$\hat{\theta} = \left(\Phi^T \Phi \right)^{-1} \Phi^T (Y - \mathbf{K} \hat{c}).$$

Using $Y - \mathbf{K} \hat{c} = \Phi \hat{\theta} + \gamma \hat{c}$ in the expression for $\hat{\theta}$, we obtain $\left(\Phi^T \Phi \right)^{-1} \Phi^T \hat{c} = 0$. Multiplying the lhs and rhs of (6.51b) by $\left(\Phi^T \Phi \right)^{-1} \Phi^T$ and using this last equality, (6.51a) is finally obtained.

The spline estimator The bias space is useful, e.g., when spline kernels are adopted. In fact, the spline space of order p contains functions all satisfying the constraints $g^{(j)}(0) = 0$ for $j = 0, \ldots, p - 1$. Then, to cope with nonzero initial conditions, one can enrich such RKHS with polynomials up to order $p - 1$. The enriched space is $\mathcal{H} \oplus \mathrm{span}\{1, x, \ldots, x^{p-1}\}$, with $\oplus$ denoting a direct sum, and enjoys the universality property mentioned at the end of Sect. 6.6.5. The resulting spline estimator becomes a notable example of (6.49): it solves

$$\min_{\substack{f \in \mathcal{H}, \\ \theta \in \mathbb{R}^p}} \sum_{i=1}^{N} \left(y_i - f(x_i) - \sum_{k=1}^{p} \theta_k x_i^{k-1} \right)^2 + \gamma \int_0^1 \left(f^{(p)}(x) \right)^2 dx, \qquad (6.52)$$

whose explicit solution is given by (6.50) setting $\phi_k(x) = x^{k-1}$ and $\Phi_{ij} = x_i^{j-1}$.

We consider a simple numerical example to illustrate the estimator (6.52) and the impact of different choices of γ on its performance. The task is the reconstruction of the function $g(x) = e^{\sin(10x)}$, with $x \in [0, 1]$, from 100 direct samples corrupted by

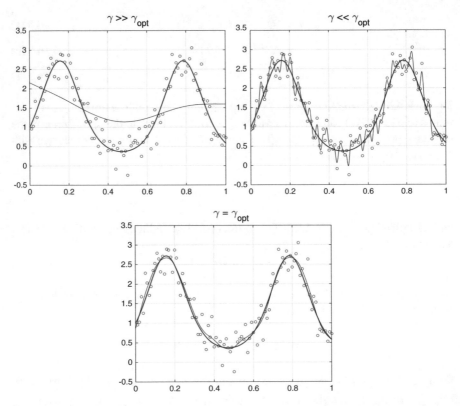

Fig. 6.5 Cubic spline estimator (6.52) with three different values of the regularization parameter: truth (red thick line), noisy data (○) and estimate (black solid line)

white and Gaussian noise with standard deviation 0.3. The estimates coming from
(6.52) with $p = 2$ and three different values of γ are displayed in the three panels
of Fig. 6.5. The cubic spline estimate plotted in the top left panel is affected by
oversmoothing: the too large value of γ overweights the norm of f in the objective
(6.52), introducing a large bias. Hence, the model is too rigid, unable to describe the
data. The top right panel displays the opposite situation obtained adopting a too low
value for γ which overweights the loss function in (6.52). This leads to a high variance
estimator: the model is overly flexible and overfits the measurements. Finally, the
estimate in the bottom panel of Fig. 6.5 is obtained using the regularization parameter
optimal in the MSE sense. The good trade-off between bias and variance leads to an
estimate close to truth. As already pointed out in the previous chapters, the choice of
γ can thus be interpreted as the counterpart of model order selection in the classical
parametric paradigm.

6.7   Asymptotic Properties ⋆

6.7.1   The Regression Function/Optimal Predictor

In what follows, we use μ to indicate a probability measure on the input space $\mathscr{X}$.
For simplicity, we assume that it admits a probability density function (pdf) denoted
by p_x. The input locations x_i are now seen as random quantities and p_x models
the stochastic mechanism through which they are drawn from $\mathscr{X}$. For instance, in
the system identification scenario treated in Sect. 6.6.1, each input location contains
system input values, e.g., see (6.40). If we assume that the input is a stationary
stochastic process, all the x_i indeed follow the same pdf p_x.

Let also $\mathscr{Y}$ indicate the output space. Then, p_{yx} denotes the joint pdf on $\mathscr{X} \times \mathscr{Y}$
which factorizes into $\mathrm{p}_{y|x}(y|x)\mathrm{p}_x(x)$. Here, $\mathrm{p}_{y|x}$ is the pdf of the output y conditional
on a particular realization x.

Let us now introduce some important quantities function of $\mathscr{X}$, $\mathscr{Y}$ and p_{yx}. Given
a function f, the least squares error associated to f is defined by

$$\mathrm{Err}(f) = \mathscr{E}(y - f(x))^2 = \int_{\mathscr{X} \times \mathscr{Y}} (y - f(x))^2 \mathrm{p}_{yx}(y, x) dx dy. \qquad (6.53)$$

The following result, also discussed in [33], characterizes the minimizer of $\mathrm{Err}(f)$
and has connections with Theorem 4.1.

Theorem 6.19 (The regression function, based on [33]) *We have*

$$f_\rho = \arg\min_f \mathrm{Err}(f),$$

where f_ρ is the so-called regression function *defined by*

$$f_\rho(x) = \int_{\mathcal{Y}} y p_{y|x}(y|x)dy, \quad x \in \mathcal{X}. \tag{6.54}$$

One can see that the regression function does not depend on the marginal density p_x but only on the conditional $p_{y|x}$. For any given x, it corresponds to the posterior mean (Bayes estimate) of the output y conditional on x. The proof of this fact is easily obtained by first using the following decomposition

$$\begin{aligned}
\text{Err}(f) &= \int_{\mathcal{X} \times \mathcal{Y}} (y - f_\rho(x) + f_\rho(x) - f(x))^2 p_{yx}(y, x)dxdy \\
&= \mathscr{E}(f_\rho(x) - f(x))^2 + \mathscr{E}(y - f_\rho(x))^2 \\
&\quad + 2 \int_{\mathcal{X}} (f_\rho(x) - f(x)) \underbrace{\left(\int_{\mathcal{Y}} (y - f_\rho(x)) p_{y|x}(y|x)dy \right)}_{0} p_x(x)dx \\
&= \mathscr{E}(f_\rho(x) - f(x))^2 + \mathscr{E}(y - f_\rho(x))^2,
\end{aligned}$$

and then noticing that the very last term is independent of f.

Theorem 6.19 shows that f_ρ is the best output predictor in the sense that it minimizes the expected quadratic loss (MSE) on a new output drawn from p_{yx}. Now, we will consider a scenario where $p_{y|x}$ (and possibly also p_x) is unknown and only N samples $\{x_i, y_i\}_{i=1}^N$ from p_{yx} are available. We will study the asymptotic properties (N growing to infinity) of the regularized approaches previously described. The regularization network case is treated in the following subsection.

6.7.2 Regularization Networks: Statistical Consistency

Consider the following regularization network

$$\hat{g}_N = \underset{f \in \mathscr{H}}{\arg\min} \frac{\sum_{i=1}^N (y_i - f(x_i))^2}{N} + \gamma \|f\|_{\mathscr{H}}^2, \tag{6.55}$$

which coincides with (6.32) except for the introduction of the scale factor $1/N$ in the quadratic loss. We have also stressed the dependence of the estimate on the data set size N. Our goal is to assess whether $\hat{g}_N$ converges to f_ρ as $N \to \infty$ using the norm $\| \cdot \|_{\mathscr{L}_2^\mu}$ defined by the pdf p_x as follows

$$\|f\|_{\mathscr{L}_2^\mu}^2 = \int_{\mathcal{X}} f^2(x) p_x(x)dx.$$

First, details on the data generation process are provided.

Data generation assumptions The probability measure μ on $\mathscr{X}$ is assumed to be Borel non degenerate. As already recalled, this means that realizations from p_x can cover entirely $\mathscr{X}$, without holes. This happens, e.g., when $p_x(x) > 0 \ \forall x \in \mathscr{X}$. The stochastic processes x_i and y_i are jointly stationary, with joint pdf p_{yx}.

The study is not limited to the i.i.d. case. This is important, e.g., in system identification where, as visible in (6.40), input locations contain past input values shifted in time, hence introducing correlation among the x_i. Let a, b indicate two integers with $a \leq b$. Then, $\mathscr{M}_a^b$ denotes the σ-algebra generated by $(x_a, y_a), \ldots, (x_b, y_b)$. The process (x, y) is said to satisfy a strong mixing condition if there exists a sequence of real numbers ψ_m such that, $\forall k, m \geq 1$, one has

$$|P(A \cap B) - P(A)P(B)| \leq \psi_i \quad \forall A \in \mathscr{M}_1^k, B \in \mathscr{M}_{k+i}^\infty$$

with

$$\lim_{i \to \infty} \psi_i = 0.$$

Intuitively, if a, b represent different time instants, this means that the random variables tend to become independent as their temporal distance increases.

Assumption 6.20 *(Data generation and strong mixing condition)* The probability measure μ on the input space (having pdf p_x) is nondegenerate. In addition, the random variables x_i and y_i form two jointly stationary stochastic processes, with finite moments up to the third order and satisfy a strong mixing condition. Finally, denoting with ψ_i the mixing coefficients, one has

$$\sum_{i=1}^\infty \left(|\psi_i|^{1/3} \right) < \infty.$$

Consistency Result
The following theorem, whose proof is in Sect. 6.9.6, illustrates the convergence in probability of (6.55) to the best output predictor.

Theorem 6.21 *(Statistical consistency of the regularization networks)* *Let $\mathscr{H}$ be a RKHS of functions $f : \mathscr{X} \to \mathbb{R}$ induced by the Mercer kernel K, with $\mathscr{X}$ a compact metric space. Assume that $f_\rho \in \mathscr{H}$ and that Assumption 6.20 holds. In addition, let*

$$\gamma \propto \frac{1}{N^\alpha}, \tag{6.56}$$

where α is any scalar in $(0, \frac{1}{2})$. Then, as N goes to infinity, one has

$$\|\hat{g}_N - f_\rho\|_{\mathscr{L}_2^\mu} \longrightarrow_p 0, \tag{6.57}$$

where $\longrightarrow_p$ denotes convergence in probability.

The meaning of (6.56) is the following one. The regularizer $\| \cdot \|_{\mathscr{H}}^2$ in (6.55) restores the well-posedness of the problem by introducing some bias in the estimation process. Intuitively, to have consistency, the amount of regularization should decay to zero as N goes to ∞, but not too rapidly in order to keep the variance term under control. This can be obtained making the regularization parameter γ go to zero with the rate suggested by (6.56).

6.7.3   Connection with Statistical Learning Theory

We now discuss the class of estimators (6.21) within the framework of statistical learning theory.

Learning problem Let us consider the problem of *learning from examples* as defined in statistical learning. The starting point is that described in Sect. 6.7.1. There is an unknown probabilistic relationship between the variables x and y described by the joint pdf p_{yx} on $\mathscr{X} \times \mathscr{Y}$. We are given examples $\{x_i, y_i\}_{i=1}^N$ of this relationship, called *training data*, which are independently drawn from p_{yx}.   The aim of the learning process is to obtain an estimator $\hat{g}_N$ (a map from the training set to a space of functions) able to predict the output y given any $x \in \mathscr{X}$.

Generalization and consistency In the statistical learning scenario, the two funda-mental properties of an estimator are *generalization* and *consistency*. To introduce them, first we introduce a loss function $\mathscr{V}(y, f(x))$, called *risk functional*. Then, the mean error associated to a function f is the *expected risk* given by

$$I(f) = \int_{\mathscr{X} \times \mathscr{Y}} \mathscr{V}(y, f(x))\mathrm{p}_{yx}(y, x)dxdy. \tag{6.58}$$

Note that, in the quadratic loss case, the expected risk coincides with the error already introduced in (6.53). Given a function f, the *empirical risk* is instead defined by

$$I_N(f) = \frac{1}{N} \sum_{i=1}^{N} \mathscr{V}(y_i, f(x_i)). \tag{6.59}$$

Then, we introduce a class of functions forming the hypothesis space $\mathscr{F}$ where the predictor is searched for. The ideal predictor, also called the *target function*, is given by[3]

$$f_0 = \underset{f \in \mathscr{F}}{\arg\min}\; I(f). \tag{6.60}$$

---

[3] Here, and also when introducing empirical risk minimization (ERM), we assume that all the introduced minimizers exist. If this does not hold true, all the concepts remain valid by resorting to the concept of almost minimizers and almost ERM, with $I(f_0) := \inf_{f \in \mathscr{F}}\; I(f)$.

In general, even when a quadratic loss is chosen, f_0 does not coincide with the regression function f_ρ introduced in (6.54) since $\mathscr{F}$ could not contain f_ρ.

The concepts of generalization and consistency trace back to [97, 99–101]. Below, recall that $\hat{g}_N$ is stochastic since it is function of the training set which contains the random variables $\{x_i, y_i\}_{i=1}^N$.

Definition 6.3 *(Generalization and consistency, based on* [102]*)* The estimator $\hat{g}_N$ (uniformly) generalizes if $\forall \varepsilon > 0$:

$$\lim_{N \to \infty} \sup_{p_{yx}} \mathbb{P}\left\{|I_N(\hat{g}_N) - I(\hat{g}_N)| > \varepsilon\right\} = 0. \tag{6.61}$$

The estimator is instead (universally) consistent if $\forall \varepsilon > 0$:

$$\lim_{N \to \infty} \sup_{p_{yx}} \mathbb{P}\left\{I(\hat{g}_N) > I(f_0) + \varepsilon\right\} = 0. \tag{6.62}$$

From (6.61), one can see that generalization implies that the performance on the training set (the empirical error) must converge to the "true" performance on future outputs (the expected error). The presence of the $\sup_{p_{yx}}$ is then to indicate that this property must hold uniformly w.r.t. all the possible stochastic mechanisms which generate the data. Consistency, as defined in (6.62), instead requires the expected error of $\hat{g}_N$ to converge to the expected error achieved by the best predictor in $\mathscr{F}$. Note that the reconstruction of f_0 is not required. The goal is that $\hat{g}_N$ be able to mimic the prediction performance of f_0 asymptotically. Key issues in statistical learning theory are the understanding of the conditions on $\hat{g}_N$, the function class $\mathscr{F}$ and the loss $\mathscr{V}$ which ensure such properties.

Empirical Risk Minimization

The most natural technique to determine f_0 from data is the *empirical risk minimization* (ERM) approach where the empirical risk is optimized:

$$\hat{g}_N = \arg\min_{f \in \mathscr{F}} I_N(f) = \arg\min_{f \in \mathscr{F}} \frac{1}{N} \sum_{i=1}^N \mathscr{V}(y_i, f(x_i)). \tag{6.63}$$

The study of ERM has provided a full characterization of the necessary and sufficient conditions for its generalization and consistency. To introduce them, we first need to provide further details on the data generation assumptions.

Assumption 6.22 *(Data generation assumptions)* It holds that

- the $\{x_i, y_i\}_{i=1}^N$ are i.i.d. and each couple has joint pdf p_{yx};
- the input space $\mathscr{X}$ is a compact set in the Euclidean space;
- $y \in \mathscr{Y}$ almost surely with $\mathscr{Y}$ a bounded real set;
- the class of functions $\mathscr{F}$ is bounded, e.g., under the sup-norm;
- $A \leq \mathscr{V}(y, f(x)) \leq B$, for $f \in \mathscr{F}$, $y \in \mathscr{Y}$, with A, B finite and independent of f and y. ◻

Note that, if the first four points hold true, in practice any loss function of interest, such as quadratic, Huber or Vapnik, satisfies the last requirement.

We now introduce the concept of V_γ-dimension [5]. It is a complexity measure which extends the concept of Vapnik–Chervonenkis (VC) dimension originally introduced for the indicator functions.

Definition 6.4 *(V_γ-dimension, based on [5])* Let Assumption 6.22 hold. The V_γ-dimension of $\mathcal{V}$ in $\mathcal{F}$, i.e., of the set $\mathcal{V}(y, f(x))$, $f \in \mathcal{F}$, is defined as the maximum number h of vectors $(x_1, y_1), \ldots, (x_h, y_h)$ that can be separated in all 2^h possible way using rules

$$\text{Class 1: if } \mathcal{V}(y_i, f(x_i)) \geq s + \gamma,$$
$$\text{Class 0: if } \mathcal{V}(y_i, f(x_i)) \leq s - \gamma$$

for $f \in \mathcal{F}$ and some $s \geq 0$. If, for any h, it is possible to find h pairs $(x_1, y_1), \ldots,$ (x_h, y_h) that can be separated in all the 2^h possible ways, the V_γ-dimension of $\mathcal{V}$ in $\mathcal{F}$ is infinite.

So, the V_γ-dimension is infinite if, for any data set size, one can always find a function f and a set of points which can be separated by f in any possible way. Note that the required margin to distinguish the classes increases as γ augments. This means that the V_γ-dimension is a monotonically decreasing function of γ.

The following definition deals with the uniform, distribution-free convergence of empirical means to expectations for classes of real-valued functions. It is related to the so-called *uniform laws of large numbers*.

Definition 6.5 *(Uniform Glivenko Cantelli class, based on [5])* Let $\mathcal{G}$ denote a space of functions $\mathcal{Z} \to \mathcal{R}$, where $\mathcal{R}$ is a bounded real set, and let p_z denote a generic pdf on $\mathcal{Z}$. Then, $\mathcal{G}$ is said to be a Uniform Glivenko Cantelli (uGC) class[4] if

$$\forall \varepsilon > 0 \quad \lim_{N \to \infty} \sup_{p_z} \mathbb{P} \left\{ \sup_{g \in \mathcal{G}} \left| \frac{1}{N} \sum_{i=1}^{N} g(z_i) - \int_{\mathcal{X}} g(z) p_z(z) dz \right| > \varepsilon \right\} = 0.$$

It turns out that, under the ERM framework, generalization and consistency are equivalent concepts. Moreover, the finiteness of the V_γ-dimension coincides with the concept of uGC class relative to the adopted losses and turns out the necessary and sufficient condition for generalization and consistency [5]. This is formalized below.

Theorem 6.23 (ERM and V_γ-dimension, based on [5]) *Let Assumption 6.22 hold. The following facts are then equivalent:*

• *ERM (uniformly) generalizes.*

---

[4] Sometimes, the class defined by (6.5) in terms of convergence in probability is called weak uGC while almost sure convergence leads to a strong uGC. However, it can be proved that, if Assumption 6.22 holds true and the function class is the composition of the losses with $\mathcal{F}$, the two concepts become equivalent.

- *ERM is (uniformly) consistent.*
- *The V_γ-dimension of $\mathcal{V}$ in $\mathcal{F}$ is finite for any $\gamma > 0$.*
- *The class of functions $\mathcal{V}(y, f(x))$ with $f \in \mathcal{F}$ is uGC.*

In the last point regarding the uGC class, one can follow Definition 6.5 using the correspondences $\mathcal{Z} = \mathcal{X} \times \mathcal{Y}$, $z = (x, y)$, $\mathrm{p}_z = \mathrm{p}_{yx}$ and $\mathcal{R} = [A, B]$.

Connection with Regularization in RKHS

The connection between statistical learning theory and the class of kernel-based estimators (6.21) is obtained using as function space $\mathcal{F}$ a ball $\mathcal{B}_r$ in a RKHS $\mathcal{H}$, i.e.,

$$\mathcal{F} = \mathcal{B}_r := \left\{ f \in \mathcal{H} \mid \|f\|_{\mathcal{H}} \leq r \right\}. \tag{6.64}$$

The ERM method (6.63) becomes

$$\hat{g}_N = \arg\min_f \frac{1}{N} \sum_{i=1}^{N} \mathcal{V}(y_i, f(x_i)) \quad \text{s.t.} \quad \|f\|_{\mathcal{H}} \leq r, \tag{6.65}$$

which is an inequality constrained optimization problem. Exploiting the Lagrangian theory, we can find a positive scalar γ, function of r and of the data set size N, which makes (6.65) equivalent to

$$\hat{g}_N = \arg\min_{f \in \mathcal{H}} \frac{1}{N} \sum_{i=1}^{N} \mathcal{V}(y_i, f(x_i)) + \gamma \left(\|f\|_{\mathcal{H}}^2 - r^2 \right),$$

which, apart from constants, coincides with (6.21). The question now is whether (6.65) is consistent in the sense of the statistical learning theory. The answer is positive. In fact, under Assumption 6.22, it can be proved that the class of functions $\mathcal{V}$ in $\mathcal{F}$ is uGC if $\mathcal{F}$ is uGC. In addition, one sufficient (but not necessary) condition for $\mathcal{F}$ to be uGC is that $\mathcal{F}$ be a compact set in the space of continuous functions. The following important result then holds.

Theorem 6.24 (Generalization and consistency of the kernel-based approaches, based on [33, 65]) *Let $\mathcal{H}$ be any RKHS induced by a Mercer kernel containing functions $f : \mathcal{X} \to \mathbb{R}$, with $\mathcal{X}$ a compact metric space. Then, for any r, the ball $\mathcal{B}_r$ is compact in the space of continuous functions equipped with the sup-norm. It then comes that $\mathcal{B}_r$ is uGC and, if Assumption 6.22 holds, the regularized estimator (6.65) generalizes and is consistent.*

Theorem 6.24 thus shows that kernel-based approaches permit to exploit flexible infinite-dimensional models with the guarantee that the best prediction performance (achievable inside the chosen class) will be asymptotically reached.

6.8   Further Topics and Advanced Reading

Basic functional analysis principles can be found, e.g., in [59, 79, 112]. The concept of RKHS was developed in 1950 in the seminal works [13, 20]. Classical books on the subject are [6, 82, 84]. RKHSs have been introduced within the machine learning community in [46, 47] leading, in conjunction with Tikhonov regularization theory [21, 96], to the development of many powerful kernel-based algorithms [42, 86].

Extensions of the theory to vector-valued RKHSs is described in [62]. This is connected to the so-called multi-task learning problem [18, 29] which deals with the simultaneous reconstruction of several functions. Here, the key point is that measurements taken on a function (task) may be informative w.r.t. the other ones, see [16, 40, 68, 95] for illustrations of the advantages of this approach. Multi-task learning will be illustrated in Chap. 9 using also a numerical example based on real pharmacokinetics data.

Mercer theorem dates back to [60] which discusses also the connection with integral equations, see also the book [50]. Extensions of the theorem to non compact domains are discussed in [94]. The first version of the representer theorem appears in [52]. It has been then subject of many generalizations which can be found in [11, 36, 83, 103, 110]. Recent works have also extended the classical formulation to the context of vector-valued functions (multi-task learning and collaborative filtering), matrix regularization problems (with penalty given by spectral functions of matrices), matricizations of tensors, see, e.g., [1, 7, 12, 54, 87]. These different types of representer theorems are cast in a general framework in [10].

The term regularization network traces back to [71] where it is illustrated that a particular regularized scheme is equal to a radial basis function network. Support vector regression and classification were introduced in [24, 31, 37, 98], see also the classical book [102]. Robust statistics are described in [51].

The term "kernel trick" was used in [83] while interpretation of kernels as inner products in a feature space was first described in [4]. Sobolev spaces are illustrated, e.g., in [2] while classical works on smoothing splines are [32, 104]. The important spline interpolation properties are described in [3, 14, 22].

Polynomial kernels were used for the first time in [70] while an application to Wiener system identification can be found in [44], as also discussed later on in Chap. 8 devoted to nonlinear system identification. An explicit (spectral) characterization of the RKHS induced by the Gaussian kernel can be found in [91, 92], while the more general case of radial basis kernels is treated in [85]. The concept of universal kernel is discussed, e.g., in [61, 90].

The strong mixing condition is discussed, e.g., in [107] and [34].

The convergence proof for the regularization network relies upon the integral operator approach described in [88] in an i.i.d. setting and its extension to the dependent case developed in [66] in the Wiener system identification context. For other works on statistical consistency and learning rates of regularized least squares in RKHS see, e.g., [48, 93, 105, 109, 111].

Statistical learning theory and the concepts of generalization and consistency, in connection with the uniform law of large numbers, date back to the works of Vapnik and Chervonenkis [97, 99–101]. Other related works on convergence of empirical processes are [38, 39, 73]. The concept of V_γ dimension and its equivalence with the Glivenko–Cantelli class is proved in [5], see also [41] for links with RKHS. Relationships between the concept of stability of estimates (continuous dependence on the data) and generalization/consistency can be found in [63, 72], see also [26] for previous work on this subject. Numerical computation of the regularized estimate (6.21) is discussed in the literature studying the relationship between machine learning and convex optimization [19, 25, 77]. In the regularization network case (quadratic loss), if the data set size N is large, plain application of a solver with computational cost $O(N^3)$ can be highly inefficient. Then, one can use approximate representations of the kernel function [15, 53], based, e.g., on the Nyström method or greedy strategies [89, 106, 113]. One can also exploit the Mercer theorem by just using an mth-order approximation of K given by $\sum_{i=1}^{m} \zeta_i \rho_i(x)\rho_i(y)$. The solution obtained with this kernel may provide accurate approximations also when $m \ll N$, see [28, 43, 67, 114, 115]. Training of kernel machines can be also accelerated by using randomized low-dimensional feature spaces [74], see also [78] for insights on learning rates.

In the case of generic convex loss (different from the quadratic), one problem is that the objective is not differentiable everywhere. In this circumstance, the powerful interior point (IP) methods [64, 108] can be employed which applies damped Newton iterations to a relaxed version of the Karush–Kuhn–Tucker (KKT) equations for the objective [27]. A statistical and computational framework that allows their broad application to the problem (6.21) for a wide class of piecewise linear quadratic losses can be found in [8, 9]. In practice, IP methods exhibit a relatively fast convergence behaviour. However, as in the quadratic case, a difficulty can arise if N is very large, i.e., it may not be possible to store the entire kernel matrix in memory and this fact can hinder the application of second-order optimization techniques such as the (damped) Newton method. A way to circumvent this problem is given by the so-called decomposition methods where a subset of the coefficients c_i, called working set, is selected, and the associated low-dimensional sub-problem is solved. In this way, only the corresponding entries of the output kernel matrix need to be loaded into the memory, e.g., see [30, 56–58]. An extreme case of decomposition method is coordinate descent, where the working set contains only one coefficient [35, 45, 49].

6.9 Appendix

6.9.1 Fundamentals of Functional Analysis

We gather some basic functional analysis definitions and results.

Vector Spaces

We will assume that the reader is familiar with the concept of real vector space V (the field is given by the real numbers). Here, we just recall that this is a set whose elements are called vectors. The space is closed w.r.t. two operations, called addition and scalar multiplication, which satisfy the usual algebraic properties. This means that any linear and finite combination of vectors still falls in V. When the vector space contains functions $g : \mathcal{X} \to \mathbb{R}$, for any $f, g \in V$ and $\alpha \in \mathbb{R}$ the two operations are defined as follows:

$$f + g = h \text{ where } h(x) = f(x) + g(x) \ \forall x \in \mathcal{X}$$

and

$$\alpha f = h \text{ where } h(x) = \alpha f(x) \ \forall x \in \mathcal{X}.$$

Inner Products and Norms

An inner product on V is the function

$$\langle \cdot, \cdot \rangle : V \times V \to \mathbb{R}$$

which is

1. linear in the first argument

$$\langle \alpha v + \beta y, z \rangle = \alpha \langle v, z \rangle + \beta \langle y, z \rangle, \quad v, y, z \in V \ \ \alpha, \beta \in \mathbb{R};$$

2. symmetric (and so also linear in the second argument)

$$\langle v, y \rangle = \langle y, v \rangle;$$

3. positive, in the sense that

$$\langle v, v \rangle \geq 0 \ \ \forall v$$

   with

$$\langle v, v \rangle = 0 \iff v = 0,$$

   where in the r.h.s. 0 denotes the null vector.

Recall also that a norm on V is the nonnegative function

$$\| \cdot \| : V \to \mathbb{R}^+$$

which satisfies

1. absolute homogeneity

$$\|\alpha v\| = |\alpha| \|v\|, \quad v \in V \ \ \alpha \in \mathbb{R};$$

2. the triangle inequality

$$\|v + y\| \le \|v\| + \|y\|;$$

3. null vector condition

$$\|v\| = 0 \iff v = 0.$$

The norm induced by the inner product $\langle \cdot, \cdot \rangle$ is given by

$$\|v\|^2 = \langle v, v \rangle,$$

and it is easy to check that this function indeed satisfies all the three norm axioms listed above. One also has the Cauchy–Schwarz inequality

$$|\langle v, y \rangle| \le \|v\| \|y\|.$$

Finally, recall that both $\langle \cdot, x \rangle$ with $x \in V$ and $\|\cdot\|$ are examples of continuous functionals $V \to \mathbb{R}$, i.e., if $\lim_{j \to \infty} \|v - v_j\| = 0$, then

$$\lim_{j \to \infty} \|v_j\| = \|v\|, \quad \lim_{j \to \infty} \langle v_j, x \rangle = \langle v, x \rangle \ \forall x \in V.$$

Hilbert and Banach Spaces

A Hilbert space $\mathcal{H}$ is a vector space equipped with an inner product $\langle \cdot, \cdot \rangle$ which is complete w.r.t. to the norm $\|\cdot\|$ induced by such inner product. This means that, given any Cauchy sequence, i.e., a sequence of vectors $\{g_j\}_{j=1}^{\infty}$ such that

$$\lim_{m,n \to \infty} \|g_m - g_n\| = 0,$$

there exists $g \in \mathcal{H}$ such that

$$\lim_{j \to \infty} \|g - g_j\| = 0.$$

In other words, every Cauchy sequence is convergent. Examples of Hilbert spaces used in this book are

- the classical Euclidean space $\mathbb{R}^m$ of vectors $a = [a_1 \ \dots \ a_m]$ equipped with the classical Euclidean inner product

$$\langle a, b \rangle_2 = \sum_{i=1}^{m} a_i b_i$$

sometimes denoted just by $\langle \cdot, \cdot \rangle$ in the book;
- the space ℓ_2 of squared summable real sequences $a = [a_1 \ a_2 \dots]$, i.e., such that

$$\sum_{i=1}^{\infty} a_i^2 < \infty,$$

equipped with the inner product

$$\langle a, b \rangle_2 = \sum_{i=1}^{\infty} a_i b_i;$$

- the classical Lebesgue space $\mathscr{L}_2$ of functions (where the measure μ is here omitted to simplify notation) $g : \mathscr{X} \to \mathbb{R}$ which are squared summable w.r.t. the measure μ, i.e., such that

$$\int_{\mathscr{X}} g^2(x) d\mu(x) < \infty,$$

equipped with the inner product still denoted by $\langle \cdot, \cdot \rangle_2$ but now given by

$$\langle g, f \rangle_2 = \int_{\mathscr{X}} g(x) f(x) d\mu(x).$$

The spaces reported above are also instances of metric spaces where, for every couple of vectors f, g, there is a notion of distance defined by $\| f - g \|$. Other metric spaces are the Banach spaces. They are normed vector spaces complete w.r.t. the metric induced by their norm. Hence, every Hilbert space is a Banach space but the converse is not true: this happens when $\| \cdot \|$ does not derive from an inner product. Examples of Banach spaces (whose norm does not derive from an inner product) are

- the space ℓ_1 of absolutely summable real sequences $a = [a_1 \ a_2 \ldots]$, i.e., such that

$$\sum_{i=1}^{\infty} |a_i| < \infty,$$

equipped with the norm

$$\|a\|_1 = \sum_{i=1}^{\infty} |a_i|;$$

- the Lebesgue space $\mathscr{L}_1$ of functions $g : \mathscr{X} \to \mathbb{R}$ absolutely integrable w.r.t. the measure μ, i.e., such that

$$\int_{\mathscr{X}} |g(x)| d\mu(x) < \infty,$$

equipped with the norm

$$\|g\|_1 = \int_{\mathscr{X}} |g(x)| d\mu(x);$$

- the space ℓ_∞ of bounded real sequences $a = [a_1 \; a_2 \dots]$, i.e., such that

$$\sup_i |a_i| < \infty,$$

equipped with the norm

$$\|a\|_\infty = \sup_i |a_i|;$$

- the space $\mathscr{C}$ of continuous functions $g : \mathscr{X} \to \mathbb{R}$. where $\mathscr{X}$ is a compact set typically in $\mathbb{R}^m$, equipped with the sup-norm (also called uniform norm)

$$\|g\|_\infty = \max_{x \in \mathscr{X}} |g(x)|;$$

- the Lebesgue space $\mathscr{L}_\infty$ of functions $g : \mathscr{X} \to \mathbb{R}$ which are essentially bounded w.r.t. the measure μ, i.e., for any g there exists M such that

$$|g(x)| \le M \text{ almost everywhere in } \mathscr{X} \text{ w.r.t. the measure } \mu,$$

equipped with the norm

$$\|g\|_\infty = \inf \{M \mid |g(x)| \le M \text{ almost everywhere in } \mathscr{X} \text{ w.r.t. the measure } \mu\}.$$

An infinite-dimensional Hilbert (or Banach) space is said to be separable if it admits a countable basis $\{\rho_j\}_{j=1}^\infty$, i.e., for any g in the space we can find scalars c_j such that

$$\lim_{j \to \infty} \left\| g - \sum_{j=1}^\infty c_j \rho_j \right\| = 0.$$

When such vectors $\{\rho_j\}$ satisfy also the conditions

$$\|\rho_j\| = 1 \; \forall j, \quad \langle \rho_j, \rho_i \rangle = 0 \; j \ne i,$$

then the basis is said to be orthonormal.

Subspaces, Projections and Compact Sets

A subset S of the vector space V is said to be a subspace if S is itself a vector space with the same addition and multiplication operations defined in V. The symbol

$$\mathrm{span}(\{\rho_j\}_{j \in A})$$

denotes the subspace containing all the finite linear combinations of vectors taken from the (possibly uncountable) family $\{\rho_j\}_{j \in A}$.

Given a subspace (or simply a set) S contained in a Hilbert (or Banach) space, we define

$$\bar{S} = S \cup \{\text{all the limits of Cauchy sequences built using vectors in S}\}.$$

Then, S is said to be closed if

$$\bar{S} = S.$$

The orthogonal to a subspace S of a Hilbert space is denoted by $S^{\perp}$ and defined by

$$S^{\perp} = \{g \mid \langle g, f \rangle = 0 \; \forall f \in S\}.$$

It is easy to prove that $S^{\perp}$ is always a closed subspace.

The following fundamental theorem holds.

Theorem 6.25 (Projection theorem) *Let S be a closed subspace of a Hilbert space with norm $\| \cdot \|_{\mathscr{H}}$. Then, one has*

- *any $g \in \mathscr{H}$ has a unique decomposition*

$$g = g_S + g_{S^{\perp}}, \quad g_S \in S, \; g_{S^{\perp}} \in S^{\perp};$$

- *g_S is the projection of g onto S, i.e.,*

$$g_S = \arg\min_{f \in S} \|g - f\|_{\mathscr{H}};$$

- *it holds that*

$$\|g\|_{\mathscr{H}}^2 = \|g_S\|_{\mathscr{H}}^2 + \|g_{S^{\perp}}\|_{\mathscr{H}}^2.$$

A set A contained in a Hilbert (or Banach) space with norm $\| \cdot \|$ is said to be *compact* if, given any sequence $\{g_j\}$ of vectors all contained in A, it is possible to extract a subsequence $\{g_{k_j}\}$ convergent in A, i.e., there exists $g \in A$ such that

$$\lim_{j \to \infty} \|g - g_{k_j}\| = 0.$$

When the space is finite-dimensional, a set is compact iff it is closed and bounded.

Linear and Bounded Functionals

Given a Hilbert space $\mathscr{H}$ with norm $\| \cdot \|_{\mathscr{H}}$, a functional $L : \mathscr{H} \to \mathbb{R}$ is said to be bounded (or, equivalently, continuous) if there exists a positive scalar C such that

$$|L[g]| \leq C\|g\|_{\mathscr{H}}, \quad \forall g \in \mathscr{H}. \tag{6.66}$$

The following classical theorem holds.

Theorem 6.26 (Closed graph theorem) *Let $\mathscr{H}$ be a Hilbert (or Banach) space. Then $L : \mathscr{H} \to \mathbb{R}$ is linear and bounded if and only if the graph of L, i.e.,*

$$Gr(L) = \{(f, L[f]) \text{ with } f \in \mathcal{H}\},$$

is a closed set in the product space $\mathcal{H} \times \mathbb{R}$. This means that if $\{f_i\}_{i=1}^{+\infty}$ is a sequence converging to $f \in \mathcal{H}$ and $\{L[f_i]\}_{i=1}^{+\infty}$ converges to $y \in \mathbb{R}$, then $L[f] = y$.

This other fundamental theorem asserts that every linear and bounded functional over $\mathcal{H}$ is in one-to-one correspondence with a vector in $\mathcal{H}$.

Theorem 6.27 (Riesz representation theorem, based on [76]) *Let $\mathcal{H}$ be a Hilbert space and let $L : \mathcal{H} \to \mathbb{R}$. Then L is linear and bounded if and only there is a unique $f \in \mathcal{H}$ such that*

$$L[g] = \langle g, f \rangle_{\mathcal{H}}, \quad \forall g \in \mathcal{H}. \tag{6.67}$$

6.9.2 Proof of Theorem 6.1

First, we derive two lemmas which are instrumental to the main proof.

Lemma 6.1 *Let*

$$S = span(\{K_x\}_{x \in \mathcal{X}}).$$

If there exists a Hilbert space $\mathcal{H}$ satisfying conditions (6.2) and (6.3), then $\mathcal{H}$ is the closure of S, i.e., $\mathcal{H} = \bar{S}$.

Proof It comes from condition (6.2) that $\bar{S}$ is a closed subspace which must belong to $\mathcal{H}$. Theorem 6.25 (Projection theorem) then ensures that any function $f \in \mathcal{H}$ can be written as

$$f = f_{\bar{S}} + f_{\bar{S}^\perp}, \quad f_{\bar{S}} \in \bar{S}, \ f_{\bar{S}^\perp} \in \bar{S}^\perp.$$

As for the component $f_{\bar{S}^\perp}$, using condition (6.3) (reproducing property) we obtain

$$f_{\bar{S}^\perp}(x) = \langle f_{\bar{S}^\perp}, K_x \rangle_{\mathcal{H}} = 0, \ \forall x.$$

In fact, every kernel section belongs to S and is thus orthogonal to every function in $\bar{S}^\perp$. Hence, $f_{\bar{S}^\perp}$ is the null vector and this concludes the proof. $\square$

Lemma 6.2 *Let $S = span(\{K_x\}_{x \in \mathcal{X}})$ and define*

$$\|f\|_{\mathcal{H}}^2 = \sum_{i=1}^{m} \sum_{j=1}^{m} c_i c_j K(x_i, x_j), \tag{6.68}$$

where f is a generic element in S, hence admitting representation

$$f(\cdot) = \sum_{i=1}^{m} c_i K_{x_i}(\cdot).$$

Then, $\| \cdot \|_{\mathscr{H}}$ is a well-defined norm in S.

Proof The reader can easily check that absolute homogeneity and the triangle inequality are satisfied by $\| \cdot \|_{\mathscr{H}}$. We only need to prove the null vector condition, i.e., that for every $f \in S$ one has

$$\|f\|_{\mathscr{H}} = 0 \iff f = 0.$$

Now, assume that $\|f\|_{\mathscr{H}} = 0$ where $f(\cdot) = \sum_{i=1}^{m} c_i K_{x_i}(\cdot)$. While the coefficients $\{c_i\}_{i=1}^{m}$ and the input locations $\{x_i\}_{i=1}^{m}$ are fixed and define f, let also c_{m+1} and x_{m+1} be an arbitrary scalar and input location, respectively. Define $\mathbf{K} \in \mathbb{R}^{m \times m}$ and $\mathbf{K}_+ \in \mathbb{R}^{m+1 \times m+1}$ two matrices with (i, j)-entry given by $K(x_i, x_j)$. Let also $c = [c_1 \ \ldots \ c_m]^T$ and $c_+ = [c_1 \ \ldots \ c_m \ c_{m+1}]^T$. Note that $\mathbf{K}c$ is the vector which contains the values of f on the input locations $\{x_i\}_{i=1}^{m}$.

Since K is positive definite, it holds that

$$c_+^T \mathbf{K}_+ c_+ \geq 0 \quad \forall \ (c_{m+1}, x_{m+1}) \in (\mathbb{R} \times \mathscr{X}).$$

In addition, since by assumption

$$\|f\|_{\mathscr{H}}^2 = c^T \mathbf{K} c = 0,$$

it comes that the components of the vector $\mathbf{K}c$, which are the values of f on $\{x_i\}_{i=1}^{m}$, are all null. Now, we show that $f(x) = 0$ holds everywhere, also on the generic input location $x_{m+1} \in \mathscr{X}$. In fact, after simple calculations, one obtains

$$c_+^T \mathbf{K}_+ c_+ = c^T \mathbf{K} c + 2 \left[\sum_{i=1}^{m} c_i K(x_i, x_{m+1}) \right] c_{m+1} + K(x_{m+1}, x_{m+1}) c_{m+1}^2$$

$$= 2 \left[\sum_{i=1}^{m} c_i K(x_i, x_{m+1}) \right] c_{m+1} + K(x_{m+1}, x_{m+1}) c_{m+1}^2$$

$$= 2 f(x_{m+1}) c_{m+1} + K(x_{m+1}, x_{m+1}) c_{m+1}^2.$$

Now, assume that $f(x_{m+1}) > 0$. Then, since the last term on the r.h.s. is infinitesimal of order two w.r.t. c_{m+1} we can find a negative value for c_{m+1} sufficiently close to zero such that $c_+^T \mathbf{K}_+ c_+ < 0$ which contradicts the fact that K is positive definite. If $f(x_{m+1}) < 0$ we can instead find a positive value for c_{m+1} sufficiently close to zero such that $c_+^T \mathbf{K}_+ c_+ < 0$, which is still a contradiction. Hence, we must have $f(x_{m+1}) = 0$. Since x_{m+1} was arbitrary, we conclude that f must be the null function. $\square$

We now prove Theorem 6.1. Let $S = \text{span}(\{K_x\}_{x\in\mathscr{X}})$ and, for any $f, g \in S$ having representations

$$f(\cdot) = \sum_{i=1}^{m} c_i K_{x_i}(\cdot), \quad g(\cdot) = \sum_{i=1}^{p} d_i K_{y_i}(\cdot)$$

define

$$\langle f, g \rangle_{\mathscr{H}} = \sum_{i=1}^{m}\sum_{j=1}^{p} c_i d_j K(x_i, y_j).$$

By Lemma 6.2, it is immediate to check that $\langle \cdot, \cdot \rangle_{\mathscr{H}}$ is a well-defined inner product on S. Then, we now show that the desired Hilbert space is $\mathscr{H} = \bar{S}$, where $\bar{S}$ is the completion of S w.r.t. the norm induced by $\langle \cdot, \cdot \rangle_{\mathscr{H}}$.

Condition (6.2) is trivially satisfied since, by construction, all the kernel sections belong to $\mathscr{H}$.

As for the condition (6.3), we start checking that it holds over S. Introducing the couple of functions in S given by

$$f(\cdot) = \sum_{i=1}^{m} c_i K_{x_i}(\cdot), \quad g(\cdot) = K_x(\cdot),$$

we have

$$\langle f, K_x \rangle_{\mathscr{H}} = \langle f, g \rangle_{\mathscr{H}} = \sum_{i=1}^{m} c_i K(x_i, x) = f(x),$$

showing that the reproducing property holds in S. Let us now consider the completion of S. To this aim, let $\{f_j\}$ be a Cauchy sequence with $f_j \in S \ \forall j$. We have

$$|f_i(x) - f_j(x)| = |\langle f_i - f_j, K_x \rangle_{\mathscr{H}}|$$
$$\leq \|f_i - f_j\|_{\mathscr{H}} \|K_x\|_{\mathscr{H}},$$

where we have used first the reproducing property (since it holds in S) and then the Cauchy–Schwarz inequality. We have

$$\|K_x\|_{\mathscr{H}} = |\sqrt{\langle K_x, K_x \rangle_{\mathscr{H}}}| = \sqrt{K(x, x)} \leq q < +\infty,$$

where the scalar q independent of x exists because the kernel K is continuous over the compact $\mathscr{X} \times \mathscr{X}$. Combining the last two inequalities leads to

$$|f_i(x) - f_j(x)| \leq \sup_{x\in\mathscr{X}} |f_i(x) - f_j(x)| \leq q\|f_i - f_j\|_{\mathscr{H}}, \qquad (6.69)$$

which shows that the convergence in $\mathscr{H}$ implies also uniform convergence. In other words, if $f_j \to f$ in $\mathscr{H}$ w.r.t. $\|\cdot\|_{\mathscr{H}}$, then $f_j \to f$ also in the space $\mathscr{C}$ of continuous functions w.r.t. the sup-norm $\| \cdot \|_\infty$. Since $S \subset \mathscr{C}$ and $\mathscr{C}$ is Banach, all the functions in the completion of S are continuous, i.e., $\mathscr{H} \subset \mathscr{C}$. Furthermore, if $f_j \to f$ in $\mathscr{H}$, one has that for any $x \in \mathscr{X}$

$$\lim_{j \to \infty} \langle f_j, K_x \rangle_{\mathscr{H}} = \langle f, K_x \rangle_{\mathscr{H}},$$

by the continuity of the inner product. But we also have

$$\lim_{j \to \infty} \langle f_j, K_x \rangle_{\mathscr{H}} = \lim_{j \to \infty} f_j(x) = f(x),$$

since $f_j \in S \ \forall j$, the reproducing property holds in S and convergence in $\mathscr{H}$ implies uniform (and, hence, pointwise) convergence. This shows that $\langle f, K_x \rangle_{\mathscr{H}} = f(x) \ \forall f \in \mathscr{H}$, i.e., the reproducing property holds over all the space $\mathscr{H}$.

The last point is the unicity of $\mathscr{H}$. For the sake of contradiction, assume that there exists another Hilbert space $\mathscr{G}$ which satisfies conditions (6.2) and (6.3). By Lemma 6.1, we must have $\mathscr{G} = \bar{S}$ where the completion of S is w.r.t. the norm $\| \cdot \|_{\mathscr{G}}$ deriving from the inner product $\langle \cdot, \cdot \rangle_{\mathscr{G}}$. Condition (6.3) holds both in $\mathscr{H}$ and in $\mathscr{G}$, so that we have

$$\langle K_x, K_s \rangle_{\mathscr{H}} = K(x, s) = \langle K_x, K_s \rangle_{\mathscr{G}}, \ \forall (x, s) \in \mathscr{X} \times \mathscr{X}.$$

Since the functions in S are finite linear combinations of kernel sections, by the linearity of the inner product, the above equality allows to conclude that

$$\langle f, g \rangle_{\mathscr{H}} = \langle f, g \rangle_{\mathscr{G}}, \ \forall (f, g) \in S \times S.$$

Such an equality, together with the uniqueness of limits, implies that the completion of S w.r.t. $\| \cdot \|_{\mathscr{H}}$ coincides with the completion w.r.t. $\| \cdot \|_{\mathscr{G}}$. Hence, $\mathscr{H}$ and $\mathscr{G}$ are the same Hilbert space and this completes the proof.

6.9.3   Proof of Theorem 6.10

It is not difficult to see that (6.12) with the inner product (6.13) is a Hilbert space. In addition, using the Mercer theorem, in particular the expansion (6.11), from (6.13) one has

$$\|K_x\|_{\mathscr{H}}^2 = \| \sum_{i \in \mathscr{I}} \zeta_i \rho_i(x) \rho_i(\cdot) \|_{\mathscr{H}}^2$$

$$= \sum_{i \in \mathscr{I}} \frac{\zeta_i^2 \rho_i^2(x)}{\zeta_i} = K(x, x) < \infty,$$

and, for any $f = \sum_{i \in \mathscr{I}} a_i \rho_i$, it also holds that

$$\langle K_x, f \rangle_{\mathscr{H}} = \langle \sum_{i \in \mathscr{I}} \zeta_i \rho_i(x) \rho_i(\cdot), \sum_{i \in \mathscr{I}} a_i \rho_i(\cdot) \rangle_{\mathscr{H}}$$

$$= \sum_{i \in \mathscr{I}} \frac{\zeta_i \rho_i(x) a_i}{\zeta_i} = f(x).$$

This shows that every kernel section belongs to $\mathscr{H}$ and the reproducing property holds. Theorem 6.1 then ensures that $\mathscr{H}$ is indeed the RKHS associated to K.

6.9.4 Proof of Theorem 6.13

First, let $\mathscr{H}$ be the RKHS induced by $K(x, y) = \zeta \rho(x) \rho(y)$. Any RKHS is spanned by its kernel sections, hence in this case $\mathscr{H}$ is the one-dimensional subspace generated by ρ. By the reproducing property it holds that

$$\|K_x\|_{\mathscr{H}}^2 = K(x, x) = \zeta \rho^2(x).$$

In addition, one has

$$\|K_x\|_{\mathscr{H}}^2 = \|\zeta \rho(x) \rho\|_{\mathscr{H}}^2 = \zeta^2 \rho^2(x) \|\rho\|_{\mathscr{H}}^2,$$

so that

$$\|\rho\|_{\mathscr{H}}^2 = \frac{1}{\zeta}.$$

Now, consider the kernel of interest $K(x, y) = \sum_{i=1}^{\infty} \zeta_i \rho_i(x) \rho_i(y)$ associated with $\mathscr{H}$. Define $K_j(x, y) = \zeta_j \rho_j(x) \rho_j(y)$. with $\|\cdot\|_{\mathscr{H}_j}$ to denote the norm induced by K_j. From the discussion above it holds that

$$\|\rho_j\|_{\mathscr{H}_j}^2 = \frac{1}{\zeta_j}. \tag{6.70}$$

Think of $K(x, y) = \sum_{i=1}^{\infty} \zeta_i \rho_i(x) \rho_i(y)$ as the sum of $K_j(x, y)$ and $K_{-j}(x, y) = \sum_{k \neq j}^{\infty} \zeta_k \rho_k(x) \rho_k(y)$. Then, using Theorem 6.6 and (6.70), one has

$$\|\rho_j\|_{\mathscr{H}}^2 = \min_{c_j, h} \frac{c_j^2}{\zeta_j} + \|h\|_{\mathscr{H}_{-j}}^2 \text{ s.t. } \rho_j = c_j \rho_j + h, \ c_j \in \mathbb{R}, \ h \in \mathscr{H}_{-j}$$

where $\mathscr{H}_{-j}$ is the RKHS induced by K_{-j}. Evaluating the objective at $(c_j = 1, h = 0)$, one obtains

$$\|\rho_j\|_{\mathscr{H}}^2 \leq \frac{1}{\zeta_j},$$

and this shows that $\rho_j \in \mathscr{H} \; \forall j$.

Now we prove that the functions ρ_j generate all the RKHS $\mathscr{H}$ induced by K. Using Theorem 6.25 (Projection theorem), it comes that for any $f \in \mathscr{H}$ we have

$$f = g + h \quad \text{with} \; g \in G, \; h \in G^\perp$$

where G indicates the closure in $\mathscr{H}$ of the subspace generated by all the ρ_k. Using the reproducing property, one obtains

$$
\begin{aligned}
h(x) &= <h(\cdot), K(x, \cdot) >_{\mathscr{H}} \\
&= <h(\cdot), \sum_{k=1}^{\infty} \zeta_k \rho_k(x) \rho_k(\cdot) >_{\mathscr{H}} \\
&= \sum_{k=1}^{\infty} \zeta_k \rho_k(x) <h(\cdot), \rho_k(\cdot) >_{\mathscr{H}} = 0 \quad \forall x,
\end{aligned}
$$

where the last equality exploits the relation $h \perp \rho_k \; \forall k$. This completes the first part of the proof.

As for the RKHS norm characterization, first let $\mathscr{H}_j^\infty$ be the RKHS induced by the kernel $\sum_{k=j}^{\infty} K_k$ with h_j to denote a generic element of $\mathscr{H}_j^\infty$. Then, given $f \in \mathscr{H}$, using Theorem 6.6 in an iterative fashion, we obtain

$$
\begin{aligned}
\|f\|_{\mathscr{H}}^2 &= \min_{c_1, h_2} \frac{c_1^2}{\zeta_1} + \|h_2\|_{\mathscr{H}_2^\infty}^2 \; \text{s.t.} \; f = c_1 \rho_1 + h_2 \\
&= \min_{c_1, c_2, h_3} \frac{c_1^2}{\zeta_1} + \frac{c_2^2}{\zeta_2} + \|h_3\|_{\mathscr{H}_3^\infty}^2 \; \text{s.t.} \; f = c_1 \rho_1 + c_2 \rho_2 + h_3 \\
&\;\;\vdots \\
&= \min_{c_1, \ldots, c_{n-1}, h_n} \sum_{k=1}^{n-1} \frac{c_k^2}{\zeta_k} + \|h_n\|_{\mathscr{H}_n^\infty}^2 \; \text{s.t.} \; f = \sum_{i=1}^{n-1} c_i \rho_i + h_n.
\end{aligned}
$$

In particular, every equality above is obtained thinking of the kernel $\sum_{k=j}^{\infty} K_k$ as the sum of K_j and $\sum_{k=j+1}^{\infty} K_k$. Then, h_j can be decomposed into two parts, i.e., $h_j = c_j \rho_j + h_{j+1}$, with $\|\rho_j\|_{\mathscr{H}_j}^2 = 1/\zeta_j$ where, as before, $\|\cdot\|_{\mathscr{H}_j}$ denotes the norm in the one-dimensional RKHS induced by K_j. Now, let $\hat{c}_1, \ldots, \hat{c}_{n-1}, \hat{h}_n$ be the minimizers of the last objective (the minimizer can be assumed unique without loss of generality, just to simplify the exposition) and note that $\|\hat{h}_n\|_{\mathscr{H}_n^\infty}$ must go to zero as $n \to \infty$. Then, it comes that the sequence $\hat{c}_1, \hat{c}_2, \ldots$ characterizing the norm $\|f\|_{\mathscr{H}}^2$ is indeed $\min_{\{c_k\}} \sum_{k=1}^{\infty} \frac{c_k^2}{\zeta_k}$ with the $\{c_k\}$ subject to the constraints $\lim_{n \to \infty} \| f - \sum_{k=1}^{n} c_k \rho_k \|_{\mathcal{H}} = 0$.

6.9.5 Proofs of Theorems 6.15 and 6.16

We prove the following more general result that embraces as special cases Theorems 6.15 and 6.16.

Theorem 6.28 *Let $\mathcal{H}$ be a Hilbert space. Consider the optimization problem*

$$\min_{f \in \mathcal{H}} \Phi(L_1[f], \ldots, L_N[f], \|f\|_{\mathcal{H}}) \tag{6.71}$$

and assume that

- *problem (6.71) admits at least one solution;*
- *each $L_i : \mathcal{H} \to \mathbb{R}$ is linear and bounded;*
- *the objective Φ is strictly increasing w.r.t. its last argument.*

Then, all the solutions of (6.71) admit the following expression

$$\hat{g} = \sum_{i=1}^{N} c_i \eta_i, \tag{6.72}$$

where the c_i are suitable scalar expansion coefficients and each $\eta_i \in \mathcal{H}$ is the representer of L_i, i.e.,

$$L_i[f] = \langle f, \eta_i \rangle_{\mathcal{H}}, \quad \forall f \in \mathcal{H}, \ i = 1, \ldots, N.$$

In particular, if $\mathcal{H}$ is a RKHS with kernel K, each basis function is given by

$$\eta_i(x) = L_i[K(\cdot, x)].$$

To prove the above result, let $\hat{g}$ be a solution of (6.71) and denote with S the (closed) subspace spanned by the N representers η_i of the functionals L_i, i.e.,

$$S = \text{span}\{\eta_1, \ldots, \eta_N\}.$$

Exploiting Theorem 6.25 (Projection theorem), we can write

$$\hat{g} = \hat{g}_S + \hat{g}_{S^\perp}, \quad \hat{g}_S \in S, \ \hat{g}_{S^\perp} \in S^\perp.$$

For the sake of contradiction, assume that $\hat{g}_{S^\perp}$ is different from the null function. Then, we have

$$
\begin{aligned}
\Phi(L_1[\hat{g}],\ldots,L_n[\hat{g}],\|\hat{g}\|_{\mathcal{H}}) &= \Phi(\langle\eta_1,\hat{g}\rangle_{\mathcal{H}},\ldots,\langle\eta_N,\hat{g}\rangle_{\mathcal{H}},\|\hat{g}\|_{\mathcal{H}}) \\
&= \Phi(\langle\eta_1,\hat{g}_S+\hat{g}_{S^\perp}\rangle_{\mathcal{H}},\ldots,\langle\eta_N,\hat{g}_S+\hat{g}_{S^\perp}\rangle_{\mathcal{H}},\sqrt{\|\hat{g}_S\|_{\mathcal{H}}^2+\|\hat{g}_{S^\perp}\|_{\mathcal{H}}^2}) \\
&= \Phi(\langle\eta_1,\hat{g}_S\rangle_{\mathcal{H}},\ldots,\langle\eta_N,\hat{g}_S\rangle_{\mathcal{H}},\sqrt{\|\hat{g}_S\|_{\mathcal{H}}^2+\|\hat{g}_{S^\perp}\|_{\mathcal{H}}^2}) \\
&< \Phi(\langle\eta_1,\hat{g}_S\rangle_{\mathcal{H}},\ldots,\langle\eta_N,\hat{g}_S\rangle_{\mathcal{H}},\|\hat{g}_S\|_{\mathcal{H}}),
\end{aligned}
$$

where the last equality exploits the fact that each η_i is orthogonal to all the functions in $S^\perp$ while the inequality exploits the assumption that Φ is strictly increasing w.r.t. its last argument. This contradicts the optimality of $\hat{g}$ and implies that $\hat{g}_{S^\perp}$ must be the null function, hence concluding the first part of the proof.

Finally, to prove (6.28) note that, if $\mathcal{H}$ is a RKHS, one has

$$
\eta_i(x) = \langle\eta_i,K_x\rangle_{\mathcal{H}} = L_i[K(\cdot,x)],
$$

where the first equality comes from the reproducing property, while the second one derives from the fact that η_i is the representer of L_i.

6.9.6   Proof of Theorem 6.21

Preliminary Lemmas

The first lemma, whose proof can be found in [34], states a bound on the correlation between two random variables assuming values in a Hilbert space.

Lemma 6.3 (based on [34]) *Let a and b be zero-mean random variables measurable with respect to the σ-algebras $\mathcal{M}_1$ and $\mathcal{M}_2$ and with values in the Hilbert space $\mathcal{H}$ having inner product $\langle\cdot,\cdot\rangle_{\mathcal{H}}$. Then, it holds that*

$$
|\mathcal{E}[\langle a,b\rangle_{\mathcal{H}}]| \leq 15\sqrt[3]{\psi(\mathcal{M}_1,\mathcal{M}_2)\mathcal{E}\|a\|_{\mathcal{H}}^3\mathcal{E}\|b\|_{\mathcal{H}}^3}, \tag{6.73}
$$

where all the expectations above are assumed to exist and

$$
\psi(\mathcal{M}_1,\mathcal{M}_2) = \sup_{A\in\mathcal{M}_1,B\in\mathcal{M}_2} |P(A\cap B)-P(A)P(B)|.
$$

As for the second lemma, first it is useful to introduce the following integral operator:

$$
L_K[f](\cdot) = \int_X K(\cdot,x)f(x)p_x(x)dx.
$$

Since the assumptions underlying Theorem (6.9) (Mercer Theorem) hold true, there exists a complete orthonormal basis of $\mathscr{L}_2^\mu$, denoted by $\{\rho_i\}_{i \in \mathscr{I}}$, which satisfies

$$L_K[\rho_i] = \zeta_i \rho_i, \quad i \in \mathscr{I}, \quad \zeta_1 \geq \zeta_2 \geq .$$

To simplify exposition, hereby we assume $\zeta_i > 0\ \forall i$. Then, for $r > 0$, we define the operators L_K^{-r} and L_K^r as follows

$$L_K^r[f] = \sum_{i \in \mathscr{I}} \zeta_i^r c_i \rho_i \qquad (6.74)$$

$$L_K^{-r}[f] = \sum_{i \in \mathscr{I}} \frac{c_i}{\zeta_i^r} \rho_i. \qquad (6.75)$$

The function $L_K^{-r}[f]$ is less regular than f since its expansion coefficients go to zero more slowly. Instead, L_K^r is a smoothing operator since $\zeta_i^r c_i$ goes to zero faster than c_i as i goes to infinity. When $r = 1/2$ we recover the operator $L_K^{1/2}$ already defined in (6.17) which satisfies $\mathscr{H} = L_K^{1/2} \mathscr{L}_2^\mu$. The following lemma holds.

Lemma 6.4 *If $L_K^{-r} f_\rho \in \mathscr{L}_2^\mu$ for some $0 < r \leq 1$, letting*

$$\hat{f} = \arg\min_{f \in \mathscr{H}} \|f - f_\rho\|_{\mathscr{L}_2^\mu}^2 + \gamma \|f\|_{\mathscr{H}}^2, \qquad (6.76)$$

one has

$$\|\hat{f} - f_\rho\|_{\mathscr{L}_2^\mu} \leq \gamma^{\,r} \|L_K^{-r} f_\rho\|_{\mathscr{L}_2^\mu}. \qquad (6.77)$$

Proof By assumption, there exists $g \in \mathscr{L}_2^\mu$, say $g = \sum_{i \in \mathscr{I}} d_i \rho_i$, such that $f_\rho = L_K^r g$ so that $f_\rho = \sum_{i \in \mathscr{I}} \zeta_i^r d_i \rho_i$. Now, we characterize the solution $\hat{f}$ of (6.76) using $f = \sum_{i \in \mathscr{I}} c_i \rho_i$ and optimizing w.r.t. the c_i. The objective becomes

$$\sum_{i \in \mathscr{I}} (c_i - \zeta_i^r d_i)^2 + \gamma \sum_{i \in \mathscr{I}} \frac{c_i^2}{\zeta_i},$$

and setting the partial derivatives w.r.t. each c_i to zero, we obtain

$$\hat{f} = \sum_{i \in \mathscr{I}} \hat{c}_i \rho_i, \quad \hat{c}_i = \frac{\zeta_i^{r+1} d_i}{\zeta_i + \gamma}. \qquad (6.78)$$

This implies

$$\hat{f} - f_\rho = -\sum_{i \in \mathscr{I}} \frac{\gamma}{\zeta_i + \gamma} \zeta_i^r d_i \rho_i.$$

If $0 < r \le 1$, it follows that

$$\|\hat{f} - f_\rho\|_{\mathscr{L}_2^\mu} = \left\{ \sum_{i \in \mathscr{I}} \left(\frac{\gamma}{\zeta_i + \gamma} \zeta_i^r d_i \right)^2 \right\}^{1/2}$$

$$= \gamma^r \left\{ \sum_{i \in \mathscr{I}} \left(\frac{\gamma}{\zeta_i + \gamma} \right)^{2(1-r)} \left(\frac{\zeta_i}{\zeta_i + \gamma} \right)^{2r} d_i^2 \right\}^{1/2}$$

$$\le \gamma^r \sum_{i \in \mathscr{I}} \left(\frac{\gamma}{\zeta_i + \gamma} \right)^{(1-r)} \left(\frac{\zeta_i}{\zeta_i + \gamma} \right)^r |d_i|$$

$$\le \gamma^r \left\{ \sum_{i \in \mathscr{I}} d_i^2 \right\}^{1/2} = \gamma^r \|g\|_{\mathscr{L}_2^\mu} = \gamma^r \|L_K^{-r} f_\rho\|_{\mathscr{L}_2^\mu}$$

and this proves (6.77).   □

In the proof of the third lemma reported below, the notation $S_x : \mathscr{H} \to \mathbb{R}^N$ indicates the sampling operator defined by $S_x f = [f(x_1) \dots f(x_N)]$. In addition, S_x^T denotes its adjoint, i.e., for any $c \in \mathbb{R}^N$, it satisfies

$$\langle f, S_x^T c \rangle_{\mathscr{H}} = \langle S_x f, c \rangle = \sum_{i=1}^N c_i f(x_i) = \langle f, \sum_{i=1}^N c_i K_{x_i} \rangle_{\mathscr{H}},$$

where $\langle \cdot, \cdot \rangle$ is the Euclidean inner product. Hence, one has

$$S_x^T c = \sum_{i=1}^N c_i K_{x_i} \quad \forall c \in \mathbb{R}^N.$$

Lemma 6.5 *Define*

$$\eta_i(\cdot) = \left[y_i - \hat{f}(x_i) \right] K(x_i, \cdot) \tag{6.79}$$

with $\hat{f}$ defined by (6.76). Then, if $\hat{g}_N$ is given by (6.55), one has

$$\|\hat{g}_N - \hat{f}\|_{\mathscr{H}} \le \frac{1}{\gamma} \left\| \frac{1}{N} \sum_{i=1}^N (\eta_i - \mathscr{E}[\eta_i]) \right\|_{\mathscr{H}}.$$

Proof First, it is useful to derive two useful equalities involving $\hat{f}$ and $\hat{g}_N$. The first one is

$$\gamma \hat{f} = L_K(\hat{f}_\rho - \hat{f}) = \mathscr{E} \eta_i. \tag{6.80}$$

The last equality in (6.80) follows from the definition of L_k and η_i. The first equality can be obtained using the representation $f_\rho = \sum_{i \in \mathscr{I}} d_i \rho_i$, then following the same passages contained in the first part of the previous lemma's proof to obtain

$$\hat{f} = \sum_{i \in \mathscr{I}} \frac{\zeta_i}{\zeta_i + \gamma} d_i \rho_i, \quad f_\rho - \hat{f} = \sum_{i \in \mathscr{I}} \frac{\gamma}{\zeta_i + \gamma} d_i \rho_i.$$

The second result consists of the following alternative expression for $\hat{g}_N$:

$$\hat{g}_N = \left(\frac{S_x^T S_x}{N} + \gamma I \right)^{-1} \frac{S_x^T}{N} Y, \tag{6.81}$$

where I denotes the identity operator. To prove it, we will use the equality $\hat{g}_N = S_x^T (\mathbf{K} + N\gamma I_N)^{-1} Y$ which derives from the representer theorem and also the fact that, for any vector $c \in \mathbb{R}^N$, it holds that $S_x S_x^T c = \mathbf{K}c$ with $\mathbf{K}$ the kernel matrix built using $[x_1 \ldots x_N]$. Then, we have

$$\left(\frac{S_x^T S_x}{N} + \gamma I \right) \hat{g}_N = \frac{S_x^T}{N} \left(\mathbf{K} (\mathbf{K} + N\gamma I_N)^{-1} + N\gamma (\mathbf{K} + N\gamma I_N)^{-1} \right) Y$$

$$= \frac{S_x^T}{N} Y.$$

Now, it is also useful to obtain a bound on the inverse of the operator $\frac{S_x^T S_x}{N} + \gamma I$. Assume that $v \in \mathscr{H}$ and let u satisfy

$$\left(\frac{S_x^T S_x}{N} + \gamma I \right) u = v.$$

We take inner products on both sides with u and use the equality $\langle S_x S_x^T u, u \rangle_{\mathscr{H}} = \langle S_x u, S_x u \rangle$ to obtain

$$\frac{1}{N} \langle S_x u, S_x u \rangle + \gamma \|u\|_{\mathscr{H}}^2 = \langle v, u \rangle_{\mathscr{H}} \leq \|v\|_{\mathscr{H}} \|u\|_{\mathscr{H}}.$$

One has

$$\lambda_x := \inf_{f \in \mathscr{H}} \frac{\|S_x f\|}{\sqrt{N} \|f\|_{\mathscr{H}}} \implies (\lambda_x^2 + \gamma) \|u\|_{\mathscr{H}}^2 \leq \|v\|_{\mathscr{H}} \|u\|_{\mathscr{H}}.$$

Thus, we have shown that

$$\left(\frac{S_x^T S_x}{N} + \gamma I \right) u = v \implies \|u\|_{\mathscr{H}} \leq \frac{\|v\|_{\mathscr{H}}}{\lambda_x^2 + \gamma} \leq \frac{1}{\gamma} \|v\|_{\mathscr{H}}, \quad \forall v \in \mathscr{H}. \tag{6.82}$$

Now, it comes from (6.81) that

$$\hat{g}_N - \hat{f} = \left(\frac{S_x^T S_x}{N} + \gamma I\right)^{-1} \left(\frac{S_x^T Y}{N} - \frac{S_x^T S_x \hat{f}}{N} - \gamma \hat{f}\right).$$

Exploiting the equalities

$$\frac{S_x^T Y}{N} - \frac{S_x^T S_x \hat{f}}{N} = \frac{1}{N} \sum_{i=1}^{N} \eta_i, \quad \gamma \hat{f} = \mathscr{E} \eta_i,$$

which derive from (6.79) and (6.80), respectively, we obtain

$$\hat{g}_N - \hat{f} = \left(\frac{S_x^T S_x}{N} + \gamma I\right)^{-1} \frac{1}{N} \sum_{i=1}^{N} (\eta_i - \mathscr{E}[\eta_i]).$$

The use of (6.82) then completes the proof.   □

Proof of Statistical Consistency
Let $\hat{f}$ be defined by (6.76), i.e.,

$$\hat{f} = \arg \min_{f \in \mathscr{H}} \|f - f_\rho\|_{\mathscr{L}_2^\mu}^2 + \gamma \|f\|_{\mathscr{H}}^2.$$

Then, consider the following error decomposition

$$\|\hat{g}_N - f_\rho\|_{\mathscr{L}_2^\mu} \leq \|\hat{f} - f_\rho\|_{\mathscr{L}_2^\mu} + \|\hat{g}_N - \hat{f}\|_{\mathscr{L}_2^\mu}. \tag{6.83}$$

The first term $\|\hat{f} - f_\rho\|_{\mathscr{L}_2^\mu}$ on the r.h.s. is not stochastic. The assumption $f_\rho \in \mathscr{H}$ ensures that $\|L_K^{-r} f_\rho\|_{\mathscr{L}_2^\mu} < \infty$ for $0 \leq r \leq 1/2$. It thus comes from Lemma 6.4 that, at least for $0 < r \leq 1/2$, it holds that

$$\|\hat{f} - f_\rho\|_{\mathscr{L}_2^\mu} \leq \gamma^r \|L_K^{-r} f_\rho\|_{\mathscr{L}_2^\mu} < \infty. \tag{6.84}$$

Now, consider the second term $\|\hat{g}_N - \hat{f}\|_{\mathscr{L}_2^\mu}$. Since the input space (the function domain) is compact, and recalling also (6.69), there exists a constant A such that

$$\|\hat{g}_N - \hat{f}\|_{\mathscr{L}_2^\mu} \leq A \|\hat{g}_N - \hat{f}\|_{\mathscr{H}}. \tag{6.85}$$

To obtain a bound for the r.h.s. involving the RKHS norm, consider the stochastic function

$$\eta_i(\cdot) = \left[y_i - \hat{f}(x_i)\right] K(x_i, \cdot),$$

already introduced in (6.79). Using the reproducing property, one has

$$\|\eta_i\|_{\mathcal{H}}^2 = \left[y_i - \hat{f}(x_i)\right]^2 K(x_i, x_i). \tag{6.86}$$

The function $\hat{f}$ belongs to $\mathcal{H}$ and is thus continuous on the compact $\mathcal{X}$. In addition, the kernel K is continuous on the compact $\mathcal{X} \times \mathcal{X}$ and the process $\{x_i, y_i\}$ has finite moments up to the third order by assumption. Hence, there exists a constant B independent of i such that

$$\mathcal{E}\left[\|\eta_i\|_{\mathcal{H}}^k\right] \le B, \quad k = 1, 2, 3. \tag{6.87}$$

We can now come back to $\|\hat{g}_N - \hat{f}\|_{\mathcal{H}}$. From Lemma 6.5, $\forall \gamma > 0$ it holds that

$$\|\hat{g}_N - \hat{f}\|_{\mathcal{H}} \le \frac{1}{\gamma} \left\| \frac{1}{N} \sum_{i=1}^N (\eta_i - \mathcal{E}[\eta_i]) \right\|_{\mathcal{H}}. \tag{6.88}$$

Now, using first Jensen's inequality and then (6.87), (6.88), Assumption 6.20 and (6.73) in Lemma 6.3 (with a and b replaced by $\eta_i - \mathcal{E}[\eta_i]$ and $\eta_j - \mathcal{E}[\eta_j]$) one obtains constants C and D such that

$$\left(\mathcal{E}\left[\left\| \frac{1}{N} \sum_{i=1}^N (\eta_i - \mathcal{E}[\eta_i]) \right\|_{\mathcal{H}} \right] \right)^2 \le \mathcal{E}\left[\left\| \frac{1}{N} \sum_{i=1}^N (\eta_i - \mathcal{E}[\eta_i]) \right\|_{\mathcal{H}}^2 \right]$$

$$\le \frac{15}{N^2} \sum_{i=1}^N \sum_{j=1}^N \sqrt[3]{|\psi_{|i-j|}|} \left(\mathcal{E}[\|\eta - \mathcal{E}[\eta]\|_{\mathcal{H}}^3] \right)^{\frac{2}{3}}$$

$$\le \frac{C}{N} \left(\mathcal{E}[(\|\eta\|_{\mathcal{H}} + \|\mathcal{E}[\eta]\|_{\mathcal{H}})^3] \right)^{\frac{2}{3}} \le \frac{D}{N},$$

where η replaces η_i or η_j when the expectation is independent of i and j. This latter result, combined with (6.85) and (6.88), leads to the existence of a constant E such that

$$\mathcal{E}\|\hat{g}_N - \hat{f}\|_{\mathcal{L}_2^\mu} \le A \mathcal{E}\|\hat{g}_N - \hat{f}\|_{\mathcal{H}} \le \frac{E}{\gamma\sqrt{N}} \tag{6.89}$$

that, combined with (6.83) and (6.84), implies that for any $0 < r \le 1/2$

$$\mathcal{E}\|\hat{g}_N - f_\rho\|_{\mathcal{L}_2^\mu} \le \gamma^r \|L_K^{-r} f_\rho\|_{\mathcal{L}_2^\mu} + \frac{E}{\gamma\sqrt{N}}. \tag{6.90}$$

Hence, when γ is chosen according to (6.56), $\mathcal{E}\|\hat{g}_N - f_\rho\|_{\mathcal{L}_2^\mu}$ converges to zero as N grows to ∞. Using the Markov inequality, (6.57) is finally obtained.

References

1. Abernethy J, Bach F, Evgeniou T, Vert JP (2009) A new approach to collaborative filtering: operator estimation with spectral regularization. J Mach Learn Res 10:803–826
2. Adams RA, Fournier J (2003) Sobolev spaces. Academic Press
3. Ahlberg JH, Nilson EH (1963) Convergence properties of the spline fit. J Soc Indust Appl Math 11:95–104
4. Aizerman A, Braverman EM, Rozoner LI (1964) Theoretical foundations of the potential function method in pattern recognition learning. Autom Remote Control 25:821–837
5. Alon N, Ben-David S, Cesa-Bianchi N, Haussler D (1997) Scale-sensitive dimensions, uniform convergence, and learnability. J ACM 44(4):615–631
6. Alpay D (2003) Reproducing kernel Hilbert spaces and applications. Springer
7. Amit Y, Fink M, Srebro N, Ullman S (2007) Uncovering shared structures in multiclass classification. In: Proceedings of the 24th international conference on machine learning, ICML '07, New York, NY, USA. ACM, pp 17–24
8. Aravkin A, Burke J, Pillonetto G (2012) Nonsmooth regression and state estimation using piecewise quadratic log-concave densities. In: Proceedings of the 51st IEEE conference on decision and control (CDC 2012)
9. Aravkin A, Burke JV, Pillonetto G (2013) Sparse/robust estimation and Kalman smoothing with nonsmooth log-concave densities: modeling, computation, and theory. J Mach Learn Res 14:2689–2728
10. Argyriou A, Dinuzzo F (2014) A unifying view of representer theorems. In: Proceedings of the 31th international conference on machine learning, vol 32, pp 748–756
11. Argyriou A, Micchelli CA, Pontil M (2009) When is there a representer theorem? vector versus matrix regularizers. J Mach Learn Res 10:2507–2529
12. Argyriou A, Micchelli CA, Pontil M (2010) On spectral learning. J Mach Learn Res 11:935–953
13. Aronszajn N (1950) Theory of reproducing kernels. Trans Am Math Soc 68:337–404
14. Atkinson KE (1968) On the order of convergence of natural cubic spline interpolation. SIAM J Numer Anal 5(1):89–101
15. Bach FR, Jordan MI (2005) Predictive low-rank decomposition for kernel methods. In: Proceedings of the 22nd international conference on Machine learning, ICML '05, New York, NY, USA. ACM, pp 33–40
16. Bakker B, Heskes T (2003) Task clustering and gating for Bayesian multitask learning. J Mach Learn Res 4:83–99
17. Bartlett PL, Long PM, Lugosi G, Tsigler A (2020) Benign overfitting in linear regression. PNAS 117:30063–30070
18. Baxter J (1997) A Bayesian/information theoretic model of learning to learn via multiple task sampling. Mach Learn 28:7–39
19. Bennett KP, Parrado-Hernandez E (2006) The interplay of optimization and machine learning research. J Mach Learn Res 7:1265–1281
20. Bergman S (1950) The kernel function and conformal mapping. Mathematical surveys and monographs. AMS
21. Bertero M (1989) Linear inverse and ill-posed problems. Adv Electron Electron Phys 75:1–120
22. Birkhoff G, De Boor C (1964) Error bounds for spline interpolation. J Math Mech 13:827–835
23. Bochner S. Monotone Funktionen, Stieltjessche Integrale, und harmonische Analyse. Math Ann 108:378–410
24. Boser BE, Guyon IM, Vapnik VN (1992) A training algorithm for optimal margin classifiers. In: Proceedings of the 5th annual ACM workshop on computational learning theory. ACM Press, pp 144–152
25. Bottou L, Chapelle O, DeCoste D, Weston J (eds) (2007) Large scale kernel machines. MIT Press, Cambridge, MA, USA

26. Bousquet O, Elisseeff A (2002) Stability and generalization. J Mach Learn Res 2:499–526
27. Boyd S, Vandenberghe L (2004) Convex optimization. Cambridge University Press
28. Carli FP, Chiuso A, Pillonetto G (2012) Efficient algorithms for large scale linear system identification using stable spline estimators. In: IFAC symposium on system identification
29. Caruana R (1997) Multitask learning. Mach Learn 28(1):41–75
30. Collobert R, Bengio S (2001) SVMTorch: support vector machines for large-scale regression problems. J Mach Learn Res 1:143–160
31. Cortes C, Vapnik V (1995) Support-vector networks. Mach Learn 20(3):273–297
32. Craven P, Wahba G (1979) Smoothing noisy data with spline functions. Numer Math 31:377–403
33. Cucker F, Smale S (2001) On the mathematical foundations of learning. Bull Am Math Soc 39:1–49
34. Dehling H, Philipp W (1982) Almost sure invariance principles for weakly dependent vector-valued random variables. Ann Probab 10(3):689–701
35. Dinuzzo F (2011) Analysis of fixed-point and coordinate descent algorithms for regularized kernel methods. IEEE Trans Neural Netw 22(10):1576–1587
36. Dinuzzo F, Scholkopf B (2012) The representer theorem for Hilbert spaces: a necessary and sufficient condition. In: Bartlett P, Pereira FCN, Burges CJC, Bottou L, Weinberger KQ (eds) Advances in neural information processing systems, vol 25, pp 189–196
37. Drucker H, Burges CJC, Kaufman L, Smola A, Vapnik V (1997) Support vector regression machines. In: Advances in neural information processing systems
38. Dudley RM, Giné E, Zinn J (1991) Uniform and universal Glivenko-Cantelli classes. J Theor Probab 4(3):485–510
39. Dudley RM (1984) École d'Été de Probabilités de Saint-Flour XII - 1982, chapter A course on empirical processes. Springer, Berlin, Heidelberg, pp 1–142
40. Evgeniou T, Micchelli CA, Pontil M (2005) Learning multiple tasks with kernel methods. J Mach Learn Res 6:615–637
41. Evgeniou T, Pontil M (1999) On the V_γ dimension for regression in reproducing kernel Hilbert spaces. In: Algorithmic learning theory, 10th international conference, ALT '99, Tokyo, Japan, Dec 1999, Proceedings, lecture notes in artificial intelligence, vol 1720. Springer, pp 106–117
42. Evgeniou T, Pontil M, Poggio T (2000) Regularization networks and support vector machines. Adv Comput Math 13:1–50
43. Ferrari-Trecate G, Williams CKI, Opper M (1999) Finite-dimensional approximation of Gaussian processes. In: Proceedings of the 1998 conference on advances in neural information processing systems. MIT Press, Cambridge, MA, USA, pp 218–224
44. Franz MO, Schölkopf B (2006) A unifying view of Wiener and Volterra theory and polynomial kernel regression. Neural Comput 18:3097–3118
45. Friedman J, Hastie T, Tibshirani R (2010) Regularization paths for generalized linear models via coordinate descent. J Stat Softw 33(1):1–22
46. Girosi F (1998) An equivalence between sparse approximation and support vector machines. Neural Comput 10(6):1455–1480
47. Girosi F, Jones M, Poggio T (1995) Regularization theory and neural networks architectures. Neural Comput 7(2):219–269
48. Guo ZC, Zhou DX (2013) Concentration estimates for learning with unbounded sampling. Adv Comput Math 38(1):207–223
49. Ho CH, Lin CJ (2012) Large-scale linear support vector regression. J Mach Learn Res 13:3323–3348
50. Hochstadt H (1973) Integral equations. Wiley
51. Huber PJ (1981) Robust statistics. Wiley, New York, NY, USA
52. Kimeldorf G, Wahba G (1970) A correspondence between Bayesian estimation on stochastic processes and smoothing by splines. Ann Math Stat 41(2):495–502
53. Kulis B, Sustik M, Dhillon I (2006) Learning low-rank kernel matrices. In: Proceedings of the 23rd international conference on Machine learning, ICML '06, New York, NY, USA. ACM, pp 505–512

54. Lafferty J, Zhu X, Liu Y (2004) Kernel conditional random fields: representation and clique selection. In: Proceedings of the twenty-first international conference on machine learning, ICML '04, New York, NY, USA. ACM
55. Liang T, Rakhlin A (2020) Just interpolate: kernel ridgeless regression can generalize. Ann Stat 48(3):1329–1347
56. Lin CJ (2001) On the convergence of the decomposition method for support vector machines. IEEE Trans Neural Netw 12(12):1288–1298
57. List N, Simon HU (2004) A general convergence theorem for the decomposition method. In: Proceedings of the 17th annual conference on computational learning theory, pp 363–377
58. List N, Simon HU (2007) General polynomial time decomposition algorithms. J Mach Learn Res 8:303–321
59. Megginson RE (1998) An introduction to Banach space theory. Springer
60. Mercer J (1909) Functions of positive and negative type and their connection with the theory of integral equations. Philos Trans R Soc Lond 209(3):415–446
61. Micchelli CA, Xu Y, Zhang H (2006) Universal kernels. J Mach Learn Res 7:2651–2667
62. Micchelli CA, Pontil M (2005) On learning vector-valued functions. Neural Comput 17(1):177–204
63. Mukherjee S, Niyogi P, Poggio T, Rifkin R (2006) Learning theory: stability is sufficient for generalization and necessary and sufficient for consistency of empirical risk minimization. Adv Comput Math 25(1):161–193
64. Nemirovskii A, Nesterov Y (1994) Interior-point polynomial algorithms in convex programming, vol 13. SIAM, Philadelphia, PA, USA
65. Pillonetto G (2008) Solutions of nonlinear control and estimation problems in reproducing kernel hilbert spaces: existence and numerical determination. Automatica 44(8):2135–2141
66. Pillonetto G (2013) Consistent identification of Wiener systems: a machine learning viewpoint. Automatica 49(9):2704–2712
67. Pillonetto G, Bell BM (2007) Bayes and empirical Bayes semi-blind deconvolution using eigenfunctions of a prior covariance. Automatica 43(10):1698–1712
68. Pillonetto G, Dinuzzo F, De Nicolao G (2010) Bayesian on-line multi-task learning of Gaussian processes. IEEE Trans Pattern Anal Mach Intell 32(2):193–205
69. Pillonetto G, Quang MH, Chiuso A (2011) A new kernel-based approach for nonlinear system identification. IEEE Trans Autom Control 56(12):2825–2840
70. Poggio T (1975) On optimal nonlinear associative recall. Biol Cybern 19(4):201–209
71. Poggio T, Girosi F (1990) Networks for approximation and learning. Proc IEEE 78:1481–1497
72. Poggio T, Rifkin R, Mukherjee S, Niyogi P (2004) General conditions for predictivity in learning theory. Nature 428(6981):419–422
73. Pollard D (1989) Asymptotics via empirical processes. J Stat Sci 4(4):341–354
74. Rahimi A, Recht B (2007) Random features for large-scale kernel machines. Advances in neural information processing systems, pp 1177–1184
75. Ribeiro AH, Hendriks J, Wills A, Schön TB (2021) Beyond Occam's razor in system identification: double-descent when modeling dynamics. In: Proceedings of the 19th IFAC symposium on system identification (SYSID), Online, July 2021
76. Riesz F (1909) Sur les operations fonctionnelles lineaires. Comptes rendus de l'Academie des Sciences (in French) 149:974–977
77. Rockafellar RT (1970) Convex analysis. Princeton Landmarks in Mathematics. Princeton University Press
78. Rudi A, Rosasco L (2017) Generalization properties of learning with random features. In: Advances in neural information processing systems, pp 3218–3228
79. Rudin W (1987) Real and complex analysis. McGraw-Hill, Singapore
80. Rudin W (1990) Fourier analysis on groups. Wiley
81. Runge C (1901) Uber empirische funktionen und die interpolation zwischen aquidistanten ordinaten. Zeitschrift für Mathematik und Physik 46:224–243
82. Saitoh S (1988) Theory of reproducing kernels and its applications, vol 189. Pitman research notes in mathematics series. Longman Scientific and Technical, Harlow

83. Schölkopf B, Herbrich R, Smola AJ (2001) A generalized representer theorem. Neural Netw Comput Learn Theory 81:416–426
84. Schölkopf B, Smola AJ (2001) Learning with kernels: support vector machines, regularization, optimization, and beyond. (Adaptive computation and machine learning). MIT Press
85. Scovel C, Hush D, Steinwart I, Theiler J (2010) Radial kernels and their reproducing kernel Hilbert spaces. J Complex 26(6):641–660
86. Shawe-Taylor J, Cristianini N (2004) Kernel methods for pattern analysis. Cambridge University Press
87. Signoretto M, Tran DQ, De Lathauwer L, Suykens JAK (2014) Learning with tensors: a framework based on convex optimization and spectral regularization. Mach Learn 94(3):303–351
88. Smale S, Zhou DX (2007) Learning theory estimates via integral operators and their approximations. Constr Approx 26:153–172
89. Smola A, Schölkopf B (2000) Sparse greedy matrix approximation for machine learning. In: Proceedings of the seventeenth international conference on machine learning, ICML '00, San Francisco, CA, USA. Morgan Kaufmann Publishers Inc., pp 911–918
90. Sriperumbudur BK, Fukumizu K, Lanckriet G (2011) Universality, characteristic kernels and RKHS embedding of measures. J Mach Learn Res 12:2389–2410
91. Steinwart I (2002) On the influence of the kernel on the consistency of support vector machines. J Mach Learn Res 2:67–93
92. Steinwart I, Hush D, Scovel C (2006) An explicit description of the reproducing kernel Hilbert space of Gaussian RBF kernels. IEEE Trans Inf Theory 52:4635–4643
93. Steinwart I, Hush D, Scovel C (2009) Learning from dependent observations. J Multivar Anal 100(1):175–194
94. Sun H (2005) Mercer theorem for RKHS on noncompact sets. J Complex 21(3):337–349
95. Thrun S, Pratt L (1997) Learning to learn. Kluwer
96. Tikhonov AN, Arsenin VY (1977) Solutions of ill-posed problems. Winston/Wiley, Washington, D.C
97. Vapnik V (1982) Estimation of dependences based on empirical data: springer series in statistics (Springer series in statistics). Springer, New York Inc., Secaucus, NJ, USA
98. Vapnik V (1997) The nature of statistical learning theory. Springer
99. Vapnik V, Chervonenkis A (1971) On the uniform convergence of relative frequencies of events to their probabilities. Theory Probab Appl 16(2):264–280
100. Vapnik V, Chervonenkis A (1981) Necessary and sufficient conditions for the uniform convergence of means to their expectations. Theory Probab Appl 26:532–553
101. Vapnik V, Chervonenkis A (1991) The necessary and sufficient conditions for consistency in the empirical risk minimization method. Pattern Recognit Image Anal 1(3):283–305
102. Vapnik V (1998) Statistical learning theory. Wiley, New York, NY, USA
103. De Vito E, Rosasco L, Caponnetto A, Piana M, Verri A (2004) Some properties of regularized kernel methods. J Mach Learn Res 5:1363–1390
104. Wahba G (1990) Spline models for observational data. SIAM, Philadelphia
105. Wang C, Zhou DX (2011) Optimal learning rates for least squares regularized regression with unbounded sampling. J Complex 27(1):55–67
106. Williams CKI, Seeger M (2000) Using the Nyström method to speed up kernel machines. In: Proceedings of the 2000 conference on advances in neural information processing systems, Cambridge, MA, USA. MIT Press, pp 682–688
107. Withers CS (1981) Conditions for linear processes to be strong-mixing. Zeitschrift für Wahrscheinlichkeitstheorie und Verwandte Gebiete 57(4):477–480
108. Wright SJ (1997) Primal-dual interior-point methods. Siam, Englewood Cliffs, N.J., USA
109. Wu Q, Ying Y, Zhou DX (2006) Learning rates of least-square regularized regression. Found Comput Math 6:171–192
110. Yu Y, Cheng H, Schuurmans D, Szepesvari C (2013) Characterizing the representer theorem. In: Proceedings of the 30th international conference on machine learning, pp 570–578

111. Yuan M, Tony Cai T (2010) A reproducing kernel Hilbert space approach to functional linear
     regression. Ann Stat 38:3412–3444
112. Zeidler E (1995) Applied functional analysis. Springer
113. Zhang K, Kwok JT (2010) Clustered Nyström method for large scale manifold learning and
     dimension reduction. IEEE Trans Neural Netw 21(10):1576–1587
114. Zhu H, Rohwer R (1996) Bayesian regression filters and the issue of priors. Neural Comput
     Appl 4:130–142
115. Zhu H, Williams CKI, Rohwer RJ, Morciniec M (1998) Gaussian regression and optimal
     finite dimensional linear models. In: Neural networks and machine learning. Springer, Berlin

Chapter 7
Regularization in Reproducing Kernel Hilbert Spaces for Linear System Identification

Abstract In the previous parts of the book, we have studied how to handle linear system identification by using regularized least squares (ReLS) with finite-dimensional structures given, e.g., by finite impulse response (FIR) models. In this chapter, we cast this approach in the RKHS framework developed in the previous chapter. We show that ReLS with quadratic penalties can be reformulated as a function estimation problem in the finite-dimensional RKHS induced by the regularization matrix. This leads to a new paradigm for linear system identification that provides also new insights and regularization tools to handle infinite-dimensional problems, involving, e.g., IIR and continuous-time models. For all this class of problems, we will see that the representer theorem ensures that the regularized impulse response is a linear and finite combination of basis functions given by the convolution between the system input and the kernel sections. We then consider the issue of kernel estimation and introduce several tuning methods that have close connections with those related to the regularization matrix discussed in Chap. 3. Finally, we introduce the notion of stable kernels, that induce RKHSs containing only absolutely summable impulse responses and study minimax properties of regularized impulse response estimation.

7.1 Regularized Linear System Identification in Reproducing Kernel Hilbert Spaces

7.1.1 Discrete-Time Case

We will consider linear discrete-time systems in the form of the so-called output error (OE) models. Data are generated according to the relationship

$$y(t) = G^0(q)u(t) + e(t), \quad t = 1, \ldots, N, \tag{7.1}$$

© The Author(s) 2022

G. Pillonetto et al., *Regularized System Identification*, Communications and Control Engineering, https://doi.org/10.1007/978-3-030-95860-2_7

where $y(t)$, $u(t)$ and $e(t) \in \mathbb{R}$ are the system output, the known system input and the noise at time instant $t \in \mathbb{N}$, respectively. In addition, $G^0(q)$ is the "true" system that has to be identified from the input–output samples with q being the time shift operator, i.e., $qu(t) = u(t + 1)$. Here, and also in all the remaining parts of the chapter, we assume that e is white noise (all its components are mutually uncorrelated).

In Chap. 2, we have seen that there exist different ways to parametrize $G^0(q)$. In what follows, we will start our discussions exploiting the simplest impulse response descriptions given by FIR models and then we will consider more general infinite-dimensional models also in continuous time. We will see that there is a common way to estimate them through regularization in the RKHS framework and the representer theorem.

7.1.1.1 FIR Case

The FIR case corresponds to

$$y(t) = G(q, \theta)u(t) + e(t)$$
$$= \sum_{k=1}^{m} g_k u(t - k) + e(t), \quad \theta = [g_1, \ldots, g_m]^T, \tag{7.2}$$

where m is the FIR order, $g_1, \ldots, g_m$ are the FIR coefficients and θ is the unknown vector that collects them. Model (7.2) can be rewritten in vector form as follows:

$$Y = \Phi\theta + E, \tag{7.3}$$

where

$$Y = [y(1) \ \ldots \ y(N)]^T, \quad E = [e(1) \ \ldots \ e(N)]^T$$

and

$$\Phi = [\varphi(1) \ \ldots \ \varphi(N)]^T$$

with

$$\varphi^T(t) = [u(t - 1) \ \ldots u(t - m)].$$

Instead of describing FIR model estimation directly in the regularized RKHS framework, let us first recall the ReLS method with quadratic penalty term introduced in Chap. 3. It gives the estimate of θ by solving the following problem:

$$\hat{\theta} = \arg\min \sum_{t=1}^{N}(y(t) - \sum_{k=1}^{m} g_k u(t-k))^2 + \gamma \theta^T P^{-1}\theta \qquad (7.4a)$$

$$= \arg\min_{\theta} \|Y - \Phi\theta\|^2 + \gamma\theta^T P^{-1}\theta \qquad (7.4b)$$

$$= (\Phi^T\Phi + \gamma P^{-1})^{-1}\Phi^T Y \qquad (7.4c)$$

$$= P\Phi^T(\Phi P\Phi^T + \gamma I_N)^{-1}Y, \qquad (7.4d)$$

where the regularization matrix $P \in \mathbb{R}^{m \times m}$ is positive semidefinite, assumed invertible for simplicity. The regularization parameter γ is a positive scalar that, as already seen, has to balance adherence to experimental data and strength of regularization.

Now we show that (7.4) can be reformulated as a function estimation problem with regularization in the RKHS framework. For this aim, we will see that the key is to use the $m \times m$ matrix P to define the kernel over the domain $\{1, 2, \ldots, m\} \times \{1, 2, \ldots, m\}$. This in turn will define a RKHS of functions $g : \{1, 2, \ldots, m\} \to \mathbb{R}$. Such functions are connected with the components g_i of the m-dimensional vector θ by the relation $g(i) = g_i$. So, the functional view is obtained replacing the vector θ with the function that maps i into the ith component of θ.

Let us define a positive semidefinite kernel $K : \mathscr{X} \times \mathscr{X} \to \mathbb{R}$ as follows:

$$K(i, j) = P_{ij}, \quad i, j \in \mathscr{X} = \{1, 2, \ldots, m\}, \qquad (7.5)$$

where P_{ij} is the (i, j)th entry of the regularization matrix P. It is obvious that K is positive semidefinite because P is positive semidefinite. Its kernel sections will be denoted by K_i with $i = 1, \ldots, m$ and are the columns of P seen as functions mapping $\mathscr{X}$ into $\mathbb{R}$.

Now, using the Moore–Aronszajn Theorem, illustrated in Theorem 6.2, the kernel K reported in (7.5) defines a unique RKHS $\mathscr{H}$ such that $\langle K_i, g \rangle_{\mathscr{H}} = g(i), \forall (i, g) \in (\mathscr{X}, \mathscr{H})$. This is the function space where we will search for the estimate of the FIR coefficients. According to the discussion following Theorem 6.2, since there are just m kernel sections K_i associated to the m columns of P, for any impulse response candidate $g \in \mathscr{H}$, there exist m scalars a_j such that

$$g(i) = \sum_{j=1}^{m} a_j K(i, j) = P(i, :)a \qquad (7.6)$$

where $P(i, :)$ is the ith row of P. Since $g(i)$ is the ith component of θ, one has

$$\theta = Pa.$$

By the reproducing property, we also have

$$\|g\|_{\mathcal{H}}^2 = \langle \sum_{j=1}^m a_j K_j, \sum_{l=1}^m a_l K_l \rangle_{\mathcal{H}} = \sum_{j=1}^m \sum_{l=1}^m a_j a_l K(j, l)$$

$$= \sum_{j=1}^m \sum_{l=1}^m a_j a_l P_{jl} = a^T P a$$

and this implies

$$\|g\|_{\mathcal{H}}^2 = \theta^T P^{-1} \theta.$$

As a result, the ReLS method (7.4) can be reformulated as follows:

$$\hat{g} = \arg\min_{g \in \mathcal{H}} \sum_{t=1}^N (y(t) - \sum_{k=1}^m g(k)u(t-k))^2 + \gamma \|g\|_{\mathcal{H}}^2 \qquad (7.7)$$

which is a regularized function estimation problem in the RKHS $\mathcal{H}$.

In view of the equivalence between (7.4) and (7.7), the FIR function estimate $\hat{g}$ has the closed-form expression given by (7.4d). The correspondence is established by $\hat{g}(i) = \hat{\theta}_i$. We will show later that such closed-form expression can be derived/interpreted by exploiting the representer theorem.

Remark 7.1 ⋆ Besides (7.7), there is also an alternative way to reformulate the ReLS method (7.4) as a function estimation problem with regularization in the RKHS framework. This has been sketched in the discussions on linear kernels in Sect. 6.6.1. The difference lies in the choice of the function to be estimated and the choice of the corresponding kernel. In particular, in this chapter, we have obtained (7.7) choosing the function and the corresponding kernel to be the FIR g and (7.5), respectively. In contrast, in Sect. 6.6.1, the RKHS is defined by the kernel

$$K(x, y) = x^T P y, \quad x, y \in \mathcal{X} = \mathbb{R}^m \qquad (7.8)$$

and contains the linear functions $x^T \theta$, where the input locations x incapsulate m past input values. So, using (7.8), the corresponding RKHS does not contain impulse responses but functions that represent directly linear systems mapping regressors (built with input values) into outputs.

7.1.1.2   IIR Case

The infinite impulse response (IIR) case corresponds to

$$y(t) = G(q, \theta)u(t) + e(t) = \sum_{k=1}^{\infty} g_k u(t-k) + e(t), \quad t = 1, \ldots, N \qquad (7.9)$$

where $\theta = [g_1, \ldots, g_\infty]^T$. So, model order m is set to ∞ and we have to handle infinite-dimensional objects. To face the intrinsic ill-posedness of the estimation problem, one could think to introduce an infinite-dimensional regularization matrix P. But the penalty $\theta^T P^{-1}\theta$, adopted in (7.4) for the FIR case, would turn out to be undefined. So, the RKHS setting is needed to define regularized IIR estimates. The first step is to choose a positive semidefinite kernel $K : \mathbb{N} \times \mathbb{N} \to \mathbb{R}$. Then, let $\mathcal{H}$ be the RKHS associated with K and $g \in \mathcal{H}$ be the IIR function with $g(k) = g_k$ for $k \in \mathbb{N}$. Finally, the estimate is given by

$$\hat{g} = \arg\min_{g \in \mathcal{H}} \sum_{t=1}^{N} (y(t) - \sum_{k=1}^{\infty} g(k)u(t-k))^2 + \gamma \|g\|_{\mathcal{H}}^2. \qquad (7.10)$$

One may wonder whether it is possible to obtain a closed-form expression of the IIR estimate $\hat{g}$ as in the FIR case. The answer is positive and given by the following representer theorem. It derives from Theorem 6.16 reported in the previous chapter applied to the case of quadratic loss functions, as discussed in Example 6.17, that allows to recover the expansion coefficients of the estimate just solving a linear system of equations, see (6.29) and (6.31). Before stating in a formal way the result, it is useful to point out the following two facts:

- in the dynamic systems context treated in this chapter any functional L_i present in Theorem 6.16 is now applied to discrete-time impulse responses g which lives in the RKHS $\mathcal{H}$. Hence, it represents the discrete-time convolution with the input, i.e., L_i maps $g \in \mathcal{H}$ into the system output evaluated at the time instant $t - i$;
- from the discussion after Theorem 6.16, recall also that a linear functional L is linear and bounded in $\mathcal{H}$ if and only if the function f, defined for any x by $f(x) = L[K(x, \cdot)]$, belongs to $\mathcal{H}$. Hence, the condition (7.11) reported below is equivalent to assume that the system input defines linear and bounded functionals over the RKHS induced by K.

Theorem 7.1 (Representer theorem for discrete-time linear system identification, based on [73, 90]). *Consider the function estimation problem (7.10). Assume that $\mathcal{H}$ is the RKHS induced by a positive semidefinite kernel $K : \mathbb{N} \times \mathbb{N} \to \mathbb{R}$ and that, for $t = 1, \ldots, N$, the functions η_t defined by*

$$\eta_t(i) = \sum_{k=1}^{\infty} K(i, k)u(t-k), \quad i \in \mathbb{N} \qquad (7.11)$$

are all well defined in $\mathcal{H}$. Then, the solution of (7.10) is

$$\hat{g}(i) = \sum_{t=1}^{N} \hat{c}_t \eta_t(i), \quad i \in \mathbb{N}, \qquad (7.12)$$

where $\hat{c}_t$ is the tth entry of the vector

$$\hat{c} = (O + \gamma I_N)^{-1} Y \tag{7.13}$$

with $Y = [y(1), \dots y(N)]^T$ and with the (t, s)th entry of O given by

$$O_{ts} = \sum_{i=1}^{\infty} \sum_{k=1}^{\infty} K(i, k) u(t - k) u(s - i), \quad t, s = 1, \dots, N. \tag{7.14}$$

Theorem 7.1 discloses an important feature of regularized impulse response estimation in RKHS. The function estimate $\hat{g}$ has a finite dimensional representation that does not depend on the dimension of the RKHS $\mathcal{H}$ induced by the kernel but only on the data set size N.

Example 7.2 (*Stable spline kernel for IIR estimation*) To estimate high-order FIR models, in the previous chapters, we have introduced some regularization matrices related to the DC, TC and stable spline kernels, see (5.40) and (5.41). Consider now the TC kernel, also called first-order stable spline, with support extended to $\mathbb{N} \times \mathbb{N}$, i.e.,

$$K(i, j) = \alpha^{\max(i, j)}, \quad 0 < \alpha < 1, \quad (i, j) \in \mathbb{N}. \tag{7.15}$$

This kernel induces a RKHS that contains IIR models and can be conveniently adopted in the estimator (7.10). An interesting question is to derive the structure of the induced regularizer $\|g\|_{\mathcal{H}}^2$. One could connect K with the matrix P entering (7.4a) but its inverse is undefined since now P is infinite dimensional. To derive the stable spline norm, it is instead necessary to resort to functional analysis arguments. In particular, in Sect. 7.7.1, it is proved that

$$\|g\|_{\mathcal{H}}^2 = \sum_{t=1}^{\infty} \frac{(g_{t+1} - g_t)^2}{(1 - \alpha)\alpha^t}, \tag{7.16}$$

an expression that well reveals how the kernel (7.15) includes information on smooth exponential decay. When used in (7.10), the resulting IIR estimate balances the data fit (sum of squared residuals) and the energy of the impulse response increments weighted by coefficients that increase exponentially with time t and thus enforce stability.

Let us now consider a simple application of the representer theorem. Assume that the system input is a causal step of unit amplitude, i.e., $u(t) = 1$ for $t \geq 0$ and $u(t) = 0$ otherwise. The functions (7.11) are given by

$$\eta_t(i) = \sum_{k=1}^{\infty} K(i, k) u(t - k), \quad i \in \mathbb{N}.$$

For instance, the first three basis functions are

$$\eta_1(i) = \sum_{k=1}^{\infty} K(i,k)u(1-k) = \alpha^{\max(i,1)}$$

$$\eta_2(i) = \sum_{k=1}^{\infty} K(i,k)u(2-k) = \alpha^{\max(i,1)} + \alpha^{\max(i,2)}$$

$$\eta_3(i) = \sum_{k=1}^{\infty} K(i,k)u(3-k) = \alpha^{\max(i,1)} + \alpha^{\max(i,2)} + \alpha^{\max(i,3)}$$

and, in general, one has

$$\eta_t(i) = \sum_{k=1}^{t} \alpha^{\max(i,k)}.$$

Hence, any η_t is a well-defined function in the RKHS induced by K, being the sum of the first t kernel sections. Then, according to Theorem 7.1, we conclude that the IIR estimate returned by (7.10) is spanned by the functions $\{\eta_t\}_{t=1}^{N}$ with coefficients then computable from (7.13). $\qquad\qquad\Box$

Although Theorem 7.1 is stated for the IIR case (7.10), the same result also holds for the FIR case (7.7). The only difference is that the series in (7.11) and (7.14) have to be replaced by finite sums up to the FIR order m. Then, interestingly, one can interpret the regularized FIR estimate (7.4d) in a different way exploiting the representer theorem perspective. In particular, one finds $O = \Phi P \Phi^T$ while the basis functions $\{\eta_t\}_{t=1}^{N}$ are in one-to-one correspondence with the N columns of $P\Phi^T$, each of dimension m.

7.1.2 Continuous-Time Case

Now, we consider linear continuous-time systems still focusing on the output error (OE) model structure. The system outputs are collected over N time instants t_i. Hence, the measurements model is

$$y(t_i) = \int_0^{\infty} g^0(\tau)u(t_i - \tau)d\tau + e(t_i), \quad i = 1, \ldots, N, \tag{7.17}$$

where $y(t)$, $u(t)$ and $e(t)$ are the system output, the known input and the noise at time instant $t \in \mathbb{R}^+$, respectively, while $g^0(t)$, $t \in \mathbb{R}^+$ is the "true" system impulse response.

Similarly to what done in the previous section, we will study how to determine from a finite set of input–output data a regularized estimate of the impulse response g^0 in the RKHS framework. The first step is to choose a positive semidefinite kernel K : $\mathbb{R}^+ \times \mathbb{R}^+ \to \mathbb{R}$. It induces the RKHS $\mathscr{H}$ containing the impulse response candidates $g \in \mathscr{H}$. Then, the linear model can be estimated by solving the following function estimation problem:

$$\hat{g} = \underset{g \in \mathcal{H}}{\arg\min} \sum_{i=1}^{N} \left(y(t_i) - \int_0^{\infty} g(\tau) u(t_i - \tau) d\tau \right)^2 + \gamma \|g\|_{\mathcal{H}}^2. \qquad (7.18)$$

The closed-form expression of the impulse response estimate $\hat{g}$ is given by the following representer theorem that again derives from Theorem 6.16 and the same discussion reported before Theorem 7.1. Note just that now any functional L_i entering Theorem 6.16 is applied to continuous-time impulse responses g in the RKHS $\mathcal{H}$. Hence, it represents the continuous-time convolution with the input, i.e., L_i maps $g \in \mathcal{H}$ into the system output evaluated at the time instant t_i.

Theorem 7.3 (Representer theorem for continuous-time linear system identification, based on [73, 90]) *Consider the function estimation problem* (7.18). *Assume that $\mathcal{H}$ is the RKHS induced by a positive semidefinite kernel $K : \mathbb{R}^+ \times \mathbb{R}^+ \to \mathbb{R}$ and that, for $i = 1, \ldots, N$, the functions η_i defined by*

$$\eta_i(s) = \int_0^{\infty} K(s, \tau) u(t_i - \tau) d\tau, \quad s \in \mathbb{R}^+ \qquad (7.19)$$

are all well defined in $\mathcal{H}$. Then, the solution of (7.18) *is*

$$\hat{g}(s) = \sum_{i=1}^{N} \hat{c}_i \eta_i(s), \quad s \in \mathbb{R}^+ \qquad (7.20)$$

where $\hat{c}_i$ is the ith entry of the vector

$$\hat{c} = (O + \gamma I_N)^{-1} Y \qquad (7.21)$$

with $Y = [y(t_1), \ldots y(t_N)]^T$ and the (i, j)th entry of O given by

$$O_{ij} = \int_0^{\infty} \int_0^{\infty} K(\tau, s) u(t_i - s) u(t_j - \tau) ds d\tau, \quad i, j = 1, \ldots, N. \qquad (7.22)$$

Example 7.4 (*Stable spline kernel for continuous-time system identification*) In Example 6.5, we introduced the first-order spline kernel $\min(x, y)$ on $[0, 1] \times [0, 1]$. It describes a RKHS of continuous functions f on the unit interval that satisfy $f(0) = 0$ whose squared norm is the energy of the first-order derivative, i.e.,

$$\int_0^1 \left(\dot{f}(x) \right)^2 dx. \qquad (7.23)$$

To describe stable impulse responses g, we instead need a kernel defined over the positive real axis $\mathbb{R}^+$ that induces the constraint $g(+\infty) = 0$. A simple way to obtain this is to exploit the composition of the spline kernel with an *exponential* change of coordinates mapping $\mathbb{R}^+$ into $[0, 1]$. The resulting kernel is called (continuous-time)

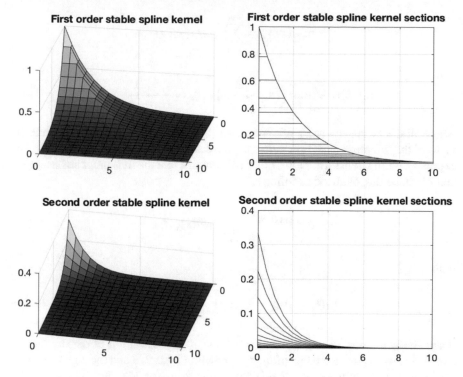

Fig. 7.1 First-order (top left) and second-order (bottom left) stable spline kernel with some kernel sections (right panels) obtained with $\beta = 0.5$ and centred on $0, 0.5, 1, \ldots, 10$ (bottom)

first-order stable spline kernel. It is given by

$$K(s, t) = \min(e^{-\beta s}, e^{-\beta t}) = e^{-\beta \max(s,t)}, \quad s, t \in \mathbb{R}^+, \quad (7.24)$$

where $\beta > 0$ regulates the change of coordinates and, hence, the impulse responses decay rate. So, β can be seen as a kernel parameter related to the dominant pole of the system.

It is interesting to note the similarity between the kernel (7.15) and the first-order stable spline kernel (7.24). By letting $\alpha = \exp(-\beta)$, the sampled version of the first-order stable spline kernel (7.24) corresponds exactly to the TC kernel (7.15). Top panel of Fig. 7.1 plots (7.24) and also some kernel sections: they are all continuous and exponentially decaying to zero. Such kernel inherits also the universality property of the splines. In fact, its kernel sections can approximate any continuous impulse response on all the compact subsets of $\mathbb{R}^+$.

The relationship with splines permits also to easily achieve one spectral decomposition of (7.24). In particular, in Example 6.11, we obtained the following expansion of the spline kernel:

$$\min(x, y) = \sum_{i=1}^{+\infty} \zeta_i \rho_i(x) \rho_i(y)$$

with

$$\rho_i(x) = \sqrt{2} \sin\left(i\pi x - \frac{\pi x}{2}\right), \quad \zeta_i = \frac{1}{(i\pi - \pi/2)^2},$$

where all the ρ_i are mutually orthogonal on $[0, 1]$ w.r.t. the Lebesque measure. In view of the simple connection between spline and stable spline kernels given by exponential time transformations, one easily obtains that the first-order stable spline kernel can be diagonalized as follows:

$$e^{-\beta \max(s,t)} = \sum_{i=1}^{\infty} \zeta_i \phi_i(s) \phi_i(t) \tag{7.25}$$

with

$$\phi_i(t) = \rho_i(e^{-\beta t}), \quad \zeta_i = \frac{1}{(i\pi - \pi/2)^2}, \tag{7.26}$$

where the ϕ_i are now orthogonal on $[0, +\infty)$ w.r.t. the measure μ of density $\beta e^{-\beta t}$. In Fig. 6.3, we reported the eigenfunctions ρ_i with $i = 1, 2, 8$ and the eigenvalues ζ_i for the first-order spline kernel (6.47). For comparison, we now show in Fig. 7.2 the corresponding eigenfunctions ϕ_i of the first-order stable spline kernel (7.24) with $\beta = 1$ and also the ζ_i. While the eigenvalues are the same, differently from the ρ_i the eigenfunctions ϕ_i now decay exponentially to zero.

Having obtained one spectral decomposition of (7.24), we can now exploit Theorem 6.10 to obtain the following representation of the RKHS induced by the first-order stable spline kernel:

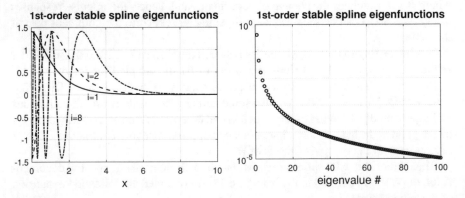

Fig. 7.2 Expansion of the continuous-time first-order stable spline kernel $e^{-\beta \max(x, y)}$ with $\beta = 1$: eigenfunctions $\rho_i(x)$ for $i = 1, 2, 8$ (left panel) and eigenvalues ζ_i (right)

$$\mathcal{H} = \left\{ g \mid g(t) = \sum_{i=1}^{\infty} c_i \phi_i(t), \; t \geq 0, \; \sum_{i=1}^{\infty} \frac{c_i^2}{\zeta_i} < \infty \right\}, \tag{7.27}$$

and the squared norm of g turns out to be

$$\|g\|_{\mathcal{H}}^2 = \sum_{i=1}^{\infty} \frac{c_i^2}{\zeta_i}. \tag{7.28}$$

Now we will exploit the above results to obtain a more useful expression for $\|g\|_{\mathcal{H}}^2$. The deep connection between spline and stable spline kernel implies that these two spaces are isometrically isomorphic, i.e., there is an one-to-one correspondence that preserves inner products. In fact, we can associate to any stable spline function $g(t)$ in $\mathcal{H}$ the spline function $f(t)$ in the space induced by (6.47) such that $g(t) = f(e^{-\beta t})$. So, $g(t) = \sum_{i=1}^{\infty} c_i \phi_i(t)$ implies $f(t) = \sum_{i=1}^{\infty} c_i \rho_i(t)$ and the two functions have indeed the same norm $\sum_{i=1}^{\infty} \frac{c_i^2}{\zeta_i}$. Now, using (7.23) and (7.28), we obtain

$$\|g\|_{\mathcal{H}}^2 = \int_0^1 \left(\dot{f}(t) \right)^2 dt = \int_0^{+\infty} (\dot{g}(t))^2 \frac{e^{\beta t}}{\beta} dt. \tag{7.29}$$

This expression gives insights into the nature of the stable spline space. Compared to the classical Sobolev space induced by the first-order spline kernel, the norm penalizes the energy of the first-order derivative of g with a weight proportional to $e^{\beta t}$. Such norm thus enforces all the function in $\mathcal{H}$ to be continuous impulse responses decaying to zero at least exponentially. Note also that (7.29) really seems the continuous-time counterpart of the norm (7.16) associated to the discrete-time stable spline kernel.

Let us see now how to generalize the kernel (7.24). In Sect. 6.6.6 of the previous chapter, we have introduced the general class of spline kernels. Here, we started our discussion using the first-order (linear) spline kernel $\min(x, y)$ but we have seen that higher-order models can be useful to reconstruct smoother functions, an important example being the second-order (cubic) spline kernel (6.48). Applying exponential time transformations to the splines, the class of the so-called *stable spline kernels* is obtained. For instance, from (6.48), one obtains the second-order stable spline kernel

$$\frac{e^{-\beta(s+t+\max(s,t))}}{2} - \frac{e^{-3\beta \max(s,t)}}{6}. \tag{7.30}$$

The bottom panels of Fig. 7.1 plots (7.30) and also some kernel sections: they exponentially decay to zero and are more regular than those associated to (7.24). □

7.1.3 More General Use of the Representer Theorem for Linear System Identification ⋆

Theorems 7.1 and 7.3 are special cases of the more general representer theorem involving function estimation from sparse and noisy data. It was reported as Theorem 6.16 in the previous chapter. Let us briefly recall it. Its starting point was the optimization problem

$$\hat{g} = \arg\min_{g \in \mathcal{H}} \sum_{i=1}^{N} \mathcal{V}_i(y_i, L_i[g]) + \gamma \|g\|_{\mathcal{H}}^2, \tag{7.31}$$

where $\mathcal{V}_i$ is a loss function, e.g., the quadratic loss adopted in this chapter, and each functional $L_i : \mathcal{H} \to \mathbb{R}$ is linear and bounded. Then, all the solutions of (7.31) are given by

$$\hat{g} = \sum_{i=1}^{N} c_i \eta_i, \tag{7.32}$$

where each $\eta_i \in \mathcal{H}$ is the representer of L_i given by

$$\eta_i(t) = L_i[K(\cdot, t)]. \tag{7.33}$$

How to compute the expansion coefficients c_i will then depend on the nature of the $\mathcal{V}_i$, as described in Sect. 6.5.

The estimator (7.31) can be exploited for linear system identification thinking of g as an impulse response, using e.g., a stable spline kernel to define $\mathcal{H}$. The linear functional L_i is then defined by a convolution and returns the system noiseless outputs at instant t_i. In particular, in discrete-time one has

$$L_i[g] = \sum_{k=1}^{\infty} g(k)u(t_i - k), \quad t_i = 1, \dots, N \tag{7.34}$$

while in continuous time, it holds that

$$L_i[g] = \int_0^{\infty} g(\tau)u(t_i - \tau)d\tau. \tag{7.35}$$

When quadratic losses are used, (7.31) becomes the regularization network described in Sect. 6.5.1 whose expansions coefficients are available in closed form. One has $\hat{c} = (O + \gamma I_N)^{-1} Y$ with the (t, s)-entry of the matrix O given by $O_{ts} = L_s[L_t[K]]$, as given by (7.14) in discrete time and by (7.22) in continuous time. The use of losses $\mathcal{V}_i$ different from quadratic then opens the way also to the definition of many new algorithms for impulse response estimation. For example, the use of the Vapnik's ϵ-insensitive loss described in Sect. 6.5.3 leads to support vector regression

Fig. 7.3 Discrete-time
Laguerre functions of order
$j = 1, 2, 8$ obtained with
$\alpha = 0.99$ (samples are
linearly interpolated)

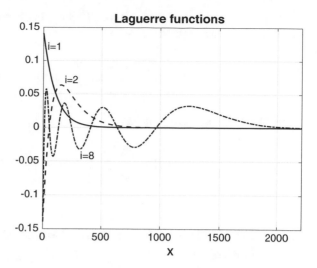

for linear system identification. Beyond promoting sparsity in the coefficients c_i, it also makes the estimator robust against outliers since penalties on large residuals grows linearly. Outliers can be tackled also by adopting the ℓ_1 or Huber loss, see Sect. 6.5.2. A general system identification framework that includes all the convex piecewise linear quadratic losses and penalties is, e.g., described in [2].

Interestingly, the estimator (7.31) can be conveniently adopted for linear system identification also giving g a different meaning from an impulse response. For instance, in system identification there are important IIR models that use Laguerre functions see e.g., [91, 92] whose z-transform is

$$\frac{\sqrt{1-\alpha^2}}{z-\alpha}\left(\frac{1-\alpha z}{z-\alpha}\right)^{j-1}, \quad j = 1, 2, \dots .$$

They form an orthonormal basis in ℓ_2 and some of them are displayed in Fig. 7.3.

Another option is given by the Kautz basis functions that allow also to include information on the presence of system resonances [46]. Using ϕ_i to denote such basis functions, the impulse response model can be written as

$$f(t) = \sum_{i=1}^{\infty} g_i \phi_i(t).$$

A problem is how to determine the coefficients g_i from data. Classical approaches use truncated expansions $f = \sum_{i=1}^{d} g_i \phi_i$, with model order d estimated using, e.g., Akaike's criterion, as discussed in Sect. 2.4.3, and then determine the g_i by least squares. An interesting alternative is to let $d = +\infty$ and to think that the g_i define the function g such that $g(i) = g_i$. One can then estimate the coefficients through (7.31) adopting a kernel, like TC and stable spline, that includes information on the expansion coefficients' decay to zero. Working in discrete time, the functionals L_i entering (7.31) are in this case defined by

$$L_i[g] = \sum_{j=1}^{\infty} g_j \sum_{k=1}^{\infty} \phi_j(k) u(t_i - k),$$

while in continuous time, one has

$$L_i[g] = \sum_{j=1}^{\infty} g_j \int_0^{\infty} \phi_j(\tau) u(t_i - \tau) d\tau.$$

7.1.4 Connection with Bayesian Estimation of Gaussian Processes

Similarly to what discussed in the finite-dimensional setting in Sect. 4.9, also the more general regularization in RKHS can be given a probabilistic interpretation in terms of Bayesian estimation. In this paradigm, the different loss functions correspond to alternative statistical models for the observation noise, while the kernel represents the covariance of the unknown random signal, assumed independent of the noise. In particular, when the loss is quadratic, all the involved distributions are Gaussian.

We now discuss the connection under the linear system identification perspective where the "true" impulse response g^0 is seen as the random signal to estimate. Consider the measurements model

$$y(t_i) = L_i[g^0] + e(t_i), \quad i = 1, \dots, N, \tag{7.36}$$

where L_i is a linear functional of the true impulse response g^0 defined by convolution with the system input evaluated at t_i. One has

$$L_i[g^0] = \sum_{k=1}^{\infty} g^0(k) u(t_i - k)$$

in discrete time and

$$L_i[g^0] = \int_0^{\infty} g^0(\tau) u(t_i - \tau) d\tau$$

in continuous time. So, the impulse response estimators discussed in this chapter can be compactly written as

$$\hat{g} = \arg \min_{g \in \mathcal{H}} \sum_{i=1}^{N} (y(t_i) - L_i[g])^2 + \gamma \|g\|_{\mathcal{H}}^2, \tag{7.37}$$

where the RKHS $\mathscr{H}$ contains functions $g : \mathscr{X} \to \mathbb{R}$ with $\mathscr{X} = \mathbb{N}$ in discrete time and $\mathscr{X} = \mathbb{R}^+$ in continuous time.

The following result (whose simple proof is in Sect. 7.7.2) shows that, under Gaussian assumptions on the impulse response and the noise, (7.37) provides the minimum variance estimate of g^0 given the measurements $Y = [y(t_1), \ldots, y(t_N)]^T$.

Proposition 7.1 *Let the following assumptions hold:*

- *the impulse response g^0 is a zero-mean Gaussian process on $\mathscr{X}$. Its covariance function is defined by*

$$\mathscr{E}(g^0(t)g^0(s)) = \lambda K(t, s),$$

  where λ is a positive scalar and K is a kernel;
- *the $e(t)$ are mutually independent zero-mean Gaussian random variables with variance σ^2. Moreover, they are independent of g^0.*

Let $\mathscr{H}$ be the RKHS induced by K, set $\gamma = \sigma^2/\lambda$ and define

$$\hat{g} = \arg\min_{g \in \mathscr{H}} \left(\sum_{i=1}^{N} (y(t_i) - L_i[g])^2 + \gamma \|g\|_{\mathscr{H}}^2 \right).$$

Then, $\hat{g}$ is the minimum variance estimator of g^0 given Y, i.e.,

$$\mathscr{E}[g^0(t)|Y] = \hat{g}(t) \quad \forall t \in \mathscr{X}.$$

Remark 7.2 The connection between regularization in RKHS and estimation of Gaussian processes was first pointed out in [51] in the context of spline regression, using quadratic losses, see also [41, 83, 90]. The connection also holds for a wide class of losses $\mathscr{V}_i$ also different from quadratic. For instance, in this statistical framework, using the absolute value loss corresponds to Laplacian noise assumptions. The statistical interpretation of an ϵ-insensitive loss in terms of Gaussians with mean and variance given by suitable random variables can be found in [79], see also [40, 67]. For all this kind of noise models, and many others, it can be shown that the RKHS estimate $\hat{g}$ includes all the possible finite-dimensional maximum a posteriori estimates of g^0, see [3] for details.

Remark 7.3 The relation between RKHSs and Gaussian stochastic processes, or more general Gaussian random fields, is stated by Proposition 7.1 in terms of minimum variance estimators. In particular, since the representer theorem ensures that such estimator is sum of a finite number of basis functions belonging to $\mathscr{H}$, it turns out that $\hat{g}$ belongs to the RKHS induced by the covariance of g^0 with probability one. Now, one may also wonder what happens a priori, before seeing the data. In other words, the question is whether realizations of a zero-mean Gaussian process of covariance K fall in the RKHS induced by K. If the kernel K is associated with an infinite-dimensional $\mathscr{H}$, the answer is negative with probability one, as graphically

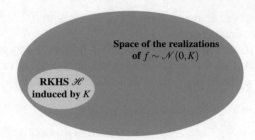

Fig. 7.4 The largest space contains all the realizations of a zero-mean Gaussian process of covariance K. The smallest space is the RKHS $\mathscr{H}$ induced by K, assumed here infinite dimensional. The probability that realizations of f fall in the RKHS is zero. Instead, when the assumptions underlying the representer theorem hold, the realizations of the minimum variance estimator $\mathscr{E}[f|Y]$ are contained in $\mathscr{H}$ with probability one

illustrated in Fig. 7.4. While deep discussions can be found in [9, 34, 59, 68], here we give just a hint on this fact. Assume that the kernel admits the decomposition

$$K(s, t) = \sum_{i=1}^{M} \zeta_i \phi_i(s) \phi_i(t)$$

inducing an M-dimensional RKHS $\mathscr{H}$. Let the deterministic functions ϕ_i be independent. Then, we know from Theorem 6.13 that, if $f(t) = \sum_{i=1}^{M} a_i \phi_i(t)$, then

$$\|f\|_{\mathscr{H}}^2 = \sum_{i=1}^{M} \frac{a_i^2}{\zeta_i}.$$

Now, think of K as a covariance and let a_i be zero-mean Gaussian and independent random variables of variance ζ_i, i.e.,

$$a_i \sim \mathcal{N}(0, \zeta_i).$$

Then, the so-called Karhunen–Loève expansion of the Gaussian random field $f \sim \mathcal{N}(0, K)$, also discussed in Sect. 5.6 to connect regularization and basis expansion in finite dimension, is given by

$$f(t) = \sum_{i=1}^{M} a_i \phi_i(t)$$

with M possibly infinite and convergence in quadratic mean. The RKHS norm of f is now a random variable and, since the a_i are mutually independent with $\mathscr{E} a_i^2 = \zeta_i$, one has

$$\mathscr{E}\|f\|_{\mathscr{H}}^2 = \mathscr{E}\sum_{i=1}^{M}\frac{a_i^2}{\zeta_i} = \sum_{i=1}^{M}\frac{\mathscr{E}a_i^2}{\zeta_i} = M.$$

So, if the RKHS is infinite dimensional, one has $M = \infty$ and the expected (squared) RKHS norm of the process f diverges to infinity.

7.1.5   A Numerical Example

Our goal now is to illustrate the influence of the choice of the kernel on the quality of the impulse response estimate using also the Bayesian interpretation of regularization. The example is a simple linear discrete time system in the form of (7.1). Using the z-transform, its transfer function is

$$y(t) = \frac{1}{z(z - 0.85)}u(t) + e(t), \quad t = 1, \ldots, 20. \tag{7.38}$$

The system's impulse response is reported in Fig. 7.5. The disturbances $e(t)$ are independent and Gaussian random variables with mean zero and variance 0.05^2. For ease of visualization, we let the input $u(t)$ be an impulsive signal, i.e., $u(0) = 1$ and $u(t) = 0$ elsewhere. Thus, the impulse response have to be estimated from 20 direct and noisy impulse response measurements.

We consider a Monte Carlo simulation of 200 runs. At any run, the outputs are obtained by generating mutually independent measurement noises. One data set is shown in Fig. 7.5. For each of the 200 data sets, we use the regularized IIR estimator (7.10). For what regards $K : \mathbb{N} \times \mathbb{N} \to \mathbb{R}$,, we will compare the performance of three kernels: the Gaussian (6.43), the cubic spline (6.48) and the stable spline (7.15) defined, respectively, by

$$\exp\left(-\frac{(i-j)^2}{\rho}\right), \quad \frac{ij\min\{i,j\}}{2} - \frac{(\min\{i,j\})^3}{6}, \quad \alpha^{\max(i,j)}.$$

Recall that the Gaussian and the cubic spline kernel are the most used in machine learning to include information on smoothness. The cubic spline estimator could be also complemented with a bias space given, e.g., by a linear function, as described in Sect. 6.6.7. However, one would obtain results very similar to those described in what follows.

To adopt the estimator (7.10), we need to find a suitable value for the regularization parameter γ and also for the unknown kernel parameters, i.e., the kernel width ρ in the Gaussian kernel and the stability parameter α for stable spline. As already done, e.g., in Sect. 1.2 for ridge regression, an oracle-based procedure is adopted to optimally balance bias and variance. The unknown parameters are achieved by maximizing the measure of fit defined as follows:

Fig. 7.5 The true impulse
response (thick line) and one
out of the 200 data sets (○)

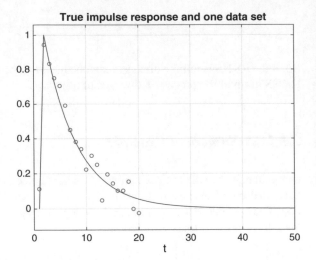

$$100\left(1-\left[\frac{\sum_{k=1}^{50}|g_k^0-\hat{g}(k)|^2}{\sum_{k=1}^{50}|g_k^0-\bar{g}^0|^2}\right]^{\frac{1}{2}}\right), \quad \bar{g}^0=\frac{1}{50}\sum_{k=1}^{50}g_k^0, \quad (7.39)$$

where computation is restricted only to the first 50 samples where, in practice, the
impulse response is different from zero. This tuning procedure is ideal since it exploits
the true function g^0. It is useful here since it excludes the uncertainty brought by the
kernel tuning procedure and will fully reveal the influence of the kernel choice on
the quality of the impulse response estimate.

The impulse response estimates obtained by the cubic spline, the Gaussian and
the stable spline kernel are reported in Fig. 7.6. When the cubic spline kernel (6.48) is
chosen, the impulse response estimates diverge as time goes. This result can be also
given a Bayesian interpretation where (6.48) becomes the covariance of the stochastic
process g^0. Specifically, the cubic spline kernel models the impulse response as
double integration of white noise. So, impulse responses coefficients are correlated
but the prior variance increases in time. For stable systems, variability is instead
expected to decay to zero as t progresses. When the Gaussian kernel (6.43) is chosen,
quality of the impulse response estimates much improves, but many of them exhibit
oscillations and the variance of the impulse response estimator is still large. Bayesian
arguments here show that the Gaussian kernel models g^0 as a stationary stochastic
process. Smoothness information is encoded but not the fact that that one expects
the prior variance to decay to zero. Finally, the impulse response estimates returned
by the stable spline kernel (7.15) are all very close to the truth. These outcomes
are similar to those described, e.g., in Example 5.4 in Sect. 5.5. In particular, even
if this example is rather simple, it shows clearly that a straightforward application
of standard kernels from machine learning and smoothing splines literature may
give unsatisfactory results. Inclusion of dynamic systems features in the regularizer,

Fig. 7.6 True impulse response (thick line) and 200 impulse response estimates obtained using the cubic spline kernel (6.48) (top panel), the Gaussian kernel (6.43) (middle) and the stable spline kernel (bottom). The unknown parameters are estimated by an oracle that maximizes the fit (7.39) for each data set

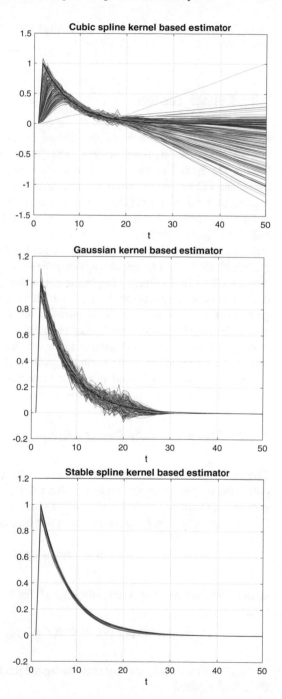

like smooth exponential decay, greatly enhances the quality of the impulse response estimates.

7.2 Kernel Tuning

As we have seen in the previous parts of the book, the kernels depend on some unknown parameters, the so-called hyperparameters. They can, e.g., include scale factors, the kernel width of the Gaussian kernel or the impulse response's decay rate in the TC and stable spline kernels. In real-world applications, the oracle-based procedure used in the previous section cannot be used. The kernels need instead to be tuned from data. Such procedure is referred to as hyperparameter estimation and is the counterpart of model order selection in the classical paradigm of system identification. It determines model complexity within the new paradigm where system identification is seen as regularized function estimation in RKHSs. This calibration step will thus have a major impact on model's performance, e.g., in terms of predictive capability on new data. Due to the connection with the ReLS methods in quadratic form, the tuning methods introduced in Chaps. 3 and 4 can be easily applied also in the RKHS framework. In particular, let $K(\eta)$ denote a kernel, where η is the hyperparameter vector belonging to the set Γ. Such vector could also include other parameters not present in the kernel, e.g., the noise variance σ^2. Some calibration methods to estimate η from data are then reported below.

7.2.1 Marginal Likelihood Maximization

The first approach we describe is marginal likelihood maximization (MLM), also called the empirical Bayes method in Sect. 4.4. MLM relies on the Bayesian interpretation of function estimation in RKHS discussed in Sect. 7.1.4. Under the same assumptions stated in Proposition 7.1, η can be estimated by maximum likelihood

$$\hat{\eta} = \arg\max_{\eta \in \Gamma} \mathrm{p}(Y|\eta), \tag{7.40}$$

with $\mathrm{p}(Y|\eta)$ obtained by integrating out g^0 from the joint density $\mathrm{p}(Y|g^0)\mathrm{p}(g^0|\eta)$, i.e.,

$$\mathrm{p}(Y|\eta) = \int \mathrm{p}(Y|g^0)\mathrm{p}(g^0|\eta)dg^0. \tag{7.41}$$

The probability density $\mathrm{p}(Y|\eta)$ is the marginal likelihood and, hence, (7.40) is called the MLM method.

Computation of (7.41) is especially simple in our case since our measurements model is linear and Gaussian. In fact, in the Bayesian interpretation of regularized linear system identification in RKHS, the impulse response g^0 is a zero-mean Gaussian process with covariance λK, where λ is a positive scale factor. The impulse response is also assumed independent of the noises $e(t)$ which are white and Gaussian of variance σ^2. Recall also the definition of the matrix O, now possibly function of η, reported in (7.14) for the discrete-time case, i.e., when $\mathscr{X} = \mathbb{N}$, and in (7.22) for the continuous-time case, i.e., when $\mathscr{X} = \mathbb{R}^+$. The matrix $\lambda O(\eta)$ plays an important role in the MLM method since it corresponds to the covariance matrix of the noise-free output vector $[L_1[g^0], \ldots, L_N[g^0]]^T$ and is thus often called the output kernel matrix. Then, as also discussed in Sect. 7.7.2, it comes that the vector Y turns out to be Gaussian with zero mean, i.e.,

$$Y \sim \mathscr{N}(0, Z(\eta)),$$

where the covariance matrix $Z(\eta)$ is given by

$$Z(\eta) = \lambda O + \sigma^2 I_N$$

with I_N the $N \times N$ identity matrix. Here, the vector η could, e.g., contain both λ and σ^2. One then obtains that the empirical Bayes estimate of η in (7.40) becomes

$$\hat{\eta} = \arg \min_{\eta \in \Gamma} \ Y^T Z(\eta)^{-1} Y + \log \det(Z(\eta)), \tag{7.42}$$

where the objective is proportional to the minus log of the marginal likelihood.

As discussed in Chap. 4, the MLM method includes the Occam's razor principle, i.e., unnecessarily complex models are automatically penalized, see e.g., [83]. In particular, the Occam's factor arises thanks to the marginalization and it manifests itself in the term $\log \det(Z(\eta))$ in (7.42). A simple example can be obtained thinking of the behaviour of the objective for different values of the kernel scale factor λ. When λ increases, the model becomes more complex since, under a stochastic viewpoint, the prior variance of the impulse response g^0 increases. In fact, the term $Y^T Z(\eta)^{-1} Y$, related to the data fit, decreases since the inverse of $Z(\eta)$ tends to the null matrix (the model has infinite variance and can describe any kind of data). But the Occam's factor increases since $\det(Z(\eta))$ grows to infinity. In this way, $\hat{\eta}$ will balance data fit and model complexity.

7.2.1.1 Numerical Example

To illustrate the effectiveness of MLM, we revisit the example reported in Sect. 1.2. The problem is to reconstruct the impulse response reported in Fig. 7.7 (red line) from the 1000 input–output data displayed in Fig. 1.2. System input is low pass and this makes estimation hard due to ill-conditioning.

Fig. 7.7 True impulse response (red thick line) and impulse response estimates obtained by ridge regression with hyperparameters estimated by an oracle that optimizes the fit (top panel), and by the stable spline kernels of order 1 (middle) and 2 (bottom) with hyperparameters estimated by marginal likelihood maximization

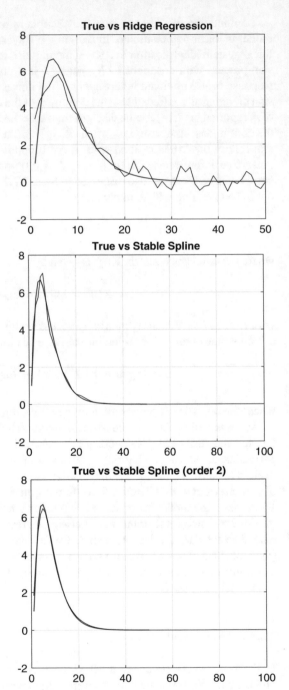

We will adopt three kernels. Using δ to denote the Kronecker delta, the value $K(i, j)$ is defined, respectively, by

$$\delta_{ij}, \quad \alpha^{\max(i,j)}, \quad \frac{\alpha^{i+j+\max(i,j)}}{2} - \frac{\alpha^{3\max(i,j)}}{6}.$$

The first choice corresponds to ridge regression with the regularizer given by the sum of squared impulse response coefficients. The other two are the first- and second-order stable spline kernel reported in (7.15) and in (7.30), respectively. More specifically, the last kernel corresponds to the discrete-time version of (7.30) with $\alpha = e^{-\beta}$.

In Fig. 1.5, we reported the ridge regularized estimate with γ chosen by an oracle to maximize the fit. To ease comparison with other approaches, such a figure is also reproduced in the top panel of Fig. 7.7. The reconstruction is not satisfactory since the regularizer does not include information on smoothness and decay. In fact, the Bayesian interpretation reveals that ridge regression describes the impulse response as realization of white noise, a poor model for stable dynamic systems. This also explains the presence of oscillations in the reconstructed profile.

The middle and bottom panel report the estimates obtained by the stable spline kernels with the noise variance and the hyperparameters γ, α tuned from data through MLM. Even if no oracle is used, the quality of the impulse response reconstruction greatly increases. This is also confirmed by a Monte Carlo study where 200 data sets are obtained using the same kind of input but generating new independent noise realizations. MATLAB boxplots of the 200 fits for all the three estimators are in Fig. 7.8. Here, the median is given by the central mark while the box edges are the 25th and 75th percentiles. Then, the whiskers extend to the most extreme fits not seen as outliers. Finally, the outliers are plotted individually. Average fits are 73.7% for ridge, 83.9% for first-order and 90.2% for second-order stable spline.

In this example, one can see that it is preferable to use the second-order stable spline kernel. This is easily explained by the fact that the true impulse response is quite regular so that increasing our expected smoothness improves the performance.

Fig. 7.8 Boxplot of the fits over the 200 data sets achieved by ridge regression with oracle (left) and by the stable spline kernels of order 1 (middle) and 2 (right) with hyperparameters estimated via marginal likelihood maximization

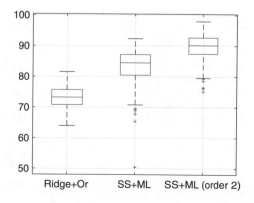

Interestingly, the selection between different kernels, like first- and second-order stable spline, can be also automatically performed by MLM, so addressing the problem of model comparison described in Sect. 2.6.2. In fact, let s denote an additional hyperparameter that may assume only value 0 or 1. Then, we can consider the combined kernel

$$s\alpha^{\max(i,j)} + (1-s)\left(\frac{\alpha^{i+j+\max(i,j)}}{2} - \frac{\alpha^{3\max(i,j)}}{6}\right)$$

and optimize the hyperparameters s, α and γ by MLM. Clearly, the role of s is to select one of the two kernels, e.g., if the estimate $\hat{s}$ is 0, then the impulse response estimate will be given by a second-order stable spline. Applying this procedure to our problem, one finds that the second-order stable spline kernel is selected 177 times out of the 200 Monte Carlo runs. Obtained fits are shown in Fig. 7.9, their mean is 88.8%.

Remark 7.4 Kernel choice via MLM has also connections with selection through the concept of Bayesian model probability discussed in Sect. 4.11, see also [50]. In fact, assume we are given different competitive kernels (covariances) K^i and, for a while, assume also that all the hyperparameter vectors η^i are known. We can then interpret each kernel as a different model. We can also assign a priori probabilities that data have been generated by the ith covariance K^i, hence thinking of any model as a random variable itself. If all the kernels are given the same probability, the marginal likelihood computed using K^i becomes proportional to the posterior probability of the ith model. This permits to exploit the marginal likelihood to select the "best" kernel-based estimate among those generated by the K^i. When hyperparameters are unknown, the marginal likelihoods can be evaluated with each η^i set to its estimate $\hat{\eta}^i$. In this case, care is needed since maximized likelihoods define model posterior probabilities that do not account for hyperparameters uncertainty. For example, if the dimensions of η^i change with i, the risk is to select a kernel that have many parameters and overfits. This problem can be mitigated, e.g., by adopting the criteria described in Sect. 2.4.3, e.g., using BIC, we compute

Fig. 7.9 Boxplot of the fits over the 200 data sets achieved by a stable spline estimator where, beyond hyperparameters, also the kernel order (1 or 2) is estimated by marginal likelihood maximization

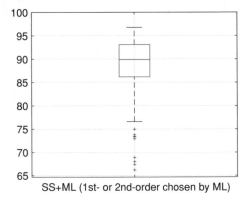

SS+ML (1st- or 2nd-order chosen by ML)

$$\hat{i} = \arg\min_{i} \; -2\log \mathrm{p}(Y|\hat{\eta}^i) + (\dim \eta^i)\log N,$$

where N is the number of available output measurements and $\dim \eta^i$ is the number of hyperparameters contained in the ith model. Note that, when using stable spline kernels as in the above example, the BIC penalty is irrelevant since the first- and the second-order stable spline estimator contain the same number of unknown hyperparameters.

7.2.2 Stein's Unbiased Risk Estimator

The second method is the Stein's unbiased risk estimator (SURE) method introduced in Sect. 3.5.3.2. The idea of SURE is to minimize an unbiased estimator of the risk, which is the expected in-sample validation error of the model estimate. In what follows, g^0 is no more stochastic as in the previous subsection but corresponds to a deterministic impulse response. Identification data are given by

$$y(t_i) = L_i[g^0] + e(t_i), \quad i = 1, \ldots, N,$$

where the $e(t_i)$ are independent, with zero mean and known variance σ^2, and each L_i is the linear functional defined by convolutions with the system input evaluated at t_i. One thus has $L_i[g^0] = \sum_{k=1}^{\infty} g^0(k)u(t_i - k)$ in discrete time, where the t_i assume integer values, and $L_i[g^0] = \int_0^{\infty} g^0(\tau)u(t_i - \tau)d\tau$ in continuous time. The N independent validation output samples $y_v(t_i)$ are then defined by using the same input that generates the identification data but an independent copy of the noises, i.e.,

$$y_v(t_i) = L_i[g^0] + e_v(t_i), \quad i = 1, \ldots, N. \tag{7.43}$$

So, all the $2N$ random variables $e_v(t_i)$ and $e(t_i)$ are mutually independent, with zero mean and noise variance σ^2. Consider the impulse response estimator

$$\hat{g} = \arg\min_{g \in \mathscr{H}} \left(\sum_{i=1}^{N} (y(t_i) - L_i[g])^2 + \gamma \|g\|_{\mathscr{H}}^2 \right)$$

as a function of the hyperparameter vector η. The predictions of the $y_v(t_i)$ are then given by $L_i[\hat{g}]$ and also depend on η. The expected in-sample validation error of the model estimate $\hat{g}$ is then given by the mean prediction error

$$\mathrm{EVE}_{\mathrm{in}}(\eta) = \frac{1}{N} \sum_{i=1}^{N} \mathscr{E}(y_v(t_i) - L_i[\hat{g}])^2, \tag{7.44}$$

where the expectation $\mathscr{E}$ is over the random noises $e_v(t_i)$ and $e(t_i)$. Note that the result not only depends on η but also on the unknown (deterministic) impulse response g^0. So, we cannot compute the prediction error. However, it is possible to derive an unbiased estimate of it. To obtain this, let $\hat{Y}(\eta)$ be the (column) vector with components $L_i[\hat{g}]$. The output kernel matrix $O(\eta)$, already introduced to describe marginal likelihood maximization, then gives the connection between the vector Y containing the measured outputs $y(t_i)$ and the predictions. In fact, using the representer theorem to obtain $\hat{g}$, and hence the $L_i[\hat{g}]$, one obtains

$$\hat{Y}(\eta) = O(\eta)(O(\eta) + \gamma I_N)^{-1}Y. \tag{7.45}$$

Following the same line of discussion developed in Sect. 3.5.3.2 to obtain (3.96), we can derive the following unbiased estimator of (7.44):

$$\widehat{\text{EVE}_{\text{in}}}(\eta) = \frac{1}{N}\|Y - \hat{Y}(\eta)\|^2 + 2\sigma^2\frac{\text{dof}(\eta)}{N}, \tag{7.46}$$

where $\text{dof}(\eta)$ are the degrees of the freedom of $\hat{Y}(\eta)$ given by

$$\text{dof}(\eta) = \text{trace}(O(\eta)(O(\eta) + \gamma I_N)^{-1}) \tag{7.47}$$

that vary from N to 0 as γ increases from 0 to ∞.

Note that (7.46) is function only of the N output measurements $y(t_i)$. Thus, we can then estimate the hyperparameter η by minimizing the unbiased estimator $\widehat{\text{EVE}_{\text{in}}}(\eta)$ of $\text{EVE}_{\text{in}}(\eta)$ to achieve

$$\hat{\eta} = \underset{\eta \in \Gamma}{\arg\min} \frac{1}{N}\|Y - \hat{Y}(\eta)\|^2 + 2\sigma^2\frac{\text{dof}(\eta)}{N}. \tag{7.48}$$

The above formula has the same form of the AIC criterion (2.33) computed assuming Gaussian noise of known variance σ^2 except that the dimension m of the model parameter θ is now replaced by the degrees of freedom $\text{dof}(\eta)$.

7.2.3 Generalized Cross-Validation

The third approach is the generalized cross-validation (GCV) method. As discussed in Sects. 2.6.3 and 3.5.2.3, cross-validation (CV) is a classical way to estimate the expected validation error by efficient reuse of the data and GCV is closely related with the N-fold CV with quadratic losses. To describe it in the RKHS framework, let $\hat{g}^k$ be the solution of the following function estimation problem:

$$\hat{g}^k = \arg\min_{g \in \mathcal{H}} \sum_{i=1, i \neq k}^{N} (y(t_i) - L_i[g])^2 + \gamma \|g\|_{\mathcal{H}}^2 . \tag{7.49}$$

So, $\hat{g}^k$ is the function estimate when the kth datum $y(t_k)$ is left out. As also described, e.g., in [90, Chap. 4], the following relation between the prediction error of $\hat{g}$ and the prediction error of $\hat{g}^k$ holds:

$$y(t_k) - L_k[\hat{g}^k] = \frac{y(t_k) - L_k[\hat{g}]}{1 - H_{kk}(\eta)}, \tag{7.50}$$

where $H_{kk}(\eta)$ is the (k, k)th element of the influence matrix

$$H(\eta) = O(\eta)(O(\eta) + \gamma I_N)^{-1}.$$

Therefore, the validation error of the N-fold CV with quadratic loss function is

$$\sum_{k=1}^{N} \left(y(t_k) - L_k[\hat{g}^k] \right)^2 = \sum_{k=1}^{N} \left(\frac{y(t_k) - L_k[\hat{g}]}{1 - H_{kk}(\eta)} \right)^2 . \tag{7.51}$$

Minimizing the above equation as a criterion to estimate the hyperparameter η leads to the predicted residual sums of squares (PRESS) method

$$\hat{\eta} = \arg\min_{\eta \in \Gamma} \sum_{k=1}^{N} \left(\frac{y(t_k) - L_k[\hat{g}]}{1 - H_{kk}(\eta)} \right)^2 . \tag{7.52}$$

The above criterion coincides with that derived in (3.80) working in the finite-dimensional setting.

GCV is a variant of (7.52) obtained by replacing each $H_{kk}(\eta)$, $k = 1, \ldots, N$, in (7.52) with their average. One obtains

$$\hat{\eta} = \arg\min_{\eta \in \Gamma} \sum_{k=1}^{N} \left(\frac{y(t_k) - L_k[\hat{g}]}{1 - \mathrm{trace}(H(\eta))/N} \right)^2 . \tag{7.53}$$

In view of (7.45), one has
$$\hat{Y}(\eta) = H(\eta)Y.$$

and, from (7.47) one can see that $\mathrm{trace}(H(\eta))$ corresponds to the degrees of freedom $\mathrm{dof}(\eta)$, i.e.,

$$\mathrm{trace}(H(\eta)) = \mathrm{dof}(\eta).$$

So, the GCV (7.53) can be rewritten as follows:

$$\hat{\eta} = \underset{\eta \in \Gamma}{\arg\min} \ \frac{\|Y - \hat{Y}(\eta)\|^2}{(1 - \text{dof}(\eta)/N)^2}.$$
(7.54)

This corresponds to the criterion (3.82) obtained in the finite-dimensional setting. Differently from SURE, a practical advantage of PRESS and GCV is that they do not require knowledge (or preliminary estimation) of the noise variance σ^2.

7.3   Theory of Stable Reproducing Kernel Hilbert Spaces

In the numerical experiments reported in this chapter, we have seen that regularized IIR models based, e.g., on TC and stable splines provide much better estimates of stable linear dynamic systems than other popular machine learning choices like the Gaussian kernel. The reading key was the inclusion in the identification process of information on the decay rate of the impulse response. This motivates the study of the class of the so-called stable kernels that enforces the stability constraint on the induced RKHS.

7.3.1   Kernel Stability: Necessary and Sufficient Conditions

The necessary and sufficient condition for a linear system to be bounded-input–bounded-output (BIBO) stable is that its impulse response $g \in \ell_1$ for the discrete-time case and $g \in \mathscr{L}_1$ for the continuous-time case. Here, ℓ_1 is the space of absolutely summable sequences, while $\mathscr{L}_1$ contains the absolutely summable functions on $\mathbb{R}^+$ (equipped with the classical Lebesque measure), i.e.,

$$\sum_{k=1}^{\infty} |g_k| < \infty \ \forall g \in \ell_1 \quad \text{and} \quad \int_0^{\infty} |g(x)| dx < \infty \ \forall g \in \mathscr{L}^1.$$
(7.55)

Therefore, for regularized identification of stable systems the impulse response should be searched within a RKHS that is a subspace of ℓ_1 in discrete time and a subspace of $\mathscr{L}_1$ in continuous time. This naturally leads to the following definition of stable kernels.

Definition 7.1 (*Stable kernel, based on* [32, 73]) Let $K : \mathscr{X} \times \mathscr{X} \to \mathbb{R}$ be a positive semidefinite kernel and $\mathscr{H} : \mathscr{X} \to \mathbb{R}$ be the RKHS induced by K. Then, K is said to be stable if

- $\mathscr{H} \subset \ell_1$ for the discrete-time case where $\mathscr{X} = \mathbb{N}$;
- $\mathscr{H} \subset \mathscr{L}_1$ for the continuous-time case where $\mathscr{X} = \mathbb{R}^+$.

If a kernel K is not stable, it is also said to be unstable. Accordingly, the RKHS $\mathscr{H}$ is said to be stable or unstable if K is stable or unstable.

Assigned a kernel, the question is now how to assess its stability. For this purpose, a direct use of the above definition is often challenging since it can be difficult to understand which functions belong to the associated RKHS. Stability conditions directly on K would be instead desirable. One first observation is that, since $\mathscr{H}$ contains all kernel sections according to Theorem 6.2, all of them must be stable. In discrete time, this means $K(i, \cdot) \in \ell_1$ for all i. However, this condition is necessary but not sufficient for stability, a fact which is not so surprising since we have seen in Sect. 6.2 that $\mathscr{H}$ contains also all the Cauchy limits of linear combinations of kernel sections. For instance, in Example 6.4, we have seen that the identity kernel $K(i, j) = \delta_{ij}$, connected with ridge regression but here defined over all $\mathbb{N} \times \mathbb{N}$, induces ℓ_2. Such space is not contained in ℓ_1. So, the identity kernel is not stable even if each kernel section is stable since it contains only one non-null element.

The following fundamental result can be found in a more general form in [16] and gives the desired charactherization of kernel stability. Maybe not surprisingly, we will see that the key test spaces are ℓ^∞, that contains bounded sequences in discrete time, and $\mathscr{L}_\infty$, that contains essentially bounded functions in continuous time. The proof is reported in Sect. 7.7.3.

Theorem 7.5 (Necessary and sufficient condition for kernel stability, based on [16, 32, 73]) *Let $K : \mathscr{X} \times \mathscr{X} \to \mathbb{R}$ be a positive semidefinite kernel with $\mathscr{X} = \mathbb{N}$ or $\mathscr{X} = \mathbb{R}^+$. Then,*

- *one has*

$$\mathscr{H} \subset \ell_1 \iff \sum_{s=1}^{\infty} \left| \sum_{t=1}^{\infty} K(s, t) l_t \right| < \infty, \ \forall l \in \ell_\infty \tag{7.56}$$

for the discrete-time case where $\mathscr{X} = \mathbb{N}$;
- *one has*

$$\mathscr{H} \subset \mathscr{L}_1 \iff \int_0^\infty \left| \int_0^\infty K(s, t) l(t) dt \right| ds < \infty, \ \forall l \in \mathscr{L}_\infty \tag{7.57}$$

for the continuous-time case where $\mathscr{X} = \mathbb{R}^+$.

Figure 7.10 illustrates the meaning of Theorem 7.5 by resorting to a simple system theory argument. In particular, a kernel can be seen as an acausal linear time-varying system. In discrete time it induces the following input–output relationship

$$y_i = \sum_{j=1}^{\infty} K_i(j) u_j, \quad i = 1, 2, \ldots, \tag{7.58}$$

where $K_i(j) = K(i, j)$, while u_i and y_i denote the system input and output at instant i. Then, the RKHS induced by K is stable iff system (7.58) maps every bounded input $\{u_i\}_{i=1}^{\infty}$ into a summable output $\{y_i\}_{i=1}^{\infty}$. Abusing notation, we can also see K as an infinite-dimensional matrix with i, j-entry given by $K_i(j)$ with u and y infinite-dimensional column vectors. Then, using ordinary algebra notation to

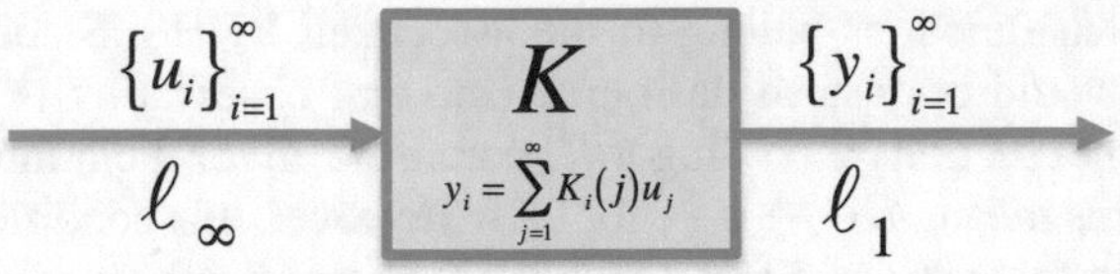

Fig. 7.10 System theoretic interpretation of RKHS stability. The kernel K is associated to an acausal linear system. In discrete time, the input–output relationship is given by $y_i = \sum_{j=1}^{\infty} K_i(j)u_j$. Then, K is stable iff every bounded input u is mapped into a summable output y

handle these objects, the input–output relationship becomes $y = K u$ and the stability condition is

$$\mathscr{H} \subseteq \ell_1 \iff K u \in \ell_1 \; \forall u \in \ell_\infty.$$

In Theorem 7.5, it is immediate to see that including the constraint $-1 \le l_t \le 1 \; \forall t$ on the test functions does not have any influence on the stability test. With this constraint, one has

$$\left| \sum_{t=1}^{\infty} K(s,t) l_t \right| \le \sum_{t=1}^{\infty} |K(s,t)| \quad \text{and} \quad \left| \int_0^{\infty} K(s,t) l(t) dt \right| \le \int_0^{\infty} |K(s,t)| dt.$$

The following result is then an immediate corollary of Theorem 7.5 obtained exploiting the above inequalities. It states that absolute summability is a sufficient condition for a kernel to be stable.

Corollary 7.1 (based on [16, 32, 73]) *Let $K : \mathscr{X} \times \mathscr{X} \to \mathbb{R}$ be a positive semidefinite kernel with $\mathscr{X} = \mathbb{N}$ or $\mathscr{X} = \mathbb{R}^+$. Then,*

• *one has*

$$\mathscr{H} \subset \ell_1 \Longleftarrow \sum_{s=1}^{\infty} \sum_{t=1}^{\infty} |K(s,t)| < \infty \tag{7.59}$$

for the discrete-time case where $\mathscr{X} = \mathbb{N}$;
• *one has*

$$\mathscr{H} \subset \mathscr{L}_1 \Longleftarrow \int_0^{\infty} \int_0^{\infty} |K(s,t)| dt ds < \infty \tag{7.60}$$

for the continuous-time case where $\mathscr{X} = \mathbb{R}^+$.

Finally, consider the class of nonnegative-valued kernels K^+, i.e., satisfying $K(s,t) \ge 0 \; \forall s, t$. If a kernel is stable, using as test function $l(t) = 1 \; \forall t$, one must have

$$\left| \sum_{t=1}^{\infty} K^+(s,t) l_t \right| = \sum_{t=1}^{\infty} K^+(s,t) < \infty$$

in discrete time, and

$$\left| \int_0^\infty K^+(s,t)l(t)dt \right| = \int_0^\infty K^+(s,t)dt < \infty$$

in continuous time. So, for nonnegative-valued kernels, stability implies (absolute) summability of the kernel. But, since we have seen in Corollary 7.1 that absolute summability implies stability, the following result holds.

Corollary 7.2 (based on [16, 32, 73]) *Let $K^+ : \mathcal{X} \times \mathcal{X} \to \mathbb{R}$ be a positive semidefinite and nonnegative-valued kernel with $\mathcal{X} = \mathbb{N}$ or $\mathcal{X} = \mathbb{R}^+$. Then,*

- *one has*

$$\mathcal{H} \subset \ell_1 \iff \sum_{s=1}^\infty \sum_{t=1}^\infty K^+(s,t) < \infty \tag{7.61}$$

for the discrete-time case where $\mathcal{X} = \mathbb{N}$;
- *one has*

$$\mathcal{H} \subset \mathcal{L}_1 \iff \int_0^\infty \int_0^\infty K^+(s,t)dtds < \infty \tag{7.62}$$

for the continuous-time case where $\mathcal{X} = \mathbb{R}^+$.

As an example, we can now show that the Gaussian kernel (6.43) defined e.g., over $\mathbb{R}^+ \times \mathbb{R}^+$ is not stable. In fact, it is nonnegative valued and one has

$$\int_0^\infty \int_0^\infty \exp\left(-(s-t)^2/\rho \right) dsdt = \infty \ \forall \rho.$$

The same holds for the spline kernels (6.45) extended to $\mathbb{R}^+ \times \mathbb{R}^+$ and also for translation invariant kernels introduced in Example 6.12, as e.g., proved in [32] using the Schoenberg representation theorem. Hence, all of these models are not suited for stable impulse response estimation.

Remark 7.5 Any unstable kernel can be made stable simply by truncation. More specifically, let $K : \mathcal{X} \times \mathcal{X} \to \mathbb{R}$ be an unstable kernel with $\mathcal{X} = \mathbb{N}$ or $\mathcal{X} = \mathbb{R}^+$. Then by setting $K(s,t) = 0$ for $s, t > T$ for any given $T \in \mathcal{X}$, a stable kernel is obtained. Care should be however taken when a FIR model is obtained through this operation. In fact, consider e.g., the use of cubic spline or Gaussian kernel in the estimation problem depicted in Fig. 7.6 setting T equal to 20 or 50. Also after truncation, such models would not give good performance: the undue oscillations affecting the estimates in the top and middle panel of Fig. 7.6 would still be present. The reason is that these two kernels do not encode the information that the variability of the impulse response decreases as time progresses, as also already discussed using the Bayesian interpretation of regularization.

7.3.2 Inclusions of Reproducing Kernel Hilbert Spaces in More General Lebesque Spaces ⋆

We now discuss the conditions for a RKHS to be contained in the spaces $\mathscr{L}_p^\mu$ equipped with a generic measure μ. The following analysis will then include both the space $\mathscr{L}_1$ (considered before with the Lebesque measure) and ℓ_1 as special cases obtained with $p = 1$. First, we need the following definition.

Definition 7.2 (*based on* [16]) Let $1 \leq p \leq \infty$ and $q = \frac{p}{p-1}$ with the convention $\frac{p}{p-1} = \infty$ if $p = 1$ and $\frac{p}{p-1} = 1$ if $p = \infty$. Moreover, let $K : \mathscr{X} \times \mathscr{X} \to \mathbb{R}$ be a positive semidefinite kernel. Then, the kernel K is said to be q-bounded if

1. the kernel section $K_s \in \mathscr{L}_p^\mu$ for almost all $s \in \mathscr{X}$, i.e., for every $s \in \mathscr{X}$ except on a set of null measure w.r.t. μ;
2. the function $\int_0^\infty K(s, t)l(t)d\mu(t) \in \mathscr{L}_p^\mu, \forall l \in \mathscr{L}_q^\mu$.

The following theorem then gives the necessary and sufficient condition for the q-boundedness of a kernel and is a special case of Proposition 4.2 in [16].

Theorem 7.6 (based on [16]) *Let $K : \mathscr{X} \times \mathscr{X} \to \mathbb{R}$ be a positive semidefinite kernel with $\mathscr{H}$ the induced RKHS. Then, $\mathscr{H}$ is a subspace of $\mathscr{L}_p^\mu$ if and only if K is q-bounded, i.e.,*

$$\mathscr{H} \subset \mathscr{L}_p^\mu \iff K \text{ is } q\text{-bounded.}$$

Theorem 7.6 permits thus to see if a RKHS is contained in $\mathscr{L}_p^\mu$ by checking the properties of the kernel. Interestingly, setting $p = 1$, that implies $q = \infty$, and μ e.g., to the Lebesque measure one can see that the concept of stable and ∞-bounded kernel are equivalent. Theorem 7.5 is then a special case of Theorem 7.6.

7.4 Further Insights into Stable Reproducing Kernel Hilbert Spaces ⋆

In this section, we provide some additional insights into the structure of the stable kernels and associated RKHSs. The analysis is focused on the discrete-time case where the kernel K can be seen as an infinite-dimensional matrix with the (i, j)-entries denoted by K_{ij}. Thus, the function domain is the set of natural numbers $\mathbb{N}$ and the RKHS contains discrete-time impulse responses of causal systems.

As discussed after (7.58) to comment Fig. 7.10, the kernel K can be also associated with an acausal linear time-varying system, often called kernel operator in the literature. It maps the infinite-dimensional input (sequence) u into the infinite-dimensional output Ku whose ith component is $\sum_{j=1}^\infty K_{ij}u_j$. Two important kernel operators will be considered. The first one maps ℓ_∞ into ℓ_1 and is key for kernel stability as pointed out in Theorem 7.5. The second one maps ℓ_2 into ℓ_2 itself and will be important to discuss spectral decompositions of stable kernels.

7.4.1 Inclusions Between Notable Kernel Classes

To state some relationships between stable kernels and other fundamental classes, we start introducing some sets of RKHSs. Define

- the set $\mathscr{S}_s$ that contains all the stable RKHSs;
- the set $\mathscr{S}_1$ with all the RKHSs induced by absolutely summable kernels, i.e., satisfying

$$\sum_{ij} |K_{ij}| < +\infty;$$

- the set $\mathscr{S}_{ft}$ of RKHSs induced by finite-trace kernels, i.e., satisfying

$$\sum_{i} K_{ii} < +\infty;$$

- the set $\mathscr{S}_2$ associated to squared summable kernels, i.e., satisfying

$$\sum_{ij} K_{ij}^2 < +\infty.$$

One has then the following result from [8] (see Sect. 7.7.4 for some details on its proof).

Theorem 7.7 (based on [8]) *It holds that*

$$\mathscr{S}_1 \subset \mathscr{S}_s \subset \mathscr{S}_{ft} \subset \mathscr{S}_2. \tag{7.63}$$

Figure 7.11 gives a graphical description of Theorem 7.7 in terms of inclusions of kernels classes. Its meaning is further discussed below.

In Corollary 7.1, we have seen that absolute summability is a sufficient condition for kernel stability. The result $\mathscr{S}_1 \subset \mathscr{S}_s$ shows also that such inclusion is strict. Hence, one cannot conclude that a kernel is unstable from the sole failure of absolute summability.

The fact that $\mathscr{S}_s \subset \mathscr{S}_{ft}$ means that the set of finite-trace kernels contains the stable class. This inclusion is strict, hence the trace analysis can be used only to show that a given RKHS is not contained in ℓ_1. There are however interesting consequences of this fact. Consider all the RKHSs induced by translation invariant kernels

$$K_{ij} = h(i - j),$$

where h satisfies the positive semidefinite constraints. The trace of these kernels is $\sum_i K_{ii} = \sum_i h(0)$ and it always diverges unless h is the null function. So, all the translation invariant kernels are unstable (as already mentioned after Corollary 7.2). Other instability results become also immediately available. For instance, all the kernels with diagonal elements satisfying $K_{ii} \propto i^{-\delta}$ are unstable if $\delta \leq 1$.

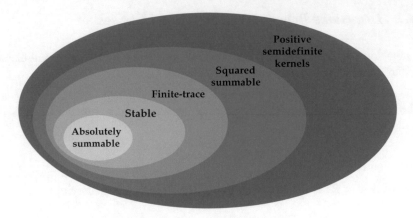

Fig. 7.11 Inclusion properties of some important kernel classes

Finally, the strict inclusion $\mathscr{S}_{ft} \subset \mathscr{S}_2$ shows that the finite-trace test is more powerful than a check of kernel squared summability.

7.4.2 Spectral Decomposition of Stable Kernels

As discussed in Sect. 6.6.3 and in Remark 6.3, kernels can define spaces rich of functions by (implicitly) mapping the space of the regressors into high-dimensional feature spaces where linear estimators can be used. This allows to reduce nonlinear algorithms even without knowing explicitly the feature map, i.e., without the exact knowledge of which functions are encoded in the kernel. In particular, in Sect. 6.3, we have seen that if the kernel admits the spectral representation

$$K(x, y) = \sum_{i=1}^{\infty} \zeta_i \rho_i(x)\rho_i(y), \tag{7.64}$$

then the $\rho_i(x)$ are the basis functions that span the RKHS induced by K. For instance, the basis functions $\rho_1(x) = 1, \rho_2(x) = x, \rho_3(x) = x^2, \dots$ describe polynomial models which are, e.g., included up to a certain degree in the polynomial kernel discussed in Sect. 6.6.4. Now, we will see that stable kernels always admit an expansion of the type (7.64) with the ρ_i forming a basis of ℓ_2. The number of ζ_i different from zero then corresponds to the dimension of the induced RKHS.

Formally, it is now necessary to consider the operator induced by a stable kernel K as a map from ℓ_2 into ℓ_2 itself. Again, it is useful to see K as an infinite-dimensional matrix so that we can think of Kv as the result of the kernel operator applied to $v \in \ell_2$. An operator is said to be compact if it maps any bounded sequence $\{v_i\}$ into a sequence $\{Kv_i\}$ from which a convergent subsequence can be extracted [85,

95]. From Theorem 7.7, we know that any stable kernel K is finite trace and, hence, squared summable. This fact ensures the compactness of the kernel operator, as discussed in [8] and stated below.

Theorem 7.8 (based on [8]) *Any operator induced by a stable kernel is self-adjoint, positive semidefinite and compact as a map from ℓ_2 into ℓ_2 itself.*

This result allows us to exploit the spectral theorem [35] to obtain an expansion of K. Now, recall that spectral decompositions were discussed in Sect. 6.3 where the Mercer's theorem was also reported. Mercer's theorem derivations exploit the spectral theorem and, as, e.g., in Theorem 6.9, they typically assume that the kernel domain is compact, see also [86] for discussions and extensions. Indeed, first formulations consider continuous kernels on compact domains (proving also uniform convergence of the expansion). However, the spectral theorem does not require the domain to be compact and, when applied to discrete-time kernels on $\mathbb{N} \times \mathbb{N}$, it guarantees pointwise convergence. It thus becomes the natural generalization of the decomposition of a symmetric matrix in terms of eigenvalues and eigenvectors, initially discussed in the finite-dimensional setting in Sect. 5.6 to link regularization and basis expansion. This is summarized in the following proposition that holds in virtue of Theorem 7.8.

Proposition 7.2 (Representation of stable kernels, based on [8]) *Assume that the kernel K is stable. Then, there always exists an orthonormal basis of ℓ_2 composed by eigenvectors $\{\rho_i\}$ of K with corresponding eigenvalues $\{\zeta_i\}$, i.e.,*

$$K\rho_i = \zeta_i \rho_i, \quad i = 1, 2, \ldots.$$

In addition, the kernel admits the following expansion:

$$K_{xy} = \sum_{i=1}^{+\infty} \zeta_i \rho_i(x) \rho_i(y), \qquad (7.65)$$

with $x, y \in \mathbb{N}$.

While in the next subsection, we will use the above theorem to discuss the representation of stable RKHSs, some numerical considerations regarding (7.65) are now in order. Under an algorithmic viewpoint, many efficient machine learning procedures use truncated Mercer expansions to approximate the kernel, see [42, 52, 75, 93, 96] for discussions on their optimality in a stochastic framework. Applications for system identification can be found in [15] where it is shown that a relatively small number of eigenfunctions (w.r.t. the data set size) can well approximate impulse responses regularized estimates. These works trace back to the so-called Nyström method where an integral equation is replaced by finite-dimensional approximations [5, 6]. However, obtaining the Mercer expansion (7.65) in closed form is often hard. Fortunately, the ℓ_2 basis and related eigenvalues of a stable RKHS can be numerically recovered (with arbitrary precision w.r.t. the ℓ_2 norm) through a sequence of SVDs applied to truncated kernels [8]. Formally, let $K^{(d)}$ denote the $d \times d$ positive

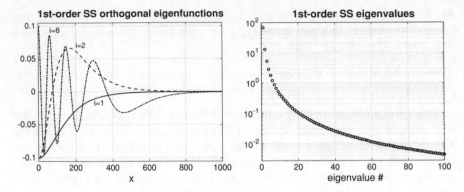

Fig. 7.12 Expansion of the first-order discrete-time stable spline kernel $K_{xy} = \alpha^{\max(x,y)}$ with $\alpha = 0.99$: eigenfunctions $\rho_i(x)$ orthogonal in ℓ_2 for $i = 1, 2, 8$ (left panel, samples are linearly interpolated) and eigenvalues ζ_i (right)

semidefinite matrix obtained by retaining only the first d rows and columns of K. Let also $\rho_i^{(d)}$ and $\zeta_i^{(d)}$ be, respectively, the eigenvectors of $K^{(d)}$, seen as elements of ℓ_2 with a tail of zeros, and the eigenvalues returned by the SVD of $K^{(d)}$. Assume, for simplicity, single multiplicity of each ζ_i. Then, for any i, as d grows to ∞ one has

$$\zeta_i^{(d)} \rightarrow \zeta_i \tag{7.66a}$$

$$\|\rho_i^{(d)} - \rho_i\|_2 \rightarrow 0, \tag{7.66b}$$

where $\|\cdot\|_2$ is the ℓ_2 norm.

In Fig. 7.12, we show some eigenvectors (left panel) and the first 100 eigenvalues (right) of the stable spline kernel $K_{xy} = \alpha^{\max(x,y)}$ with $\alpha = 0.99$. Results are obtained applying SVDs to truncated kernels of different sizes and monitoring convergence of eigenvectors and eigenvalues. The final outcome was obtained with $d = 2000$.

7.4.3 Mercer Representations of Stable Reproducing Kernel Hilbert Spaces and of Regularized Estimators

Now we exploit the representations of the RKHSs induced by a diagonalized kernel as discussed in Theorems 6.10 and 6.13 (where compactness of the input space is not even required). In view of Proposition 7.2, assuming for simplicity all the ζ_i different from zero, one obtains that the RKHS associated to a stable K always admits the representation

$$\mathcal{H} = \left\{ g = \sum_{i=1}^{\infty} a_i \rho_i \ \text{s.t.} \ \sum_{i=1}^{\infty} \frac{a_i^2}{\zeta_i} < +\infty \right\}, \tag{7.67}$$

where the ρ_i are the eigenvectors of K forming an orthonormal basis of ℓ_2.[1] If $g = \sum_{i=1}^{\infty} a_i \rho_i$, one also has

$$\|g\|_{\mathcal{H}}^2 = \sum_{i=1}^{\infty} \frac{a_i^2}{\zeta_i}. \tag{7.68}$$

The fact that any stable RKHS is generated by an ℓ_2 basis gives also a clear connection with the important impulse response estimators which adopt orthonormal functions, e.g., the Laguerre functions illustrated in Fig. 7.3 [46, 91, 92]. A classical approach used in the literature is to introduce the model $g = \sum_i a_i \rho_i$ and then to use linear least squares to determine the expansion coefficients a_i. In particular, let $L_t[g]$ be the system output, i.e., the convolution between the known input and g evaluated at the time instant t. Then, the impulse response estimate is

$$\hat{g} = \sum_{i=1}^{d} \hat{a}_i \rho_i \tag{7.69a}$$

$$\{\hat{a}_i\}_{i=1}^{d} = \arg\min_{\{a_i\}_{i=1}^{d}} \sum_{t=1}^{N} \left(y(t) - L_t \left[\sum_{i=1}^{d} a_i \rho_i \right] \right)^2, \tag{7.69b}$$

where d determines model complexity and is typically selected using AIC or cross-validation (CV) as discussed in Chap. 2.

In view of (7.67) and (7.68), the regularized estimator (7.10), equipped with a stable RKHS, is equivalent to

$$\hat{f} = \sum_{i=1}^{\infty} \hat{a}_i \rho_i \tag{7.70a}$$

$$\{\hat{a}_i\}_{i=1}^{\infty} = \arg\min_{\{a_i\}_{i=1}^{\infty}} \sum_{t=1}^{N} \left(y(t) - L_t \left[\sum_{i=1}^{\infty} a_i \rho_i \right] \right)^2 + \gamma \sum_{i=1}^{\infty} \frac{a_i^2}{\zeta_i}. \tag{7.70b}$$

---

[1] In (7.67), we have assumed that all the kernel eigenvalues are strictly positive so that $\mathcal{H}$ is infinite dimensional. If some ζ_i is null, $\mathcal{H}$ is spanned only by the eigenvectors associated to those non-null. If only a finite number of ζ_i is different from zero, K is finite rank and $\mathcal{H}$ is finite dimensional. A notable case is that of the RKHSs induced by truncated kernels, i.e., such that there exists d such that $K_{ii} = 0 \ \forall i > d$. As we have seen, this kind of kernels induce finite-dimensional RKHSs containing FIR systems of order d.

This result is connected with the kernel trick discussed in Remark 6.3 and shows that regularized least squares in a stable (infinite-dimensional) RKHS always model impulse responses using an ℓ_2 orthonormal basis, as in the classical works on linear system identification. But the key difference between (7.69) and (7.70) is that complexity is no more controlled by the model order because d is set to ∞. Complexity instead depends on the regularization parameter γ (and possibly also on other kernel parameters) that balances the data fit and the penalty term. This latter induces stability by using the kernel eigenvalues ζ_i to constrain the decay rate to zero of the expansion coefficients.

7.4.4 Necessary and Sufficient Stability Condition Using Kernel Eigenvectors and Eigenvalues

We have seen that a fruitful way to design a regularized estimator for linear system identification is to introduce a kernel by specifying its entries K_{ij}. This modelling technique translates our expected features of an impulse response into kernel properties, e.g., smooth exponential decay as described by stable spline, TC and DC kernels. This route exploits the kernel trick, i.e., the basis functions implicit encoding. In some circumstances, it could be useful to build a kernel starting from the design of eigenfunctions ρ_i and eigenvalues ζ_i. A notable example is given by the (already cited) Laguerre or Kautz functions that belong to the more general class of Takenaka–Malmquist orthogonal basis functions [46]. They can be useful to describe oscillatory behavior or presence of fast/slow poles.

Since any stable kernel can be associated with an ℓ_2 basis, the following fundamental problem then arises. Given an orthonormal basis $\{\rho_i\}$ of ℓ_2, for example, of the Takenaka–Malmquist type, which are the conditions on the eigenvalues ζ_i ensuring stability of $K_{xy} = \sum_{i=1}^{+\infty} \zeta_i \rho_i(x) \rho_i(y)$? The answer is in the following result derived from [8] that reports the necessary and sufficient condition (the proof is given in Sect. 7.7.5).

Theorem 7.9 (RKHS stability using Mercer expansions, based on [8]) *Let $\mathcal{H}$ be the RKHS induced by K with*

$$K_{xy} = \sum_{i=1}^{+\infty} \zeta_i \rho_i(x) \rho_i(y),$$

where the $\{\rho_i\}$ form an orthonormal basis of ℓ_2. Let also

$$\mathcal{U}_\infty = \left\{ u \in \ell_\infty : |u(i)| = 1, \ \forall i \geq 1 \right\}.$$

Then, one has

$$\mathcal{H} \subset \ell_1 \iff \sup_{u \in \mathcal{U}_\infty} \sum_i \zeta_i \langle \rho_i, u \rangle_2^2 < +\infty, \tag{7.71}$$

where $\langle \cdot, \cdot \rangle_2$ is the inner product in ℓ_2.

Thus, clearly, there is no stability if one function ρ_i associated to $\zeta_i > 0$ doesn't belong to ℓ_1. In fact, one can choose u containing the signs of the components of ρ_i and this leads to $\langle \rho_i, u \rangle_2 = +\infty$. Nothing is instead required for the eigenvectors associated to $\zeta_i = 0$. Theorem 7.9 permits also to derive the following sufficient stability condition.

Corollary 7.3 (based on [8]) *Let $\mathcal{H}$ be the RKHS induced by the kernel $K_{xy} = \sum_{i=1}^{+\infty} \zeta_i \rho_i(x) \rho_i(y)$ with $\{\rho_i\}$ an orthonormal basis of ℓ_2. Then, it holds that*

$$\mathcal{H} \subset \ell_1 \Longleftarrow \sum_i \zeta_i \|\rho_i\|_1^2 < +\infty. \tag{7.72}$$

Furthermore, such condition also implies kernel absolute summability and, hence, it is not necessary for RKHS stability.

It is easy to exploit the stability condition (7.72) to design models of stable impulse responses starting from an ℓ_2 basis. Let us reconsider, e.g., Laguerre or Kautz basis functions $\{\rho_i\}$ to build the impulse response model

$$g = \sum_{i=1}^{\infty} a_i \rho_i.$$

To exploit (7.70), one has to define stability constraints on the expansion coefficients a_i. This corresponds to define ζ_i in such a way that the regularizer

$$\sum_{i=1}^{\infty} \frac{a_i^2}{\zeta_i}$$

enforces absolute summability of g. Laguerre and Kautz models belong to the Takenaka–Malmquist class of functions ρ_i that all satisfy

$$\|\rho_i\|_1 \leq Mi,$$

with M a constant independent of i [46]. Then, Corollary 7.3 ensures that the choice

$$\zeta_i \propto i^{-\nu}, \quad \nu > 2$$

includes the stability contraint for the entire Takenaka–Malmquist class.

Let us now consider the class of orthonormal basis functions ρ_i all contained in a ball of ℓ_1. Then, the necessary and sufficient stability condition assumes a form especially simple as the following result shows.

Corollary 7.4 (based on [8]) *Let $\mathcal{H}$ be the RKHS induced by the kernel $K_{xy} = \sum_{i=1}^{+\infty} \zeta_i \rho_i(x) \rho_i(y)$ with $\{\rho_i\}$ an orthonormal basis of ℓ_2 and $\|\rho_i\|_1 \leq M < +\infty$ if*

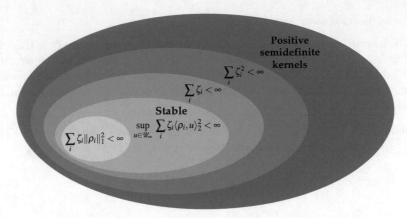

Fig. 7.13 Inclusion properties of some important kernel classes in terms of Mercer expansions. This representation is the dual of that reported in Fig. 7.11 and defines kernel sets through properties of the kernel eigenvectors ρ_i, forming an orthonormal basis in ℓ_2, and of the corresponding kernel eigenvalues ζ_i. The condition $\sum_i \zeta_i \|\rho_i\|_1^2 < \infty$ is the most restrictive since it implies kernel absolute summability. The necessary and sufficient condition for stability is $\sup_{u \in \mathcal{U}_\infty} \sum_i \zeta_i \langle \rho_i, u \rangle_2^2 < \infty$. Finally, $\sum_i \zeta_i < \infty$ and $\sum_i \zeta_i^2 < \infty$ are exactly the conditions for a kernel to be finite trace and squared summable, respectively

$\zeta_i > 0$, *with M not dependent on i. Then, one has*

$$\mathcal{H} \subset \ell_1 \iff \sum_i \zeta_i < +\infty. \tag{7.73}$$

Finally, Fig. 7.13 illustrates graphically all the stability results here obtained starting from Mercer expansions.

7.5 Minimax Properties of the Stable Spline Estimator ⋆

In this section, we will derive non-asymptotic upper bounds on the MSE of the regularized IIR estimator (7.10) valid for all the exponentially stable discrete-time systems whose poles belong to the complex circle of radius ρ. Obtained bounds can be evaluated before any data is observed. This kind of results give insight into the so-called sample complexity, i.e., the number of measurements needed to achieve a certain accuracy on impulse response reconstruction. This is an attractive feature even if, since the bounds need to hold for all the models falling in a particular class, often they are quite loose for the particular dynamic system at hand. However, they have a considerable theoretical value since permit also to assess the quality of (7.10) through nonparametric minimax concepts. Such setting considers the worst-case inside an infinite-dimensional class and has been widely studied in nonparametric regression and density estimation [88]. In particular, obtained bounds will lead to conditions which ensure the optimality in order, i.e., the best convergence rate of (7.10) in the minimax sense. We will derive them by considering system inputs given by white noises and using the TC/stable spline kernel (7.15) as regularizer. The important dependence between the convergence rate of (7.10) to the true impulse response, the stability kernel parameter α and the stability radius ρ will be elucidated.

7.5.1 Data Generator and Minimax Optimality

As in the previous part of the chapter, we use g^0 to denote the impulse response of a discrete-time linear system. The measurements are generated as follows:

$$y(t) = \sum_{k=1}^{\infty} g^0(k) u_{t-k} + e(t), \tag{7.74}$$

where $g^0(k)$ are the impulse response coefficients. We will always assume g^0 as a deterministic and exponentially stable impulse response, while the input u and the noise e are stochastic as specified below.

Assumption 7.10 The impulse response g^0 belongs to the following set:

$$\mathscr{S}(\varrho, L) = \left\{ g : |g(k)| \leq L\varrho^k \right\}, \quad 0 \leq \rho < 1. \tag{7.75}$$

The system input and the noise are discrete-time stochastic processes. One has that $\{u(t)\}_{t \in \mathbb{Z}}$ are independent and identically distributed (i.i.d.) zero-mean random variables with

$$\mathscr{E}[u(t)^2] = \sigma_u^2, \quad |u(t)| \leq C_u < \infty. \tag{7.76}$$

Finally, $\{e(t)\}_{t \in \mathbb{Z}}$ are independent random variables, independent of $\{u(t)\}_{t \in \mathbb{Z}}$, with

$$\mathscr{E}[e(t)] = 0, \quad \mathscr{E}[e(t)^2] \leq \sigma^2. \tag{7.77}$$

The available measurements are

$$\mathscr{D}_T = \{u(1), \ldots, u(N), y(1), \ldots, y(N)\}, \tag{7.78}$$

where N is the data set size.

The quality of an impulse response estimator $\hat{g}$ function of $\mathscr{D}_T$ will be measured by computing the estimation error $\mathscr{E}\|g^0 - \hat{g}\|_2$, where $\|\cdot\|_2$ is the norm in the space ℓ_2 of squared summable sequences. Note that the expectation is taken w.r.t. the randomness of the system input and the measurement noise. The worst-case error over the family $\mathscr{S}$ of exponentially stable systems reported in (7.75) will be also considered. In particular, the uniform ℓ_2-risk of $\hat{g}$ is

$$\sup_{g \in \mathscr{S}} \mathscr{E} \|g - \hat{g}\|_2.$$

An estimator g^* is then said to be *minimax* if the following equality holds for any data set size N:

$$\sup_{g \in \mathscr{S}} \mathscr{E} \|g - g^*\|_2 = \inf_{\hat{g}} \sup_{g \in \mathscr{S}} \mathscr{E} \|g - \hat{g}\|_2,$$

meaning that g^* minimizes the error w.r.t. the worst-case scenario. Building such kind of estimator is in general really difficult. For this reason, it is often convenient to consider just the asymptotic behaviour introducing the concept of optimality in order. Specifically, an estimator $\bar{g}$ is *optimal in order* if

$$\sup_{g \in \mathscr{S}} \mathscr{E} \|g - \bar{g}\|_2 \le C_N \sup_{g \in \mathscr{S}} \mathscr{E} \|g - g^*\|_2$$

with C_N is function of the data set size and satisfies $\sup_N C_N < \infty$ and g^* is minimax. In our linear system identification setting, optimality in order thus ensures that, as N grows to infinity, the convergence rate of $\bar{g}$ to the true impulse response g^0 cannot be improved by any other system identification procedure in the minimax sense.

7.5.2 Stable Spline Estimator

As anticipated, our study is focused on the following regularized estimator:

$$\hat{g} = \arg\min_{g \in \mathscr{H}} \sum_{t=1}^{N} (y(t) - \sum_{k=1}^{\infty} g(k)u(t-k))^2 + \gamma \|g\|_{\mathscr{H}}^2, \tag{7.79}$$

equipped with the stable spline kernel

$$K(i, j) = \alpha^{\max(i, j)}, \quad 0 < \alpha < 1, \quad (i, j) \in \mathbb{N}. \tag{7.80}$$

For future developments, it is important to control complexity of (7.79) not only by using the hyperparameters γ and α but also through the dimension d of the following subspace:

$$\mathscr{H}_d = \{g \in \mathscr{H} \text{ s.t. } g(d+1) = g(d+2) = \cdots = 0\}$$

over which optimization of the objective in (7.79) is performed. In particular, we will consider the estimator

$$\hat{g}^d = \arg\min_{g \in \mathscr{H}_d} \sum_{t=1}^{N} \left(y(t) - \sum_{k=1}^{d} g(k)u(t-k) \right)^2 + \gamma \|g\|_{\mathscr{H}}^2, \tag{7.81}$$

and will study how N and the choice of γ, α, d influence the estimation error and, hence, the convergence rate. This will lead to complexity control rules that are a hybrid of those seen in the classical and in the regularized framework. To obtain this, first, we rewrite (7.81) in terms of regularized FIR estimation by exploiting the structure of the stable spline norm (7.16) which shows that

$$g \in \mathscr{H}_d \Longrightarrow \|g\|_{\mathscr{H}}^2 = \left(\sum_{t=1}^{d-1} \frac{(g(t+1) - g(t))^2}{(1-\alpha)\alpha^t} \right) + \frac{g^2(d)}{(1-\alpha)\alpha^d}. \tag{7.82}$$

Let us define the matrix

$$R = \frac{1}{\alpha - \alpha^2} \begin{bmatrix} 1 & -1 & 0 & 0 & \cdots & 0 \\ -1 & 1+\frac{1}{\alpha} & -\frac{1}{\alpha} & 0 & \cdots & 0 \\ 0 & -\frac{1}{\alpha} & \frac{1}{\alpha}+\frac{1}{\alpha^2} & -\frac{1}{\alpha^2} & \cdots & 0 \\ & & \ddots & \ddots & \ddots & \vdots \\ 0 & 0 & & & & \\ 0 & 0 & \cdots & \cdots & -\frac{1}{\alpha^{d-2}} & \frac{1}{\alpha^{d-2}}+\frac{1}{\alpha^{d-1}} \end{bmatrix} \tag{7.83}$$

and the regressors

$$\varphi_d(t) = \begin{pmatrix} u(t-1) \\ \vdots \\ u(t-d) \end{pmatrix}. \tag{7.84}$$

Now, one can easily see that the first d components of $\hat{g}^d$ in (7.81) are contained in the vector

$$\arg\min_{\theta} \sum_{t=1}^{N} \left(y(t) - \varphi_d(t)^T \theta \right)^2 + \gamma \theta^T R \theta. \tag{7.85}$$

Hence, we obtain

$$\hat{g}^d = (\hat{g}(1), \dots, \hat{g}(d), 0, 0, \dots) \tag{7.86}$$

where

$$\begin{pmatrix} \hat{g}(1) \\ \vdots \\ \hat{g}(d) \end{pmatrix} = \left(\frac{1}{N} \sum_{t=1}^{N} \varphi_d(t)\varphi_d^T(t) + \frac{\gamma}{N} R \right)^{-1} \frac{1}{N} \sum_{t=1}^{N} \varphi_d(t) y(t). \tag{7.87}$$

In real applications, one cannot measure the inputs at all the time instants and our data set $\mathscr{D}_T$ in (7.78) could contain only the inputs $u(1), \dots, u(N)$. So, differently from what postulated in the above equations, in practice the regressors are never perfectly known. One solution is just to replace with zeros the unknown input values $\{u(t)\}_{t<1}$ entering (7.84). Also under this model misspecification, all the results introduced in the next sections still hold.

7.5.3   Bounds on the Estimation Error and Minimax Properties

The following theorem will report non asymptotic bounds that illustrate the dependence of $\mathscr{E}\|g^0 - \hat{g}^d\|_2$ on the following three key variables:

- the FIR order d which determines the truncation error;
- the parameter α contained in the matrix R reported in (7.83) that establishes the exponential decay of the estimated impulse response coefficients;
- the regularization parameter γ which trades-off the penalty defined by R and the adherence to experimental data.

In addition, it gives conditions on α which ensure optimality in order if some conditions on the stability radius ρ entering (7.75) and on the FIR order d (function of the data set size N) are fullfilled. Below, the notation $O(1)$ indicates an absolute constant, independent of N. Furthermore, given $x \in \mathbb{R}$, we use $\lfloor x \rfloor$ to indicate the largest integer not larger than x. The following result then holds.

Theorem 7.11 (based on [74]) *Let the FIR order d be defined by the following function of the data set size N:*

$$d^* = \left\lfloor \frac{\ln(N(1-\alpha)\sigma_u^2) - \ln(8\gamma)}{\ln(1/\alpha)} \right\rfloor, \tag{7.88}$$

with N large enough to guarantee $d^* \geq 1$.
Then, under Assumption 7.10, the estimator (7.81) satisfies

$$\mathscr{E}\|g - \hat{g}^{d^*}\|_2 \tag{7.89}$$
$$\leq O(1) \left[\frac{L\rho^{d^*+1}}{(1-\rho)} \left(\sqrt{\frac{d^*}{N}} + 1 \right) + \frac{\sigma}{\sigma_u} \sqrt{\frac{d^*}{N}} + \frac{4L\gamma}{1-\alpha} \frac{h_{d^*}}{N} \right],$$

where

$$h_{d^*} = \begin{cases} \sqrt{d^*} & \text{if } \alpha = \rho \\ \dfrac{\rho}{\sqrt{\alpha^2 - \rho^2}} & \text{if } \alpha > \rho \\ \dfrac{\rho}{\sqrt{\rho^2 - \alpha^2}} \left(\dfrac{\rho}{\alpha}\right)^{d^*} & \text{if } \alpha < \rho \end{cases} . \tag{7.90}$$

Furthermore, if the measurement noise is Gaussian and $\sqrt{\alpha} \geq \rho$, the stable spline estimator (7.81) is optimal in order.

To illustrate the meaning of Theorem 7.11, first is useful to recall a result obtained in [43] that relies on the Fano's inequality. It shows that, if a dynamic system is fed with white input and the measurement noise is Gaussian, the expected ℓ_2 error of any impulse response estimator cannot decay to zero faster than $\sqrt{\frac{\ln N}{N}}$ in a minimax sense.

Theorem 7.12 (based on [43]) *Let Assumption 7.10 hold and assume also that the measurement noise is Gaussian. Then, if $\hat{g}$ is any impulse response estimator built with $\mathcal{D}_T$, for N sufficiently large one has*

$$\sup_{g \in \mathscr{S}(\varrho, L)} \mathscr{E}\|\hat{g} - g\|_2 \geq O(1)\sqrt{\frac{\ln N}{N}}. \tag{7.91}$$

∎

To illustrate the convergence rate of the stable spline estimator, first note that the FIR dimension d^* in (7.88) scales logarithmically with N. Apart from irrelevant constants, one in fact has

$$d^* \sim \frac{\ln(N)}{\ln(1/\alpha)}. \tag{7.92}$$

We now consider the three terms on the r.h.s. of (7.89) with $d = d^*$. Since

$$\sqrt{\frac{d^*}{N}} \sim \sqrt{\frac{\ln N}{N}} \quad \text{and} \quad \rho^{d^*} \sim N^{-\frac{\ln \rho}{\ln \alpha}}, \tag{7.93}$$

the first two terms decay to zero at least as $\sqrt{\frac{\ln N}{N}}$. Regarding the third one, one has

$$\frac{h_{d^*}}{N} \sim \begin{cases} \frac{\sqrt{\ln N}}{N} & \text{if } \alpha = \rho \\ \frac{1}{N} & \text{if } \alpha > \rho \\ N^{-\frac{\ln \rho}{\ln \alpha}} & \text{if } \alpha < \rho \end{cases} \tag{7.94}$$

and this shows that the optimal convergence rate is obtained if $\alpha \geq \rho$ but the case $\alpha < \rho$ can be critical. In particular, combining (7.89) with (7.93) and (7.94), the following considerations arise:

- the convergence rate of the stable spline estimator (7.81) does not depend on γ but only on the relationship between the kernel parameter α and the stability radius ρ defining the class of dynamic systems (7.75);
- using Theorem 7.12, one can see from (7.94) that if $\alpha < \rho$ the achievement of the optimal rate is related to the term $N^{-\frac{\ln \rho}{\ln \alpha}}$ which appears as third term in (7.89). The key condition is

$$\frac{\ln \rho}{\ln \alpha} \geq 0.5 \Longrightarrow \sqrt{\alpha} \geq \rho.$$

This indeed corresponds to what was stated in the final part of Theorem 7.11: under Gaussian noise the stable spline estimator is optimal in order if $\sqrt{\alpha}$ is an upper bound on the stability radius ρ.

Relationships (7.93) and (7.94) clarify also what happens when the kernel includes a too fast exponential decay rate, i.e., when $\sqrt{\alpha} < \rho$. In this case, the error goes to

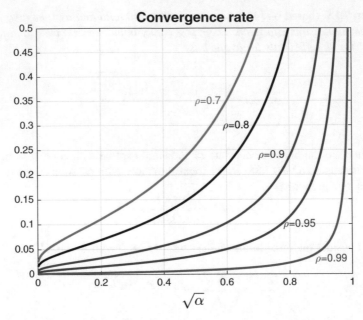

Fig. 7.14 Convergence rate $\ln \rho / \ln \alpha$ of the stable spline estimator as a function of $\sqrt{\alpha}$ for $\sqrt{\alpha} < \rho$ with ρ in the set $\{0.7, 0.8, 0.9, 0.95, 0.99\}$. When $\sqrt{\alpha} < \rho$ the estimation error converges to zero as $N^{-\frac{\ln \rho}{\ln \alpha}}$. Instead, if $\sqrt{\alpha} \geq \rho$ the error decays as $\sqrt{\frac{\ln N}{N}}$, making the stable spline estimator optimal in order when the measurement noise is Gaussian

zero as $N^{-\frac{\ln \rho}{\ln \alpha}}$, getting worse as $\sqrt{\alpha}$ drifts apart ρ. Such phenomenon has a simple explanation. A too small α enforces the impulse response estimate to decay to zero also when the true impulse response coefficients are significantly different from zero. This corresponds to a strong bias: a wrong amount of regularization is introduced in the estimation process, hence compromising the convergence rate. This is also graphically illustrated in Fig. 7.14 that plots the convergence rate $\ln \rho / \ln \alpha$ as a function of $\sqrt{\alpha}$ for five different values of ρ.

The analysis thus shows how α plays a fundamental role in controlling impulse response complexity and, hence, in establishing the properties of the regularized estimator. This is not surprising also in view of the deep connection between the decay rate and the degrees of freedom of the model. This was illustrated in Fig. 5.6 of Sect. 5.5.1 using the class of DC kernels which includes TC as special case.

7.6 Further Topics and Advanced Reading

The idea to handle linear system identification with regularization methods in the RKHS framework first appears in [72]. As already mentioned, the representer theorems introduced in this chapter are special cases of that involving linear and bounded functionals reported in the previous chapter, see Theorem 6.16. More general versions of representer theorems with, e.g., more general loss functions and/or regularization terms can be found in, e.g., [33]. Similarly to the spline smoothing problem studied in Sect. 6.6.7, it could be useful to enrich the regularized impulse response estimators here described with a parametric component. Of course, the corresponding regularized estimator will still have a closed-form finite-dimensional representation that depends on both the number of data N and the number of enriched parametric components, e.g., see [72, 90].

The stable spline kernel [72] and the diagonal correlated kernel [19] are the first two kernels introduced in the linear system identification literature. The stability of a kernel (or equivalently the stability of a RKHS) first appeared in [32, 73]. The stability of a kernel is equivalent to the ∞-boundedness of the kernel, which is a special case of the more general q-boundedness with $1 < q \leq \infty$ in [16]. The proof in [16] for the sufficiency and necessity of the q-boundedness of a kernel is quite involved and abstract. Theorem 7.5 is also discussed in [24], see also [76] where the stability analysis exploits the output kernel. The optimal kernel that minimizes the mean squared error was studied in [19, 73]. As already discussed, unfortunately, the optimal kernel cannot be applied in practice because it depends on the true impulse response to be estimated. Nevertheless, it offers a guideline to design kernels for linear system identification and more general function estimation problems. Motivated by these findings, many stable kernels have been introduced over the years, e.g., [17, 21, 77, 80, 97]. In particular, [17] proposed linear multiple kernels to handle systems with complicated dynamics, e.g., with distinct time constants and distinct resonant frequencies, and [77] further extended this idea and proposed "integral" versions of the stable spline kernels. To design kernels to embed more general prior knowledge, e.g., the overdamped/underdamped dynamics, common structure, etc., it is natural to divide the prior knowledge into different types and then develop systematic ways to design kernels accordingly, see [21, 80, 97]. In particular, the approaches proposed in [21] are based on machine learning and a system theory perspectives, those in [80] rely on the maximum entropy principle, and the method proposed in [97] uses harmonic analysis.

Along with the kernel design, many efforts have also been spent on "kernel analysis". In particular, many kernels can be given maximum entropy interpretations including the stable spline kernel, the diagonal correlated kernel and the more general simulation-induced kernel [14, 21, 23]. This can help to understand the prior knowledge embedded in the model. Many kernels have the Markov property e.g., [83]. Examples are the diagonal correlated kernel and some carefully designed simulation induced kernels [21]. Exploring this property could help to design efficient implementation. As we have seen, the spectral analysis of kernels is often not available in closed form, even is it can be numerically recovered, but exceptions include the stable spline and the diagonal correlated kernel [20, 22, 72].

The hyperparameter tuning problem has been studied for a long time in the context of function estimation problem from noisy observations, e.g., [83, 90]. The marginal likelihood maximization method depends on the connection with the Bayesian estimation of Gaussian processes, which was first studied in [51] in spline regression, see also [41, 83, 90]. More discussions on its relation to Bayesian evidence and Occam's razor principle can be found in e.g., [27, 60]. Stein's unbiased risk estimation method is also known as the C_p statistics [61]. The generalized cross-validation method is first proposed in [28] and found to be rotation invariant in [44]. The problem can also be tackled using full Bayes approaches relying on stochastic simulation techniques, e.g., Markov chain Monte Carlo [1, 39].

In the context of linear system identification, some theoretical results on the hyperparameter estimation problem have been derived. In particular, it was shown in [4] that the marginal likelihood maximization method is consistent for diagonal kernels in terms of the mean square error and asymptotically minimizes a weighted mean square error for nondiagonal kernels. In [78], the robustness of the marginal likelihood maximization is analysed with the help of the excess degrees of freedom. It is further shown in [63, 64, 66] that Stein's unbiased risk estimation as well as many cross-validation methods are asymptotically optimal in the sense of mean square error. In [4, 17, 94], the optimal hyperparameter of the marginal likelihood maximization is shown to be sparse. By exploring such property it is possible to handle various structure detection problems in system identification like sparse dynamic network identification [17, 26]. Full Bayes approaches can be found, e.g., in [69].

As also recalled in the previous chapter, straightforward implementation of the regularization method in RKHS framework has computational complexity $O(N^3)$ and thus is prohibitive to apply when N is large. Many efficient approximation methods have been proposed in machine learning, e.g., [53, 81, 82]. In the context of linear system identification, there is another practical issue that must be noted in the implementation: the ill-conditioning possibly arising from the use of stable kernels, which is unavoidable due to the nature of stability. Hence, extra care has to be taken when developing efficient implementations. Some approximation methods have been proposed to reduce the computational complexity and avoid numerical computation. The first one is to truncate the IIR at a suitable finite-order n. Then, computational complexity becomes $O(n^3)$ and one can also use the approach proposed in [18] relying on some fundamental algebraic techniques and reliable matrix factorizations. The other one is to truncate the infinite expansion of a kernel at a finite-order l. Then, computational complexity becomes $O(l^3)$, see [15]. See also [36] for efficient kernel-based regularization implementation using Alternating Direction Method of Multipliers (ADMM). Another practical issue is the difficulty caused by local minima. For kernels with few number of hyperparameters, e.g., the stable spline kernel and the diagonal correlated kernel, this difficulty can be well faced using different starting points or also some grid methods. For systems with complicated dynamics, it is suggested to apply linear multiple kernels [17] since the corresponding marginal likelihood maximization is a difference of convex programming problem and a stationary point can be found efficiently using sequential convex optimization technique, e.g., [48, 87].

We only considered single-input single-output linear systems in the chapter with white measurement noise. For multiple-input single-output linear systems, it is natural to use multi-input impulse response models and then assume that the overall system has a block diagonal kernel [73]. The regularization method can also be extended to handle linear systems with colored noise, e.g., ARMAX models. One can exploit the fact that such systems can be approximated arbitrarily well by finite-order ARX models [57]. The problem thus becomes a special case of multiple-output single-input systems where the regressors contain also past outputs [71]. This will be also illustrated in Chap. 9.

In practice, the data could be contaminated by outliers due to a failure in the measurement or transmission equipment, e.g., [56, Chap. 15]. In the presence of outliers, it is suggested to use heavy-tailed distributions instead of the commonly used Gaussian distribution for the noise in robust statistics, e.g., [49]. For regularization methods in the RKHS framework, the key difficulty is that the hyperparameter estimation criteria and the regularized estimate may not have closed-form expressions. Several methods have been proposed to overcome this difficulty. In particular, an expectation maximization (EM) method was proposed in [10] and further improved in [55] exploiting a variational expectation method.

Input design is an important issue for classical system identification and many results have been obtained, e.g., [38, 45, 47, 56]. For regularized system identification in RKHS, some results have been reported recently. The first result was given in [37] where the mutual information between the output and the impulse response was chosen as the input design criterion. Unfortunately, obtaining the optimal input involves the solution of a nonconvex optimization problem. Differently from [37], [65] adopts scalar measures of the Bayesian mean square error as input design criterion, proposing a two-step procedure to find the global optimal input through convex optimization.

For what concerns the building of uncertainty regions around the dynamic system estimates, approaches are available which return bounds that, beyond being non-asymptotic, are also exact, i.e., with the desired inclusion probability. This requires some assumptions on data generation, like the introduction of prior distributions on the impulse response. An important example, already widely discussed in this book, is the use of a Bayesian framework that interprets regularization as Gaussian regression [83]. The posterior density becomes available in closed form and Bayes intervals can be easily obtained. Another approach to compute bounds for linear regression is the sign-perturbed sums (SPS) technique [30]. Following a randomization principle, it builds guaranteed uncertainty regions for deterministic parametric models in a quasi-distribution free setup [11, 12]. Recently, there have been notable extensions to the class of models that SPS can handle. The first line of thought still sees the unknown parameters as deterministic but introduces regularization, see [29, 70, 89] and also [31] which is a first attempt to move beyond the strictly parametric nature of SPS. A second line of thought allows for the exploitation of some form of prior knowledge at a more fundamental probabilistic level [13, 70].

Finally, the interested readers are referred to the survey [73] for more references, see also [25, 58].

7.7 Appendix

7.7.1 Derivation of the First-Order Stable Spline Norm

We will exploit a representation of the RKHS induced by the first-order discrete-time stable spline kernel given by a linear transformation of the space ℓ_2 containing the squared summable sequences. This has some connections with the relationship between squared summable function spaces and RKHS discussed in Remark 6.2, even if no spectral decomposition of the kernel will be needed below.

Let $\mathscr{H}$ be the RKHS induced by the stable spline kernel (7.15) with elements denoted by $g = \{g_t\}_{t=1}^{+\infty}$. We will see that any $g \in \mathscr{H}$ can be written as

$$g_t = \sum_{j=1}^{\infty} \psi_{tj} w_j, \qquad w \in \ell_2, \tag{7.95}$$

where the scalars $\{\psi_{tj}\}$ define the linear operator mapping ℓ_2 into $\mathscr{H}$. By adopting notation of ordinary algebra to handle infinite-dimensional objects, one can see g as an infinite-dimensional column vector. In addition, (7.95) can be rewritten as $g = \Psi w$, where Ψ is an infinite-dimensional matrix with (t, j)-entry given by ψ_{tj}. We will now obtain the expression of Ψ. Let

$$\Lambda = \text{diag}\{\lambda_1, \lambda_2, \lambda_3, \dots\}, \qquad \lambda_t = \alpha^t - \alpha^{t+1}$$

$$\mathscr{M} = [v^1 \ v^2 \ v^3 \cdots], \qquad v^t = \sum_{j=1}^{t} e^j$$

where e^j is the infinite-dimensional column vector with all null elements except its jth entry which is equal to one. Let also

$$\Psi = \mathscr{M} \Lambda^{1/2}. \tag{7.96}$$

The inverse Ψ^{-1} of Ψ acts as follows: given a sequence g, it maps g into

$$\Psi^{-1} g = \begin{bmatrix} \frac{1}{\sqrt{\alpha - \alpha^2}}(g_1 - g_2) \\ \frac{1}{\sqrt{\alpha^2 - \alpha^3}}(g_2 - g_3) \\ \vdots \end{bmatrix}. \tag{7.97}$$

Then, given Ψ in (7.96), we will show that the space

$$\mathscr{H} = \left\{ \Psi w \ \middle| \ w \in \ell_2 \right\}, \tag{7.98}$$

with inner product given by

$$\langle f, g \rangle_{\mathscr{H}} = \langle \Psi^{-1} f, \Psi^{-1} g \rangle_2, \tag{7.99}$$

is the RKHS induced by the stable spline kernel. First, it is easy to see that the null space of Ψ contains only the null vector. Then, since ℓ_2 is Hilbert, one obtains that $\mathscr{H}$ is a Hilbert space. We can now exploit Theorem 6.2, i.e., the Moore–Aronszajn theorem, to prove that it is also the desired RKHS. To obtain this, the two conditions described below have to be checked.

The first condition says that any kernel section must belong to the space $\mathscr{H}$ in (7.98). Thanks to the algebraic view, we can see the stable spline kernel K as an infinite-dimensional matrix. Hence, the kernel sections are the infinite-dimensional columns of K and, in particular, we use K_t to indicate the tth column. Now, one has to assess that $\|K_t\|^2_{\mathscr{H}} < \infty \; \forall t$. Note that

$$\Psi^{-1} K_t = \begin{bmatrix} 0 \\ \vdots \\ 0 \\ \sqrt{\alpha^t - \alpha^{t+1}} \\ \sqrt{\alpha^{t+1} - \alpha^{t+2}} \\ \vdots \end{bmatrix} \quad \leftarrow t\text{th row.} \tag{7.100}$$

Then, we have

$$\langle K_t, K_t \rangle_{\mathscr{H}} = \langle \Psi^{-1} K_t, \Psi^{-1} K_t \rangle_2$$
$$= \sum_{j=t}^{+\infty} (\alpha^j - \alpha^{j+1}) = \alpha^t < \infty$$

and the first condition is so satisfied.

The second condition is the reproducing property, i.e., one has to assess that

$$\langle K_t, g \rangle_{\mathscr{H}} = g_t \quad \forall g \in \mathscr{H}, \; \forall t.$$

This holds true since

$$\langle K_t, g \rangle_{\mathscr{H}} = \langle \Psi^{-1} K_t, \Psi^{-1} g \rangle_2$$
$$\sum_{j=t}^{+\infty} (g_j - g_{j+1}) = g_t,$$

showing that the second condition is also satisfied.

Using (7.99), one has

$$\|g\|_{\mathcal{H}}^2 = \langle \Psi^{-1}g, \Psi^{-1}g \rangle_2 = \sum_{t=1}^{\infty} \frac{(g_{t+1} - g_t)^2}{(1-\alpha)\alpha^t}$$

and this confirms the norm's structure reported in (7.16).

7.7.2 Proof of Proposition 7.1

We will exploit the results on estimation of Gaussian vectors reported in Sect. 4.2.2. Let $\text{Cov}[u, v]$ denote the covariance matrix of two random vectors u and v, i.e.,

$$\text{Cov}[u, v] := \mathscr{E}[(u - \mathscr{E}[u])(v - \mathscr{E}[v])^T].$$

First, we consider the distribution of Y. Note that $L_i[g^0]$ is a linear functional of the stochastic process g^0. Hence, since linear transformations of normal processes preserve Gaussianity, the noise-free output $[L_1[g^0], \dots, L_N[g^0]]^T$ is a multivariate zero-mean Gaussian random vector. Furthermore, since

$$\text{Cov}(L_i[g^0], L_j[g^0]) = \lambda L_i[L_j[K]],$$

the covariance matrix of $[L_1[g^0], \dots, L_N[g^0]]^T$, apart from the scale factor λ, is indeed defined by the output kernel matrix O reported in (7.14) for the discrete-time case, i.e., when $\mathscr{X} = \mathbb{N}$, and in (7.22) for the continuous-time case, i.e., when $\mathscr{X} = \mathbb{R}^+$. Now, recall that the $e(t)$, where $t = 1, \dots, N$, are assumed to be mutually independently Gaussian distributed with mean zero and variance σ^2. Moreover, they are also assumed independent of g^0. One then obtains that g^0 and Y are jointly Gaussians, with the mean and covariance matrix of Y given by

$$\mathscr{E}(Y) = 0, \quad \text{Cov}(Y, Y) = \lambda O + \sigma^2 I_N.$$

For what regards the covariance matrix of g^0 and Y, the independence assumptions imply that

$$\text{Cov}(g^0(x), Y) = \lambda[L_1[K_x], \dots, L_N[K_x]].$$

Then, using also the correspondence $\gamma = \sigma^2/\lambda$, we have

$$\begin{aligned}
\mathscr{E}[g^0(x)|Y] &= \lambda[L_1[K_x] \dots L_N[K_x]] \left(\lambda O + \sigma^2 I_N\right)^{-1} Y \\
&= [L_1[K_x] \dots L_N[K_x]] \left(O + \gamma I_N\right)^{-1} Y \\
&= \sum_{t=1}^{N} \hat{c}_t L_t[K_x]
\end{aligned}$$

where $\hat{c}_t$ is the tth entry of vector $\hat{c}$ defined in (7.13) for the continuous-time case or in (7.21) for the discrete-time case. This completes the proof.

7.7.3 Proof of Theorem 7.5

We only consider the proof for the discrete-time case (7.56). The continuous-time case (7.57) can be proved in a similar way. To prove (7.56), we first need a lemma.

Lemma 7.1 *Consider the linear operator L_K defined by*

$$L_K[l](\cdot) = \sum_{t=1}^{\infty} K(\cdot, t)l_t, \tag{7.101}$$

where $K : \mathbb{N} \times \mathbb{N} \to \mathbb{R}$ is a positive semidefinite kernel. Assume that L_K satisfies the following property: for any $l \in \ell_{\infty}$, one has $L_K[l] \in \ell_1$. Then, L_k is a continuous (bounded) linear operator, i.e., there exists a scalar $b > 0$, independent of l, such that

$$\|L_K[l]\|_1 \leq b\|l\|_{\infty}, \ \forall l \in \ell_{\infty}. \tag{7.102}$$

Proof First, we show that for any $s \in \mathbb{N}$, the kernel section $K_s(\cdot)$ belongs to ℓ_1. To show this, for any $s \in \mathbb{N}$, we can define a sequence $l \in \ell_{\infty}$ in the following way:

$$l_t = \begin{cases} 1 & \text{if } K(s, t) \geq 0 \\ -1 & \text{otherwise.} \end{cases}$$

Then plugging this l into (7.101) yields $L_K[l] = \sum_{t=1}^{\infty} |K(s, t)|$. Since $L_K[l] \in \ell_1$ for every $l \in \ell_{\infty}$, then we obtain

$$\sum_{t=1}^{\infty} |K(s, t)| < \infty, \quad \forall s \in \mathbb{N}. \tag{7.103}$$

Now, for any $l, a \in \ell_{\infty}$, it holds that

$$|L_K[l](s) - L_K[a](s)| = \left| \sum_{t=1}^{\infty} K(s, t)(l_t - a_t) \right| \leq \|l - a\|_{\infty} \sum_{t=1}^{\infty} |K(s, t)|, \tag{7.104}$$

where both $\|l - a\|_{\infty}$ and $\sum_{t=1}^{\infty} |K(s, t)|$ are finite for any $s \in \mathbb{N}$ since $l, a \in \ell_{\infty}$ and in view of (7.103). Following (7.104), the remaining proof is a simple application of the closed graph theorem, see Theorem 6.26. In fact, let $l \to a$ in ℓ_{∞} and $L_K[l] \to g$

in ℓ_1. Then (7.104) shows that $L_K[l](s) \to L_K[a](s)$ for every $s \in \mathbb{N}$, implying that $g_s = L_K[a](s)$ for every $s \in \mathbb{N}$. As a result, the graph $(l, L_K[l])$ is closed and thus L_K is continuous (bounded) by the closed graph theorem. $\qquad\square$

Now let us consider (7.56) in Theorem 7.5. We first prove the sufficient part, i.e.,

$$\sum_{s=1}^{\infty}\left|\sum_{t=1}^{\infty} K(s,t)l_t\right| < \infty, \ \forall l \in \ell_\infty \implies \mathscr{H} \subset \ell_1.$$

We start by introducing some definitions. For any $f \in \mathscr{H}$, we let $l \in \ell_\infty$ be a sequence defined by the signs of f, i.e.,

$$l_t = \begin{cases} 1 & \text{if } f_t \geq 0 \\ -1 & \text{otherwise} \end{cases}$$

and let also l^n be a sequence defined by

$$l_t^n = \begin{cases} l_t & \text{for } t = 1, \dots, n \\ 0 & \text{otherwise.} \end{cases}$$

Then we have

$$\sum_{t=1}^{n} |f_t| = \sum_{t=1}^{\infty} f_t l_t^n = \sum_{t=1}^{\infty} \langle f(\cdot), l_t^n K_t(\cdot)\rangle_{\mathscr{H}},$$

where the last identity is due to the reproducing property of K. Moreover, by the Cauchy–Schwarz inequality, we have

$$\sum_{t=1}^{n} |f_t| \leq \|f\|_{\mathscr{H}} \left\|\sum_{t=1}^{\infty} l_t^n K_t(\cdot)\right\|_{\mathscr{H}}. \qquad (7.105)$$

Now we show that $\left\|\sum_{t=1}^{\infty} l_t^n K_t(\cdot)\right\|_{\mathscr{H}}$ is finite. First, we note that

$$\left\|\sum_{t=1}^{\infty} l_t^n K_t(\cdot)\right\|_{\mathscr{H}}^2 = \langle \sum_{s=1}^{\infty} l_s^n K_s(\cdot), \sum_{t=1}^{\infty} l_t^n K_t(\cdot)\rangle_{\mathscr{H}}$$

$$= \sum_{s=1}^{\infty}\left(\sum_{t=1}^{\infty} l_t^n K(s,t)\right) l_s^n$$

$$\leq \sum_{s=1}^{\infty}\left|\sum_{t=1}^{\infty} l_t^n K(s,t))\right| \|l^n\|_\infty,$$

and then from the linear operator L_K defined in (7.101) and its boundedness property (7.102) proved in Lemma 7.1, we obtain

$$\left\|\sum_{t=1}^{\infty} l_t^n K_t(\cdot)\right\|_{\mathcal{H}}^2 \leq \|L_K[l^n]\|_1 \|l^n\|_\infty \leq b\|l^n\|_\infty^2 = b,$$

where we have used the fact that $\|l^n\|_\infty = 1$ for any $n \in \mathbb{N}$. Noting the above equation and (7.105) yields

$$\sum_{t=1}^{n} |f_t| \leq \|f\|_{\mathcal{H}} \sqrt{b}, \quad \forall n \in \mathbb{N}.$$

Since $f \in \mathcal{H}$ and thus $\|f\|_{\mathcal{H}}$ is finite, $\sum_{t=1}^{n} |f_t|$ is bounded above for any $n \in \mathbb{N}$. Further note that the partial sum $\sum_{t=1}^{n} |f_t|$ is an increasing sequence and bounded above, therefore by monotone convergence theorem, the limit of $\sum_{t=1}^{n} |f_t|$, i.e., $\lim_{n \to \infty} \sum_{t=1}^{n} |f_t|$ exists, and is denoted by $\sum_{t=1}^{\infty} |f_t|$, which shows that $f \in \ell_1$. Since f was chosen arbitrarily, this implies $\mathcal{H} \subset \ell_1$ and thus completes the proof for the sufficient part.

Now, we prove the necessary part, i.e.,

$$\mathcal{H} \subset \ell_1 \implies \sum_{s=1}^{\infty} \left|\sum_{t=1}^{\infty} l_t K(s,t)\right| < \infty \ \forall l \in \ell_\infty.$$

Again, we start by introducing some definitions. For any $f \in \mathcal{H}$ and $l \in \ell_\infty$, we define a new sequence lf by letting $[lf]_t = l_t f_t$, $\forall t \in \mathbb{N}$, where $[lf]_t$ is the tth entry in the sequence lf. Then we have $lf \in \ell_1$, because $l \in \ell_\infty$ and $f \in \ell_1$ due to $\mathcal{H} \subset \ell_1$. Moreover, we define $g^n(\cdot) = \sum_{t=1}^{n} l_t K_t(\cdot)$ with $n \in \mathbb{N}$. Now we show that the sequence of functions $g^n(\cdot)$ with $n \in \mathbb{N}$ is a *weak Cauchy sequence* in $\mathcal{H}$. To show this, we take without loss of generality $m \leq n$ and $m \in \mathbb{N}$, and then we have

$$g^n(\cdot) - g^m(\cdot) = \sum_{t=m+1}^{n} l_t K_t(\cdot). \tag{7.106}$$

Moreover, we have

$$\langle g^n(\cdot) - g^m(\cdot), f(\cdot)\rangle_{\mathcal{H}} = \langle \sum_{t=m+1}^{n} l_t K_t(\cdot), f(\cdot)\rangle_{\mathcal{H}} = \sum_{t=m+1}^{n} l_t f_t, \ \forall f \in \mathcal{H}.$$

Since $lf \in \ell_1$, i.e., $\sum_{t=1}^{\infty} |l_t f_t| < \infty$, the Cauchy criterion ensures that

$$\lim_{m,n \to \infty} \sum_{t=m+1}^{n} |l_t f_t| = 0, \tag{7.107}$$

which implies

$$\lim_{m,n\to\infty} \sum_{t=m+1}^{n} l_t f_t = 0.$$

Noting the above equation and (7.106) yields that the sequence of functions $g^n(\cdot) = \sum_{t=1}^{n} l_t K_t(\cdot)$ with $n \in \mathbb{N}$ is a weak Cauchy sequence. Recall that every Hilbert space, beyond being complete, is also *weakly sequentially complete*, which is because every Hilbert space is reflexive, see Definition 2.5.23 along with Corollaries 2.8.10 and 2.8.11 in [62]. Hence, the sequence of functions $g^n(\cdot) = \sum_{t=1}^{n} l_t K_t(\cdot)$ with $n \in \mathbb{N}$ is also a *weakly convergent sequence*, i.e., there exists an $h \in \mathcal{H}$ such that

$$\lim_{n\to\infty} \langle g^n(\cdot), f(\cdot)\rangle_{\mathcal{H}} = \langle h(\cdot), f(\cdot)\rangle_{\mathcal{H}}, \ \forall f \in \mathcal{H}.$$

Now, we take $f(\cdot) = K_s(\cdot)$ in the above equation. Using the reproducing property of K, the left-hand side becomes

$$\lim_{n\to\infty} \langle g^n(\cdot), K_s(\cdot)\rangle_{\mathcal{H}} = \sum_{t=1}^{\infty} l_t K(s, t),$$

while the right-hand side becomes

$$\langle h(\cdot), K_s(\cdot)\rangle_{\mathcal{H}} = h(s).$$

This implies that

$$\sum_{t=1}^{\infty} l_t K(s, t) = h(s) \quad \forall s \in \mathbb{N}.$$

Finally, note that $h \in \mathcal{H} \subset \ell_1$, therefore

$$\sum_{s=1}^{\infty} \left| \sum_{t=1}^{\infty} l_t K(s, t) \right| < \infty, \ \forall l \in \ell_\infty,$$

which completes also the necessary part and, hence, concludes the proof.

7.7.4 Proof of Theorem 7.7

First, it is useful to set up some notation. Let r be an integer or $r = \infty$. Then, we define the set $\mathcal{U}_r$ as follows:

$$\mathcal{U}_r := \{ x \in \mathbb{R}^r : x(i) = \pm 1, \forall i = 1, \ldots, r \}. \tag{7.108}$$

Let p be another integer associated with the odd number $m = 2p + 1$ and with $n = 2^m$. We also use $x_i \in \mathscr{U}_m$, with $i = 1, 2, \ldots, n$, to indicate distinct vectors containing exactly m elements ± 1 (their ordering is irrelevant). Then, for any $n = 2^3, 2^5, 2^7, \ldots$, the $n \times m$ matrix $V^{(n)}$ is given by

$$V^{(n)} = \begin{bmatrix} x_1 \, x_2 \, \ldots \, x_n \end{bmatrix}^T \qquad (7.109)$$

and its rows contain all the possible permutations of ± 1. We now discuss the inclusions stated in the theorem.

The inclusion $\mathscr{S}_1 \subseteq \mathscr{S}_s$ derives from Corollary 7.1 where we have seen that absolute summability is a sufficient condition for kernel stability. The proof of the strict inclusion $\mathscr{S}_1 \subset \mathscr{S}_s$ is not trivial and is reported in [7] where one can find a particular kernel, function of the matrices $V^{(n)}$ in (7.109), that is stable but non-absolutely summable.

For what concerns the inclusion $\mathscr{S}_s \subset \mathscr{S}_{ft}$, let M_m denote a positive semidefinite matrix of size $m \times m$. Consider also the linear operator $M_m : \mathbb{R}^m \to \mathbb{R}^m$ with domain and co-domain equipped, respectively, with the ℓ_∞ and the ℓ_1 norms. Its operator norm is then given by

$$\|M_m\|_{\infty,1} := \max_{\|u\|_\infty = 1} \|M_m u\|_1 = \max_{x \in \mathscr{U}_m} \|M_m x\|_1, \qquad (7.110)$$

where the last equality follows from the so-called Bauer's maximum principle for convex functions. First, we prove that

$$\operatorname{trace}(M_m) \leq \|M_m\|_{\infty,1} \leq n \operatorname{trace}(M_m). \qquad (7.111)$$

For this aim, since $V^{(n)T}$ contains all the vectors in $\mathscr{U}_m$ as columns, the problem is equal to evaluating

$$M_m V^{(n)T}$$

and to find the column with maximum ℓ_1 norm. The ℓ_1 norm of each column can be obtained as the scalar product of the column with a suitable $x \in \mathscr{U}_m$ containing the signs of the column entries. Hence, the n^2 entries of

$$V^{(n)} M_m V^{(n)T}$$

surely contain these n ℓ_1 norms. Furthermore, the maximum ℓ_1 norm which needs to be found is the maximum of all these n^2 entries since $x_1^T c \leq x_2^T c$, $\forall x_1 \in \mathscr{U}_m$ if $x_2 = \operatorname{sign}(c)$, where the function sign returns, for each entry of c, value 1 if such entry is larger than zero and -1 otherwise. Also, since $V^{(n)} M_m V^{(n)T}$ is positive semidefinite, the maximum is found along its diagonal, i.e.,

$$\|M_m\|_{\infty,1} = \max_{i=1,\ldots,n} [V^{(n)} M_m V^{(n)T}]_{ii}.$$

We now note that the trace of $V^{(n)} M_m V^{(n)T}$ satisfies

$$\text{trace}[V^{(n)} M_m V^{(n)T}] \geq \|M_m\|_{\infty,1} \geq \frac{1}{n} \text{ trace}[V^{(n)} M_m V^{(n)T}].$$

Finally,

$$\begin{aligned}
\text{trace}[V^{(n)} M_m V^{(n)T}] &= \text{trace}[M_m V^{(n)T} V^{(n)}] \\
&= \text{trace}[M_m(n I_m)] = n \text{ trace}[M_m]
\end{aligned}$$

and this proves (7.111).

Now, think of M_k as the $k \times k$ submatrix of the stable kernel represented by the infinite-dimensional matrix K. We also use L_K to denote the associated kernel operator mapping ℓ_∞ into ℓ_1. So, it holds that

$$\|M_k\|_{\infty,1} \leq \|L_K\|_{\infty,1} < +\infty, \ \forall k = 1, 2, \ldots,$$

where $\|L_K\|_{\infty,1}$ indicates the operator norm of L_K, i.e.,

$$\|L_K\|_{\infty,1} = \max_{x \in \mathscr{U}_\infty} \|Kx\|_1.$$

Using (7.111), we obtain

$$\text{trace}[M_k] \leq \|L_K\|_{\infty,1}, \ \forall k = 1, 2, \ldots$$

and, since $\text{trace}[M_k]$ is a monotone non-decreasing sequence upper-bounded by $\|L_K\|_{\infty,1} < +\infty$, one also has

$$\sum_i K_{ii} \leq \|L_K\|_{\infty,1} < +\infty.$$

This shows that the trace of any stable kernel is finite. Such inclusion is strict as the following example shows. Let the vector v s.t. $v \in \ell_2$ and $v \notin \ell_1$. Consider the kernel

$$K = vv^T.$$

One has $\text{trace}(K) = \|v\|_2^2 < +\infty$. If $w = sign(v) \in \ell_\infty$ one has $Kw = v\|v\|_1$ and this implies $\|Kw\|_1 = \infty$. So, the kernel K has finite trace but is unstable.

The inclusion $\mathscr{S}_{ft} \subset \mathscr{S}_2$ relies on the important relation between nuclear and Hilbert–Schmidt (HS) operators, e.g., see [35, 54, 84]. In particular, let K be a kernel, seen as an infinite-dimensional matrix, and let L_K be the induced kernel operator as a map from ℓ_2 into ℓ_2 itself. Given any orthonormal basis $\{v_i\}$ in ℓ_2, the nuclear norm of L_K is

$$\sum_{i=1}^{\infty} \langle v_i, K v_i \rangle_2, \tag{7.112}$$

and is independent of the chosen basis. Then, L_K is said to be nuclear if (7.112) is finite. Its (squared) Hilbert–Schmidt (HS) norm is instead

$$\sum_{i=1}^{\infty} \|K v_i\|_2^2 \tag{7.113}$$

and is also independent of the chosen basis. Then, L_K is said to be HS if (7.113) is finite. It is also known that any nuclear operator is HS and can be written as the composition of two HS operators.

For our purposes, we now exploit the fact that any finite-trace kernel induces a nuclear operator, as shown in [8]. So, one also has that (7.113) is finite and, choosing as $\{v_i\}$ the canonical basis $\{e_i\}$ of ℓ_2, one obtains

$$\sum_{i=1}^{\infty} \|K e_i\|_2^2 = \sum_{ij} K_{ij}^2 < \infty. \tag{7.114}$$

Such inclusion is also strict as illustrated via the example

$$K = \text{diag}\{1, 1/2, 1/3, \ldots, 1/k, \ldots\}.$$

Finally, $\mathscr{S}_2$ is contained in the set of all the positive semidefinite infinite matrices. Furthermore, the inclusion is strict: this can be seen just considering the example $K = v v^T$, where v is the infinite-dimensional column vector with all components equal to 1.

7.7.5 Proof of Theorem 7.9

The notation L_K is still used to denote the operator induced by the kernel K and mapping ℓ_∞ into ℓ_1. Its operator norm is $\|L_K\|_{\infty,1}$ while (ζ_i, ρ_i) are its eigenvalues and eigenvectors orthogonal in ℓ_2. From Theorem 7.5 and Lemma 7.1, one has

$$\mathscr{H} \subset \ell_1 \iff \|L_K\|_{\infty,1} < +\infty. \tag{7.115}$$

Since the function

$$f(u) := \|y\|_1 = \sum_i |y(i)| = \sum_i \left| \sum_h K_{ih} u(h) \right|$$

is convex, the Bauer's maximum principle ensures that

$$\|L_K\|_{\infty,1} = \sup_{u \in \mathscr{U}_\infty} f(u) = \sup_{u \in \mathscr{U}_\infty} \sum_i \left| \sum_h K_{ih} u(h) \right|, \qquad (7.116)$$

where

$$\mathscr{U}_\infty = \left\{ u \in \ell_\infty : |u(i)| = 1, \ \forall i \geq 1 \right\}.$$

Using notation of ordinary algebra to deal with infinite-dimensional matrices, we can write $K = UDU^T$, where D is diagonal and contains the eigenvalues ζ_i of K while the columns of U contain the corresponding eigenvectors ρ_i. One has

$$y = Ux, \quad x = DU^T u$$

and, hence,

$$x = \left[\zeta_1 < \rho_1, u >_2 \ \zeta_2 < \rho_2, u >_2 \ ... \right]^T$$
$$y = \zeta_1 < \rho_1, u >_2 \rho_1 + \zeta_2 < \rho_2, u >_2 \rho_2 + ...$$

Letting $s(u) = \text{sign}(y)$, we obtain

$$h(u) := \|y\|_1 = \sum_h \zeta_h < \rho_h, u >_2 < \rho_h, s(u) >_2 .$$

Using (7.116), also noticing that $f(u) = h(u)$, this implies

$$\|L_K\|_{\infty,1} = \sup_{u \in \mathscr{U}_\infty} \sum_h \zeta_h < \rho_h, u >_2 < \rho_h, s(u) >_2$$
$$= \sup_{u \in \mathscr{U}_\infty} h(u).$$

Now, define

$$g(u) := \Sigma_h \zeta_h \langle \rho_h, u \rangle_2^2, \qquad A := \sup_{u \in \mathscr{U}_\infty} \sum_h \zeta_h \langle \rho_h, u \rangle_2^2 = \sup_{u \in \mathscr{U}_\infty} g(u).$$

Exploiting the definition of $s(u)$, one has

$$h(u) \geq g(u) \implies \|L_K\|_{\infty,1} \geq A.$$

On the other hand,

$$h(u) = \sum_h \zeta_h \langle \rho_h, u \rangle_2 \langle \rho_h, s(u) \rangle_2$$

$$= \sum_h \left(\sqrt{\zeta_h} \langle \rho_h, u \rangle_2 \right) \left(\sqrt{\zeta_h} \langle \rho_h, s(u) \rangle_2 \right)$$

$$\leq \sqrt{\sum_h \zeta_h \langle \rho_h, u \rangle_2^2} \sqrt{\sum_h \zeta_h \langle \rho_h, s(u) \rangle_2^2}$$

$$\leq \sqrt{g(u)} \sqrt{g(s(u))}$$

that implies

$$\|L_K\|_{\infty,1} \leq A.$$

So, one has

$$\|L_K\|_{\infty,1} = \sup_{u \in \mathscr{U}_\infty} \sum_h \zeta_h \langle \rho_h, u \rangle_2^2$$

and this concludes the proof in view of (7.115).

References

1. Andrieu C, Doucet A, Holenstein R (2010) Particle Markov Chain Monte Carlo methods. J. R. Stat. Soc. Series B 72(3):269–342
2. Aravkin A, Burke JV, Pillonetto G (2018) Generalized system identification with stable spline kernels. SIAM J. Sci. Comput. 40(5):1419–1443
3. Aravkin A, Bell BM, Burke JV, Pillonetto G (2015) The connection between Bayesian estimation of a Gaussian random field and RKHSs. IEEE Trans. Neural Netw. Learn. Syst. 26(7):1518–1524
4. Aravkin A, Burke JV, Chiuso A, Pillonetto G (2014) Convex vs non-convex estimators for regression and sparse estimation: the mean squared error properties of ARD and GLASSO. J. Mach. Learn. Res. 15(1):217–252
5. Atkinson K (1975) Convergence rates for approximate eigenvalues of compact integral operators. SIAM J. Numer. Anal. 12(2):213–222
6. Baker C (1977) The numerical treatment of integral equations. Clarendon press
7. Bisiacco M, Pillonetto G (2020) Kernel absolute summability is sufficient but not necessary for RKHS stability. SIAM J Control Optim
8. Bisiacco M, Pillonetto G (2020) On the mathematical foundations of stable RKHSs. Automatica
9. Bogachevch7 VJ (1998) Gaussian measures. AMS
10. Bottegal G, Aravkin A, Hjalmarsson H, Pillonetto G (2016) Robust EM kernel-based methods for linear system identification. Automatica 67:114–126
11. Campi MC, Weyer E (2005) Guaranteed non-asymptotic confidence regions in system identification. Automatica 41(10):1751–1764
12. Carè A, Csáji BCs, Campi MC, Weyer E (2018) Finite-sample system identification: an overview and a new correlation method. IEEE Control Syst Lett 2(1):61–66
13. Carè A, Pillonetto G, Campi MC (2018) Uncertainty bounds for kernel-based regression: a Bayesian SPS approach. In: 2018 IEEE 28th international workshop on machine learning for signal processing (MLSP), pp 1–6

14. Carli FP, Chen T, Ljung L (2017) Maximum entropy kernels for system identification. IEEE Trans. Autom. Control 62(3):1471–1477
15. Carli FP, Chiuso A, Pillonetto G (2012) Efficient algorithms for large scale linear system identification using stable spline estimators. In: IFAC symposium on system identification
16. Carmeli C, De Vito E, Toigo A (2006) Vector valued reproducing kernel Hilbert spaces of integrable functions and Mercer theorem. Anal. Appl. 4:377–408
17. Chen T, Andersen MS, Ljung L, Chiuso A, Pillonetto G (2014) System identification via sparse multiple kernel-based regularization using sequential convex optimization techniques. IEEE Trans. Autom. Control 59(11):2933–2945
18. Chen T, Ljung L (2013) Implementation of algorithms for tuning parameters in regularized least squares problems in system identification. Automatica 49:2213–2220
19. Chen T, Ohlsson H, Ljung L (2012) On the estimation of transfer functions, regularizations and Gaussian processes - Revisited. Automatica 48:1525–1535
20. Chen T, Pillonetto G, Chiuso A, Ljung L (2015) Spectral analysis of the DC kernel for regularized system identification. In: Proceedings 54th IEEE conference on decision and control (CDC), pp. 4017–4022, Osaka, Japan
21. Chen T (2018) On kernel design for regularized LTI system identification. Automatica 90:109–122
22. Chen T (2019) Continuous-time DC kernel – a stable generalized first-order spline kernel. IEEE Trans. Autom. Control 63:4442–4447
23. Chen T, Ardeshiri T, Carli FP, Chiuso A, Ljung L, Pillonetto G (2016) Maximum entropy properties of discrete-time first-order stable spline kernel. Automatica 66:34–38
24. Chen T, Pillonetto G (2018) On the stability of reproducing kernel Hilbert spaces of discrete-time impulse responses. Automatica
25. Chiuso A (2016) Regularization and Bayesian learning in dynamical systems: Past, present and future. Annu. Rev. Control 41:24–38
26. Chiuso A, Pillonetto G (2012) A Bayesian approach to sparse dynamic network identification. Automatica 48(8):1553–1565
27. Cox RT (1946) Probability, frequency, and reasonable expectation. Am. J. Phys. 14(1):1–13
28. Craven P, Wahba G (1979) Smoothing noisy data with spline functions. Numer. Math. 31:377–403
29. Csáji B (2019) Non-asymptotic confidence regions for regularized linear regression estimates. In: Faragó István, Izsák Ferenc, Simon Péter L (eds) Progress in Industrial Mathematics at ECMI 2018. pp. Springer International Publishing, Cham, pp 605–611
30. Csáji B, Campi MC, Weyer E (2015) Sign-perturbed sums: a new system identification approach for constructing exact non-asymptotic confidence regions in linear regression models. IEEE Trans Signal Process 63(1):169–181
31. Csáji B, Kis KB (2019) Distribution-free uncertainty quantification for kernel methods by gradient perturbations. Mach Learn 108(8):1677–1699
32. Dinuzzo F (2015) Kernels for linear time invariant system identification. SIAM J. Control Optim. 53(5):3299–3317
33. Dinuzzo F, Scholkopf B (2012) The representer theorem for Hilbert spaces: a necessary and sufficient condition. In: Bartlett P, Pereira FCN, Burges CJC, Bottou L, Weinberger KQ (eds) Advances in neural information processing systems, vol 25, pp 189–196
34. Driscollch7 M (1973) The reproducing kernel Hilbert space structure of the sample paths of a Gaussian process. Zeitschrift fur Wahrscheinlichkeitstheorie und verwandte Gebiete 26:309–316
35. Dunford N, Schwartz JT (1963) Linear operators. InterScience Publishers
36. Fujimoto Y (2021) Efficient implementation of kernel regularization based on ADMM. In: Proceedings of the 19th IFAC symposium on system identification (SYSID), Online, July 2021
37. Fujimoto Y, Sugie T (2018) Informative input design for kernel-based system identification. Automatica 89(3):37–43

38. Gevers M (2005) Identification for control: From the early achievements to the revival of experiment design. Eur. J. Control 11:335–352
39. Gilks WR, Richardson S, Spiegelhalter DJ (1996) Markov chain Monte Carlo in Practice. Chapman and Hall, London
40. Girosi F (1991) Models of noise and robust estimates. A.I. Memo 1287, Artificial Intelligence Laboratory, 1287, Massachusetts Institute of Technology
41. Girosi F, Jones M, Poggio T (1995) Regularization theory and neural networks architectures. Neural Comput. 7(2):219–269
42. Gittens A, Mahoney M (2016) Revisiting the Nyström method for improved large-scale machine learning. J. Mach. Learn. Res. 17(1):3977–4041
43. Goldenshluger A (1998) Nonparametric estimation of transfer functions: rates of convergence and adaptation. IEEE Trans. Inf. Theory 44(2):644–658
44. Golub GH, Heath M, Wahba G (1979) Generalized cross-validation as a method for choosing a good ridge parameter. Technometrics 21(2):215–223 May
45. Goodwin GC, Payne RL (1977) Dynamic system identification: experiment design and data analysis. Academic Press, New York
46. Heuberger P, Van den Hof P, Wahlberg B (2005) Modelling and identification with rational orthogonal basis functions. Springer
47. Hjalmarsson H (2005) From experiment design to closed loop control. Automatica 41(3):393–438
48. Horst R, Thoai NV (1999) DC programming: overview. J Optim Theory Appl 103(1):1–43
49. Huber PJ (1981) Robust Statistics. John Wiley and Sons, New York, NY, USA
50. Kass RE, Raftery AE (1995) Bayes factors. J Am Stat Assoc 90:773–795
51. Kimeldorf G, Wahba G (1970) A correspondence between Bayesian estimation on stochastic processes and smoothing by splines. Ann. Math. Stat. 41(2):495–502
52. Kumar S, Mohri M, Talwalkar A (2012) Sampling methods for the Nyström method. J. Mach. Learn. Res. 13(1):981–1006
53. Lázaro-Gredilla M, Quiñonero-Candela J, Rasmussen CE, Figueiras-Vidal AR (2010) Sparse spectrum Gaussian process regression. J Mach Learn Res 99:1865–1881
54. Lidskii VB (1959) Non-self-adjoint operators with a trace. Dokl. Akad. Nauk. 125:485–487
55. Lindfors M, Chen T (2019) Regularized LTI system identification in the presence of outliers: a variational EM approach. Automatica, under review
56. Ljung L (1999) System identification - theory for the user, 2nd edn. Prentice-Hall, Upper Saddle River
57. Ljung L, Wahlberg B (1992) Asymptotic properties of the least-squares method for estimating transfer functions and disturbance spectra. Adv. Appl. Probab. 24:412–440
58. Ljung L, Chen T, Mu B (2019) A shift in paradigm for system identification. Int J Control 1–8
59. Lukic MN, Beder JH (2001) Stochastic processes with sample paths in reproducing kernel Hilbert spaces. Trans. Am. Math. Soc. 353:3945–3969
60. MacKay DJC (1992) Bayesian interpolation. Neural Comput. 4:415–447
61. Mallows CL (1973) Some comments on CP. Technometrics 15(4):661–675
62. Megginson RE (1998) An introduction to Banach space theory. Springer
63. Mu B, Chen T, Ljung L (2018) Asymptotic properties of generalized cross validation estimators for regularized system identification. In: The 18th IFAC symposium on system identification (SYSID)
64. Mu B, Chen T, Ljung L (2018) On asymptotic properties of hyperparameter estimators for kernel-based regularization methods. Automatica 94:381–395
65. Mu B, Chen T (2018) On input design for regularized LTI system identification: Power-constrained input. Automatica 97:327–338
66. Mu B, Chen T, Ljung L (2018) Asymptotic properties of hyperparameter estimators by using cross-validations for regularized system identification. In: Proceedings of the 57th IEEE conference on decision and control, pp 644–649
67. Palmer JA, Wipf DP, Kreutz-Delgado K, Rao BD (2006) Variational EM algorithms for non-Gaussian latent variable models. Adv Neural Inf Process Syst

68. Parzen E (1963) Probability density functionals and reproducing kernel Hilbert spaces. In: Proceedings of the symposium on time series analysis. Wiley, New York
69. Pillonetto G, Bell BM (2007) Bayes and empirical Bayes semi-blind deconvolution using eigenfunctions of a prior covariance. Automatica 43(10):1698–1712
70. Pillonetto G, Carè A, Campi MC (2018) Kernel-based SPS. IFAC-PapersOnLine 51(15):31–36. 18th IFAC symposium on system identification SYSID 2018
71. Pillonetto G, Chiuso A, De Nicolao G (2011) Prediction error identification of linear systems: a nonparametric Gaussian regression approach. Automatica 47(2):291–305
72. Pillonetto G, De Nicolao G (2010) A new kernel-based approach for linear system identification. Automatica 46(1):81–93
73. Pillonetto G, Dinuzzo F, Chen T, De Nicolao G, Ljung L (2014) Kernel methods in system identification, machine learning and function estimation: a survey. Automatica 50
74. Pillonetto G, Scampicchio A (2021) Sample complexity and minimax properties of exponentially stable regularized estimators. IEEE Trans Autom Control
75. Pillonetto G, Schenato L, Varagnolo D (2019) Distributed multi-agent Gaussian regression via finite-dimensional approximations. IEEE Trans. Pattern Anal. Mach. Intell. 41(9):2098–2111
76. Pillonetto G (2018) System identification using kernel-based regularization: New insights on stability and consistency issues. Automatica 93:321–332
77. Pillonetto G, Chen T, Chiuso A, De Nicolao G, Ljung L (2016) Regularized linear system identification using atomic, nuclear and kernel-based norms: The role of the stability constraint. Automatica 69:137–149
78. Pillonetto G, Chiuso A (2015) Tuning complexity in regularized kernel-based regression and linear system identification: The robustness of the marginal likelihood estimator. Automatica 58:106–117
79. Pontil M, Mukherjee S, Girosi F (2000) On the noise model of support vector machine regression. In: Proceedings of algorithmic learning theory 11th international conference ALT 2000, Sydney
80. Prando G, Chiuso A, Pillonetto G (2017) Maximum entropy vector kernels for MIMO system identification. Automatica 79:326–339
81. Quiñonero-Candela J, Rasmussen CE, Williams CKI (2007) Approximation methods for Gaussian process regression. In: Bottou L, Chapelle O, DeCoste D, Weston J (eds) Large-Scale Kernel Machines. MIT Press, Cambridge, MA, USA, pp 203–223
82. Quiñonero-Candela J, Rasmussen CE (2005) A unifying view of sparse approximate Gaussian process regression. J Mach Learn Res 6:1939–1959
83. Rasmussen CE, Williams CKI (2006) Gaussian processes for machine learning. The MIT Press
84. Robert D (2017) On the traces of operators (from Grothendieck to Lidskii). EMS newsletter 3:26–33
85. Rudin W (1987) Real and Complex Analysis. McGraw-Hill, Singapore
86. Sun H (2005) Mercer theorem for RKHS on noncompact sets. J. Complex. 21(3):337–349
87. Tao PD, An LTH (1997) Convex analysis approach to D.C. programming: theory, algorithms and applications. ACTA Math Vietnam 22:289–355
88. Tsybakov AB (2008) Introduction to nonparametric estimation. Springer
89. Volpe V (2015) Identification of dynamical systems with finitely many data points. University of Brescia, MSc thesis
90. Wahba G (1990) Spline models for observational data. SIAM, Philadelphia
91. Wahlberg B (1991) System identification using Laguerre models. IEEE Trans Autom Control AC-36:551–562
92. Wahlberg B (1994) System identification using Kautz models. IEEE Trans. Autom. Control 39(6):1276–1282
93. Williams CKI, Seeger M (2000) Using the Nyström method to speed up kernel machines. In: Advances in neural information processing systems. MIT Press, Cambridge, pp 682–688
94. Wipf DP, Rao BD (2004) Sparse Bayesian learning for basis selection. IEEE Trans. Signal Process. 52(8):2153–2164
95. Zeidler E (1995) Applied functional analysis. Springer

96. Zhu H, Williams CKI, Rohwer RJ, Morciniec M (1998) Gaussian regression and optimal finite dimensional linear models. In: Neural networks and machine learning. Springer, Berlin
97. Zorzi M, Chiuso A (2018) The harmonic analysis of kernel functions. Automatica 94:125–137

Chapter 8
Regularization for Nonlinear System Identification

Abstract In this chapter we review some basic ideas for nonlinear system identification. This is a complex area with a vast and rich literature. One reason for the richness is that very many parameterizations of the unknown system have been suggested, each with various proposed estimation methods. We will first describe with some details nonparametric techniques based on Reproducing Kernel Hilbert Space theory and Gaussian regression. The focus will be on the use of regularized least squares, first equipped with the Gaussian or polynomial kernel. Then, we will describe a new kernel able to account for some features of nonlinear dynamic systems, including fading memory concepts. Regularized Volterra models will be also discussed. We will then provide a brief overview on neural and deep networks, hybrid systems identification, block-oriented models like Wiener and Hammerstein, parametric and nonparametric variable selection methods.

8.1 Nonlinear System Identification

In Sect. 2.2, Eq. (2.2), a model of a dynamical system was defined as a predictor function g that maps past input–output data

$$Z^{t-1} = \{y(t-1), u(t-1), y(t-2), u(t-2), \ldots\}$$

to the next output

$$\hat{y}(t|\theta) = g(t, \theta, Z^{t-1}), \tag{8.1}$$

where θ is a parameter vector that indexes the model. The predictor could possibly also be a nonparametric map belonging to some function class. If g is a nonlinear function of Z^{t-1} the model is nonlinear and the task to infer it from all the available measurements contained in the training set $\mathscr{D}_T$ is the task of *Nonlinear System Identification*. This is a complex area with a vast and rich literature. One reason for

© The Author(s) 2022

G. Pillonetto et al., *Regularized System Identification*, Communications and Control Engineering, https://doi.org/10.1007/978-3-030-95860-2_8

the richness is that very many parameterizations of g have been suggested, each with various proposed estimation methods, e.g., see the survey [36]. The different parameterizations allow various degrees of prior knowledge about the system to be accounted for, which gives *grey box models* with different shades of grey: see the section *The Palette of Nonlinear Models* in [36].

A typical element of nonlinear models is that somewhere in the structure there can be a *static nonlinearity* present, $\zeta(t) = h(\eta(t))$. Dealing with static nonlinearities is therefore an essential feature in nonlinear identification. See the sidebar "Static Nonlinearities" in [36], and Sect. 8.5.2 for some brief remarks.

If no prior physical knowledge is available, we have a *black-box model*. Then we need to employ parameterizations for g that are very flexible and can describe any reasonable function with arbitrary accuracy. A typical choice for this are *neural networks* or *deep nets*. See Sect. 8.5.1 for some comments. Alternatively one can define g non-parametrically as belonging to a certain (possibly infinite dimensional) function class. This leads to *kernel methods*, like *regularization networks*, and *Gaussian Process inference*, treated in the next section.

Both in the case of grey and black-box models, nonlinear identification is characterized by considerable structural uncertainty. This leads typically to parametric models with many parameters and regularization will be a natural and useful tool to handle that. This chapter will discuss typical use of regularization for various tasks in nonlinear system identification.

8.2   Kernel-Based Nonlinear System Identification

Consider the measurements model

$$y(t_i) = f^0(x_i) + e(t_i), \quad i = 1, \ldots, N, \tag{8.2}$$

where $y(t_i)$ is the system output at instant t_i, corrupted by the noises $e(t_i)$, and f^0 is the unknown function to reconstruct. The link with nonlinear system identification is obtained by assuming that the x_i contains past input and/or output values, i.e.,

$$x_i = [u_{t_i-1} \, u_{t_i-2} \, \cdots \, u_{t_i-m_u} \, y_{t_i-1} \, y_{t_i-2} \, \cdots \, y_{t_i-m_y}]. \tag{8.3}$$

In this way, the function f^0 represents a dynamic system. For the sake of simplicity, let $m = m_u = m_y$, where m will be called the system memory in what follows. Then, if $m < \infty$ a nonlinear ARX (NARX) model is obtained. A nonlinear FIR (NFIR) is instead obtained when x_i contains only past inputs, i.e.,

$$x_i = [u_{t_i-1} \, u_{t_i-2} \, \cdots \, u_{t_i-m}]. \tag{8.4}$$

Now, with these correspondences, we can assume that our nonlinear predictor belongs to a function class $\mathscr{H}$ given by a RKHS. Then, given the N couples $\{x_i, y(t_i)\}$, the regularization network

$$\hat{f} = \arg\min_{f \in \mathcal{H}} \sum_{i=1}^{N} (y(t_i) - f(x_i))^2 + \gamma \|f\|_{\mathcal{H}}^2 \qquad (8.5)$$

implements regularized NARX, with $f : \mathbb{R}^{2m} \to \mathbb{R}$, or NFIR, with $f : \mathbb{R}^m \to \mathbb{R}$.

To obtain the estimate $\hat{f}$ we can now exploit Theorem 6.15, i.e., the representer theorem. Since we focus on quadratic loss functions, the results in Sect. 6.5.1 ensure that our system estimate $\hat{f}$ not only exists and is unique but is also available in closed form. In particular, let $Y = [y(t_1), \ldots, y(t_N)]^T$ and $\mathbf{K} \in \mathbb{R}^{N \times N}$ be the kernel matrix such that $\mathbf{K}_{ij} = K(x_i, x_j)$. The nonlinear system estimate is then sum of the N kernel sections centred on the x_i, i.e.,

$$\hat{f} = \sum_{i=1}^{N} \hat{c}_i K_{x_i} \qquad (8.6)$$

with coefficients $\hat{c}_i$ contained in the vector

$$\hat{c} = (\mathbf{K} + \gamma I_N)^{-1} Y, \qquad (8.7)$$

with I_N the $N \times N$ identity matrix.

For future developments, in the remaining part of this section it is useful to cast the connection between regularization in RKHS and Bayesian estimation in this nonlinear setting. Some strategies for hyperparameters tuning will be also recalled.

8.2.1 Connection with Bayesian Estimation of Gaussian Random Fields

First, we recall an important result obtained in the linear setting in Sect. 7.1.4. The starting point was the measurements model

$$y(t_i) = L_i[g^0] + e(t_i), \quad i = 1, \ldots, N,$$

with g^0 denoting the system impulse response and $L_i[g^0]$ representing the convolution between g^0 and the input, evaluated at t_i. Proposition 7.1 said that, if $\mathcal{H}$ is the RKHS induced by a kernel K, then

$$\hat{g} = \arg\min_{g \in \mathcal{H}} \sum_{i=1}^{N} (y(t_i) - L_i[g])^2 + \gamma \|g\|_{\mathcal{H}}^2$$

is the minimum variance impulse response estimator when the noise e is white and Gaussian while g^0 is a zero-mean Gaussian process (independent of e) of covariance proportional to K, i.e.,

$$\mathscr{E}(g^0(t)g^0(s)) \propto K(t, s).$$

So, the choice of K ensures that the probability is concentrated on our expected impulse responses. For instance, in previous chapters we have seen that the TC/stable spline class describes time-courses that are smooth and exponential decaying with a level established by some hyperparameters. A very simple approach to understand the prior ideas introduced in the model is to simulate some curves that will thus represent some of our candidate impulse responses. As an example, some realizations from the discrete-time TC kernel (7.15), given by $K(i, j) = \alpha^{\max(i,j)}$ with $\alpha = 0.9$, are reported in the left panels of Fig. 8.1.

Consider the nonlinear scenario with measurements model given by (8.2) and input locations containing past inputs and outputs. The fundamental difference w.r.t. the linear setting is that the unknown function f^0 now represents directly the nonlinear input–output relationship. The connection with Bayesian estimation is obtained thinking of f^0 as a nonlinear stochastic surface, in particular a zero-mean Gaussian random field. This is a generalization of a stochastic process over general domains: one has that, for any set of input locations $\{x_i^*\}_{i=1}^p$, the vector $[f^0(x_1^*) \ \ldots \ f^0(x_p^*)]$ is jointly Gaussian. In particular, the covariance of such vector is assumed to be proportional to the kernel matrix $\mathbf{K}$ whose (i, j)-entry is $\mathbf{K}_{ij} = K(x_i^*, x_j^*)$. This corresponds to saying that f^0 is a zero-mean Gaussian random field with covariance λK, with λ a positive scalar, independent of the white Gaussian noises $e(t_i)$ of variance σ^2. Then,

$$\hat{f} = \arg\min_{f \in \mathscr{H}} \sum_{i=1}^N (y(t_i) - f(x_i))^2 + \gamma \|f\|_{\mathscr{H}}^2, \quad \gamma = \frac{\sigma^2}{\lambda}$$

turns out to be the minimum variance estimator of the nonlinear system f^0. In this stochastic scenario, our model assumptions can be better understood by simulating some nonlinear surfaces from the prior. They will represent some of our candidate nonlinear systems. As an example, some realizations from the Gaussian kernel (6.43), given by $K(x, a) = \exp(-\|x - a\|^2/\rho)$ with $\rho = 1000$, are reported in the right panels of Fig. 8.1. It is apparent that such covariance includes just information on the smoothness on the input–output map, i.e., the fact that similar inputs should produce similar system outputs.

8.2.2   Kernel Tuning

As already discussed, e.g., in Sect. 3.5, even when the structure of a kernel is assigned, the estimator (8.5) typically contains unknown parameters that have to be determined from data. For example, if the Gaussian kernel $\exp(-\|x - a\|^2/\rho)$ is adopted the unknown hyperparameter vector η will contain the regularization parameter γ, the kernel width ρ and possibly also the system memory m. We now briefly discuss esti-

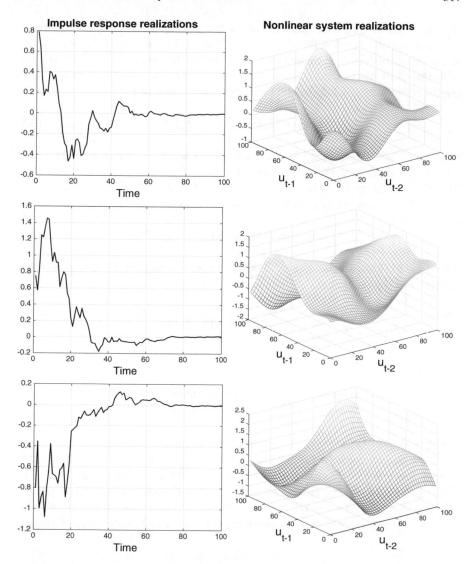

Fig. 8.1 *Left panels* Realizations of a stochastic zero-mean Gaussian process modelling discrete-time impulse response candidates. They are drawn by using the TC kernel $0.9^{\max(i,j)}$ as covariance. *Right panels* Realizations of a zero-mean Gaussian surface (random field) representing nonlinear systems candidates, in particular NFIR models with memory $m = 2$ in (8.4). They are drawn by using the Gaussian kernel $\exp(-\|x - y\|^2/1000)$ as covariance

mation of η just pointing out some natural connections with the techniques illustrated in Sect. 7.2 in the linear scenario.

An important observation is that, when a quadratic loss is adopted, even in the nonlinear setting the estimator (8.5) leads to predictors linear in the data Y. In addition, since we assume data generated according to (8.2), direct noisy measurements of f are available. Hence, the output kernel matrix O used in Sect. 7.2 just reduces to the kernel matrix $\mathbf{K}$ computed over the x_i where data are collected. In fact, from (8.6) and (8.7) one can see that the predictions $\hat{y}_i$, i.e., the estimates of the $f^0(x_i)$, are the components of $\mathbf{K}\hat{c}$. So, they are collected in the vector

$$\hat{Y}(\eta) = \mathbf{K}(\eta)(\mathbf{K}(\eta) + \gamma I_N)^{-1} Y. \tag{8.8}$$

Now, consider techniques like SURE and GCV that see f^0 as a deterministic function so that the randomness in Y derives only from the output noise. Exploiting the same line of discussion reported in Sects. 3.5.2 and 3.5.3 (see also Sect. 7.2), from (8.8) we see that the influence matrix is given by $\mathbf{K}(\eta)(\mathbf{K}(\eta) + \gamma I_N)^{-1}$. Hence, the degrees of freedom are

$$\mathrm{dof}(\eta) = \mathrm{trace}(\mathbf{K}(\eta)(\mathbf{K}(\eta) + \gamma I_N)^{-1}). \tag{8.9}$$

Then, the SURE estimate of η is obtained by minimizing the following unbiased estimator of the prediction risk

$$\hat{\eta} = \arg\min_{\eta \in \Gamma} \frac{1}{N}\|Y - \hat{Y}(\eta)\|^2 + 2\sigma^2 \frac{\mathrm{dof}(\eta)}{N} \tag{8.10}$$

while the GCV estimate is

$$\hat{\eta} = \arg\min_{\eta \in \Gamma} \frac{\|Y - \hat{Y}(\eta)\|^2}{(1 - \mathrm{dof}(\eta)/N)^2}, \tag{8.11}$$

where we have used Γ to denote the optimization domain.

If we instead consider the Bayesian framework discussed in the previous subsection, we see f^0 as a zero-mean Gaussian random field of covariance λK, with λ a positive scale factor, independent of the white Gaussian noise of variance σ^2. Since $y(t_i) = f^0(x_i) + e(t_i)$, following the same reasonings developed in the finite-dimensional context in Sect. 4.4, one obtains that the vector Y is zero-mean Gaussian, i.e.,

$$Y \sim \mathcal{N}(0, Z(\eta))$$

with covariance matrix

$$Z(\eta) = \lambda \mathbf{K}(\eta) + \sigma^2 I_N.$$

Above, the vector η could, e.g., contain λ, σ^2, m and also other parameters entering K. Then, we easily obtain that its marginal likelihood estimate is

$$\hat{\eta} = \arg\min_{\eta \in \Gamma} Y^T Z(\eta)^{-1} Y + \log \det(Z(\eta)). \tag{8.12}$$

8.3 Kernels for Nonlinear System Identification

In the previous section we have cast the kernel-based estimator (8.5) in the framework of nonlinear system identification. We have also provided its Bayesian interpretation and recalled how to estimate the hyperparameter vector η when the parametric form of K is assigned. But the crucial question is now the regularization design. This is a fundamental issue, initially discussed in Sect. 3.4.2, which in this setting consists of choosing a kernel structure suited to model nonlinear dynamic systems. Two interesting options come from machine learning literature. The first one is the (already mentioned) Gaussian kernel

$$K(x, a) = e^{\frac{-\|x-a\|^2}{\rho}}$$

that can describe input–output relationships just known to be smooth. We have also seen in Sect. 6.6.5 that this model is infinite dimensional, i.e., its induced RKHS cannot be spanned by a finite number of basis functions. It is also universal, being dense in the space of all continuous functions defined on any compact subset of the regressors' domain. These appear attractive features when little information on system dynamics are available.

A second alternative is the polynomial kernel

$$K(x, a) = (\langle x, a\rangle_2 + 1)^p, \quad p \in \mathbb{N}, \tag{8.13}$$

where $\langle \cdot, \cdot \rangle_2$ is the classical Euclidean inner product. In the NFIR case, where the input locations $x_i \in \mathbb{R}^m$ as given in (8.4), such kernel has a fundamental connection with the Volterra representations of nonlinear systems, see, e.g., [35]. In fact, we know from Sect. 6.6.4 that the induced RKHS is not universal but has dimension $\binom{m+p}{p}$ and contains all possible monomials up to the pth degree. Hence, the polynomial kernel implicitly encodes truncated discrete Volterra series of the desired order. It avoids curse of dimensionality since the possibly large number coefficients have not to be computed explicitly thanks to monomials' encoding. In fact, from (8.7) one can see that estimation complexity, even if cubic in the number N of output data turns out to be linear in the system memory m and independent of the degree p of nonlinearity.

8.3.1 A Numerical Example

We will consider a numerical example where the Gaussian and the polynomial kernel are used to estimate a nonlinear dynamic system from input–output data.

Consider the NFIR

Fig. 8.2 Coefficients g_i^0 defining the linear part of the system (8.14). They represent the impulse response of a stable linear system obtained by randomly generating a rational transfer function of order 10

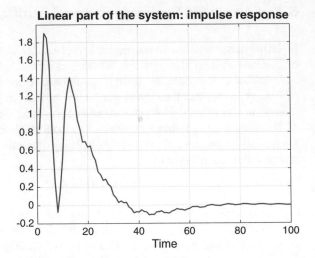

Linear part of the system: impulse response

$$f^0(x_t) = \left(\sum_{i=1}^{80} g_i^0 u_{t-i} \right) - u_{t-2} u_{t-3} - 0.25 u_{t-4}^2 + 0.25 u_{t-1} u_{t-2} +$$
$$+ 0.75 u_{t-3}^3 + 0.5 \left(u_{t-1}^2 + u_{t-1} u_{t-3} + u_{t-2} u_{t-4} \right) \tag{8.14}$$

with nonlinearities taken from [40] while the coefficients g_i^0 are reported in Fig. 8.2. The inputs are independent Gaussian random variables of variance 4. The measurements model is that reported in (8.2) with the noise e white and Gaussian of variance 4 and independent of u. Such system is strongly nonlinear: the contribution of the linear part (defined by the g_i^0) to the output variance is around 12% of the overall variance.

We generate 2000 input–output couples and display them in Fig. 8.3. The first 1000 input–output couples $\{u_k, y_k\}_{k=1}^{1000}$ are the identification data while the other 1000 $\{u_k, y_k\}_{k=1001}^{2000}$ are the test set. They are used to assess the performance of an estimator in terms of the prediction fit

$$100 \left(1 - \left[\frac{\sum_{k=1001}^{2000} |y_k - \hat{y}_k|^2}{\sum_{k=1001}^{2000} |y_k - \bar{y}|^2} \right]^{\frac{1}{2}} \right), \quad \bar{y} = \frac{1}{1000} \sum_{k=1001}^{2000} y_k, \tag{8.15}$$

where the $\hat{y}_k$ are the predictions returned by a certain estimator by assuming null initial conditions, i.e., computed by using only $\{u_k\}_{k=1001}^{2000}$ and setting to zero the inputs falling outside the test set.

First, consider the estimator (8.5) equipped with either the Gaussian or the polynomial kernel with input locations

$$x_i = [u_{t_i-1} \ u_{t_i-2} \ \cdots \ u_{t_i-m}],$$

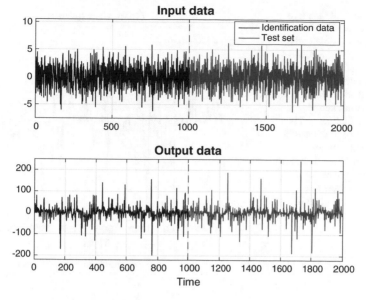

Fig. 8.3 Input and output data generated by the nonlinear system (8.14). The first 1000 couples (black line) are used as identification data while the other 1000 (red) are the test set used to assess the prediction performance of a model

where the system memory m is seen as an hyperparameter to be estimated from data. Specifically, when using the Gaussian kernel

$$K(x, a) = e^{\frac{-\|x-a\|^2}{\rho}}$$

the estimator depends on the unknown hyperparameter vector

$$\eta = [m \ \gamma \ \rho],$$

where m is the system memory, γ is the regularization parameter and ρ is the kernel width. Instead, when using the polynomial kernel

$$K(x, a) = (\langle x, a \rangle_2 + 1)^p, \quad p \in \mathbb{N},$$

we have

$$\eta = [m \ \gamma \ p],$$

where, in place of ρ, the third unknown hyperparameter is the polynomial order p. In both the cases, we estimate η by using an oracle. In particular, assigned a certain η, the estimator (8.5) determines $\hat{f}$ by using only the identification data but the oracle has access to the test set to select that hyperparameter vector that maximizes

Fig. 8.4 Test set data (red line), extracted from the last 1000 outputs visible in the right panel of Fig. 8.3, and predictions returned by (8.5) equipped with the Gaussian kernel (top panel, black) and the polynomial kernel (bottom panel, black). The estimators use the first 1000 input–output couples in Fig. 8.3 as training data, with hyperparameter vector η tuned by an oracle that maximizes the test set fit

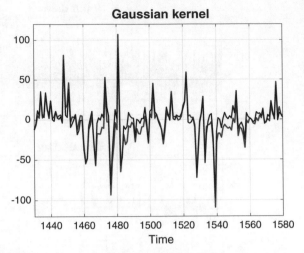

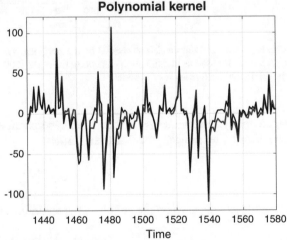

the prediction fit (8.15). Note that calibration is quite computational expensive. In fact, one has to introduce a grid to account for the discrete nature of the system memory m. The polynomial kernel requires also the introduction of another grid for the polynomial order p.

Figure 8.4 reports some test set data (red line) extracted from the last 1000 outputs displayed in the right panel of Fig. 8.3. When adopting the Gaussian kernel, the oracle chooses $m = 4$. When using the polynomial kernel it selects $m = 6$ and sets the polynomial order to $p = 3$. The top panel of Fig. 8.4 shows the predictions returned by the oracle-based Gaussian kernel (black line). The prediction fit is not so large, equal to 69.6%. The bottom panel instead plots results from the oracle-based polynomial kernel (black line). The prediction capability increases to 73.5% but does not appear so satisfactory. Figure 8.5 also reports the MATLAB boxplots of 100 prediction fits

Fig. 8.5 Boxplots of 100 predictions fits (8.15) obtained after a Monte Carlo study by the oracle-based estimators equipped with the Gaussian and polynomial kernel

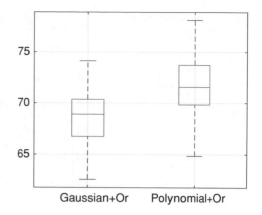

returned by the two kernel-based estimators after a Monte Carlo study. At any of the 100 runs new realizations of inputs and noises define a new identification and test set. One can see that, on average, the polynomial kernel performs a bit better than the Gaussian kernel, but its mean prediction fit is around 72%.

8.3.2 Limitations of the Gaussian and Polynomial Kernel

From (8.14) one can see that the NFIR order is $m = 80$ while the oracle sets $m = 4$ and $m = 6$ when using, respectively, the Gaussian and the polynomial kernel. This introduces a bias in the estimation process that is clearly visible in the predictions reported in Fig. 8.4. Let us try to understand the reasons of this phenomenon.

Polynomial kernel First, consider the polynomial kernel. The oracle chooses the correct polynomial order $p = 3$ to account for the highest-order term $0.75u_{t-3}^3$ present in the system. Such choice however already defines a complex model since it includes all the monomials up to order 3. In particular, with $m = 6$ and $p = 3$ the number of adopted basis functions is

$$\binom{m+p}{p} = \binom{6+3}{3} = 84,$$

that is quite large considering that 1000 outputs are available. If m is increased to 7, one would implicitly use

$$\binom{m+p}{p} = \binom{7+3}{3} = 120$$

basis functions. In general, values of m larger than 6 are not acceptable for the oracle: even a careful tuning of the regularization parameter γ does not permit to have a good control on the estimator's variance. This is illustrated in Fig. 8.6 that displays the best prediction test set fit that can be obtained by the oracle as a function of the system memory m. The maximum is indeed obtained with $m = 6$. Instead, the

Fig. 8.6 Predictions fits by
the oracle-based estimator
equipped with the
polynomial kernel as a
function of system memory
m. The optimal model
dimension is achieved for
$m = 6$

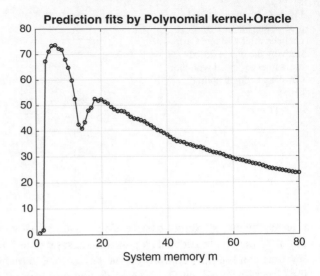

value $m = 80$ leads to a very small fit, around 25%, because this introduces an overly
complex model with

$$\binom{m + p}{p} = \binom{80 + 3}{3} = 91881$$

monomials.

Another reason that does not allow the polynomial kernel to well control model
variance is the way it regularizes (implicitly) the monomial coefficients. We describe
this point through a simple example. A quadratic polynomial kernel is considered
but similar considerations would still hold by introducing larger degrees. Let $p = 2$,
$x = [u_{t-1} \ \ldots \ u_{t-m}]$ and $a = [u_{\tau-1} \ \ldots \ u_{\tau-m}]$. Exploiting the multinomial theorem
one obtains

$$
\begin{aligned}
K(x, a) &= (\sum_{i=1}^{m} u_{t-i} u_{\tau-i} + 1)^2 \\
&= \sum_{i=1}^{m} u_{t-i}^2 u_{\tau-i}^2 + 2 \sum_{i=2}^{m} \sum_{j=1}^{i-1} (u_{t-i} u_{t-j})(u_{\tau-i} u_{\tau-j}) + 2 \sum_{i=1}^{m} u_{t-i} u_{\tau-i} + 1.
\end{aligned}
$$

This defines the following diagonalized version of the quadratic polynomial kernel

$$K(x, a) = \sum_i \zeta_i \rho_i(x) \rho_i(a),$$

where the $\rho_i(x)$ are all the monomials up to degree 3 contained in the following
vector

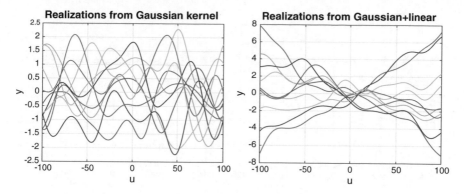

Fig. 8.7 Some realizations from a zero-mean stochastic process with covariance given by a Gaussian kernel (left panel) and by a Gaussian plus linear kernel (right)

$$\left\{ u_{t-m}^2, \ldots, u_{t-1}^2, u_{t-m}u_{t-m+1}, \ldots, u_{t-m}u_{t-1}, u_{t-m+1}u_{t-m+2}, \right.$$

$$\left. \ldots, u_{t-m+1}u_{t-1}, \ldots, u_{t-2}u_{t-1}, u_{t-m}, \ldots, u_{t-1}, 1 \right\},$$

with the corresponding ζ_i given by

$$\left\{ 1, \ldots, 1, 2, \ldots, 2, 2, \ldots, 2, \ldots, 2, 2, \ldots, 2, 1 \right\}.$$

According to the RKHS theory described in Sect. 6.3, for any f in the RKHS $\mathscr{H}$ induced by such kernel one has

$$f(x) = \sum_i c_i \rho_i(x), \quad \|f\|_{\mathscr{H}}^2 = \sum_i \frac{c_i^2}{\zeta_i},$$

where all the eigenvalues ζ_i assume value 1 or 2 (most of them are equal to 2). Hence, one can see that the regularizer $\|f\|_{\mathscr{H}}^2$ does not incorporate any fading memory concept typical of dynamic systems. In fact, the two coefficients of the monomials $\{u_{t-m}^2, u_{t-1}^2\}$ or those of the couple $\{u_{t-m}u_{t-m+1}, u_{t-2}u_{t-1}\}$ are assigned the same penalty. But, similarly to the linear case, one should instead expect that inputs u_{t-i} have less influence on y_t as the positive lag i increases.

Gaussian kernel As in the case of the polynomial model, one of the limitations of the Gaussian kernel $K(x, a) = \exp(-\|x - a\|^2/\rho)$ in modelling nonlinear systems is that it does not include any fading memory concept. Hence, the inputs $\{u_{t-1}, u_{t-2}, \ldots, u_{t-m}\}$ included in the input location are expected to have the same influence on y_t. This can be appreciated also through the Bayesian interpretation of regularization, e.g., by inspecting the system realizations generated by the Gaussian kernel reported in the right panels of Fig. 8.1.

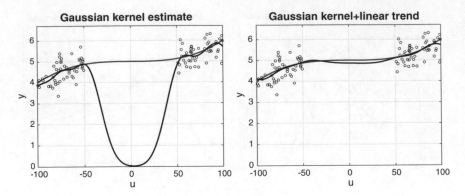

Fig. 8.8 True function (red line), noisy data and regularized estimate returned by (8.5) by using a Gaussian kernel $K(u, a) = \exp(-(u - a)^2/500)$ (left panel, black) and a Gaussian plus linear kernel $K(u, a) = \exp(-(u - a)^2/500) + 10ua$ (right panel, black). The regularization parameter γ is estimated from data via marginal likelihood optimization (8.12)

Still adopting a stochastic viewpoint, another drawback is that the covariance $\exp(-\|x - a\|^2/\rho)$ describes stationary processes and this implies that the variance of $f^0(x)$ does not depend on the input location. This is now illustrated in the one-dimensional case where $x \in \mathbb{R}$ and the kernel models a static nonlinear system $f^0(u)$, i.e., the (noiseless) output y depends only on a single input value u. The left panel of Fig. 8.7 plots some realizations from $\exp(-(u - a)^2/500)$. They can be poor nonlinear system candidates since a nonlinear system, like that reported in (8.14), often contains also a linear component. For this reason it can be useful to enrich the model with a linear kernel. Its effect can be appreciated by looking at the realizations plotted in the right panel of Fig. 8.7 that are now drawn by using $\exp(-(u - a)^2/500) + ua/400$ as covariance.

The fact that the predictive capability of a nonlinear model can much improve by adding a linear component can be understood also considering Theorem 6.15 (representer theorem). Using only a Gaussian kernel, the estimate $\hat{f}$ of the nonlinear system returned by (8.5) is the sum of N Gaussian functions centred on the x_i. Hence, in the regions where no data are available, the function $\hat{f}$ just decays to zero and this can lead to poor predictions when, e.g., a linear component is present in the system. This phenomenon is illustrated in the left panel of Fig. 8.8. In this case, the prediction performance can be greatly enhanced by adding a linear kernel, whose results are visible in the right panel of the same figure.

8.3.3 Nonlinear Stable Spline Kernel

We will build a kernel $\mathscr{K}$ for nonlinear system identification, namely the nonlinear stable spline kernel, by exploiting what has been learnt from the previous example. To simplify exposition, we consider the NFIR case but all the ideas here developed can be immediately extended to NARX models, as discussed at the end of this section.

First, it is useful to define $\mathscr{K}$ as the sum of a linear and a nonlinear kernel, i.e.,

$$\mathscr{K}(x_i, x_j) = \lambda_L x_i^T P x_j + \lambda_{NL} K(x_i, x_j), \tag{8.16}$$

where the input locations are here seen as column vectors, i.e.,

$$x_i = [u_{t_i-1}\ u_{t_i-2}\ \cdots\ u_{t_i-m}]^T,$$

$P \in \mathbb{R}^{m \times m}$ is a symmetric positive semidefinite matrix that models the impulse response of the system's linear part while K describes the nonlinear dynamics. Note that the two-scale factors λ_L and λ_{NL} are unknown hyperparameters that balance the contributions of the linear and nonlinear part to the output.

For what concerns P, such matrix can be defined by resorting to the class of stable kernels developed in the previous chapters. In particular, using the TC/stable spline kernel, the (a, b)-entry of P is

$$P_{ab} = \alpha_L^{\max(a,b)}, \quad 0 \le \alpha_L < 1, \quad a = 1, \ldots, m, \ b = 1, \ldots, m, \tag{8.17}$$

where α_L determines the decay rate of the impulse response governing the linear dynamics.

For what concerns K, we will define it by modifying the classical Gaussian kernel in order to include fading memory concepts. Following the same ideas underlying the TC kernel, we include the information that u_{t-i} is expected to have less influence on y_t as i increases by defining

$$K(x_i, x_j) = \exp\left(-\sum_{k=1}^{m} \alpha_{NL}^{k-1} \frac{(u_{t_i-k} - u_{t_j-k})^2}{\rho}\right), \quad 0 < \alpha_{NL} \le 1. \tag{8.18}$$

The additional hyperparameter α_{NL} gives the information that past inputs' influence decays exponentially to zero. To understand how this kernel models the nonlinear surface, and how different values of α_{NL} can describe different system features, we can use the Bayesian interpretation of regularization. In particular, consider an example with $m = 2$, so that the components of x_i are u_{t_i-1} and u_{t_i-2}, and let the system f^0 be a zero-mean Gaussian random field with covariance given by (8.18) with $\rho = 1000$. If $\alpha_{NL} = 1$ we recover the Gaussian kernel. Hence, before seeing any data, u_{t_i-1} and u_{t_i-2} are expected to have the same influence on the system output. This can be appreciated by drawing some realizations from such random field, e.g., see the top panel of Fig. 8.9 (or the right panels of Fig. 8.1).

With α_{NL} very close to zero, the output depends mainly on u_{t_i-1}, i.e.,

$$K(x_i, x_j) \approx \exp\left(-\frac{(u_{t_i-1} - u_{t_j-1})^2}{\rho}\right).$$

This can be appreciated by looking at the realization in the middle panel of Fig. 8.9 obtained with $\alpha_{NL} = 0.001$. One can see that, for fixed u_{t_i-1}, changes in u_{t_i-2} do not produce appreciable variations in the function value. If the value of α_{NL} is now increased, the input value u_{t_i-2} starts playing a role. This is visible in the bottom panel where the realization is now generated by using $\alpha_{NL} = 0.1$.

The nonlinear stable spline kernel enjoys also an advantage related to computational issues. Using classical machine learning kernels, like Gaussian or polynomial, the choice of the dimension m of the input space is a delicate issue. It requires discrete tuning, as encountered in classical linear system identification to estimate, e.g., FIR or ARX order, and this can be computationally expensive. In the case of the polynomial kernel, another discrete parameter is the polynomial order p that requires an additional grid. By introducing stability/fading memory hyperparameters, one can instead set m to a large value increasing the flexibility of the estimator. Then, estimation of α_L and α_{NL} from data permits to control the "effective" dimension of the regressor space in a continuous manner. In light of the continuous nature of the optimization domain, one needs to solve only one optimization problem, involving, e.g., SURE (8.10), GCV (8.11) or Empirical Bayes (8.12).

Finally, as already mentioned, the extension to NARX models is very simple. Let $x_i = [a_i^T \ b_i^T]^T$ with

$$a_i = [u_{t_i-1} \ u_{t_i-2} \ \ldots \ u_{t_i-m}]^T, \quad b_i = [y_{t_i-1} \ y_{t_i-2} \ \ldots \ y_{t_i-m}]^T.$$

Then, the kernel (8.16) can be modified as follows

$$\mathcal{K}(x_i, x_j) = \lambda_a a_i^T P_a a_j + \lambda_b b_i^T P_b b_j + \lambda_c K_c(a_i, a_j) K_d(b_i, b_j) \tag{8.19}$$

with the matrices P_a and P_b defined by the TC kernel (8.17), with possibly different decay rates α_L, and the nonlinear kernels K_c and K_d defined by (8.18), with possibly different decay rates α_{NL}. A possible variation is

$$\mathcal{K}(x_i, x_j) = \lambda_a a_i^T P_a a_j + \lambda_b b_i^T P_b b_j + \lambda_c K_c(a_i, a_j) + \lambda_d K_d(b_i, b_j), \tag{8.20}$$

where the nonlinear dynamics are no more product, as in (8.19), but instead sum of nonlinear functions which depend on either past inputs or past outputs. In fact, recall from Theorem 6.6 that sums and products of kernels induce well-defined RKHSs containing, respectively, sums and products of functions belonging to the spaces associated to the single kernels.

Nonlinear system realizations

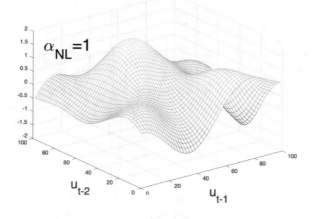

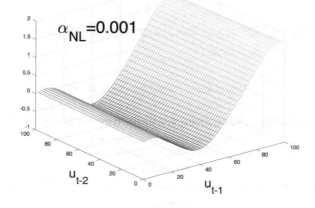

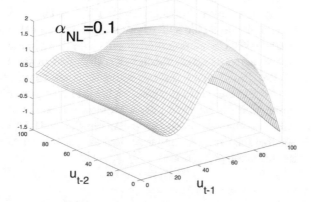

Fig. 8.9 Realizations from a zero-mean Gaussian random field having covariance $\exp\left(-\frac{1}{1000}\left((u_{t_i-1} - u_{t_j-1})^2 + \alpha_{NL}(u_{t_i-2} - u_{t_j-2})^2)\right)\right)$ for three different values of α_{NL}

Fig. 8.10 Test set data (red line), extracted from the last 1000 outputs visible in the right panel of Fig. 8.3, and predictions (black) returned by (8.5) equipped with the nonlinear stable spline kernel (8.16). The estimator uses the first 1000 input–output couples in Fig. 8.3 as training data, with kernel and regularization parameters tuned by marginal likelihood optimization

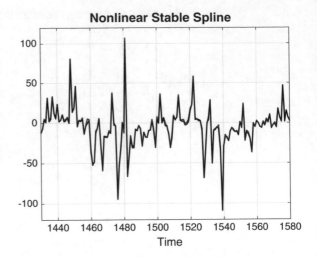

8.3.4   Numerical Example Revisited: Use of the Nonlinear Stable Spline Kernel

Let us now reconsider the numerical example where the nonlinear system (8.14) is used to generate the identification and test data reported in Fig. 8.3. Now, we use the estimator (8.5) equipped with the nonlinear stable spline kernel (8.16). System memory is set to $m = 100$. Hence, we let α_L and α_{NL} determine from data which past inputs mostly influence the output due to the linear and nonlinear system part, respectively. In particular, the hyperparameter vector $\eta = [\lambda_L\ \lambda_{NL}\ \alpha_L\ \alpha_{NL}\ \rho]$ is estimated via marginal likelihood maximization using the 1000 input–output training data.

Figure 8.10 shows the same test set data (red line) reported in Fig. 8.4 and extracted from the last 1000 outputs visible in the right panel of Fig. 8.3. The predictions (black line) returned by the nonlinear stable spline kernel are now very close to truth. The prediction fit is around 90%. Comparing these results with those in Fig. 8.4, one can see that the prediction performance is much better than that of the Gaussian and polynomial kernel. Recall also that these two estimators tune complexity by using an oracle that is not implementable in practice. Figure 8.11 also plots the MATLAB boxplots of 100 prediction fits returned after a Monte Carlo study of 100 runs by these two oracle-based estimators, already present in Fig. 8.5, and by nonlinear stable spline. One can see that the use of a regularizer that accounts for dynamic systems features largely improves the prediction fits.

Fig. 8.11 Boxplots of 100 predictions fits. The first two on the left are obtained by the oracle-based estimators equipped with the Gaussian and polynomial kernel. The boxplot on the right is obtained by the nonlinear stable spline kernel with hyperparameters estimated by marginal likelihood maximization (which exploits only the identification data)

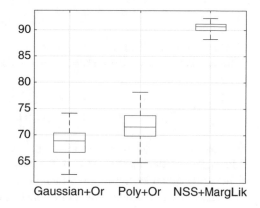

8.4 Explicit Regularization of Volterra Models

In what follows, we use $C(k, m)$ to indicate the number of ways one can form the nonnegative integer k as the sum of m nonnegative integers. This is the same problem as distributing k objects to m groups (some groups may get zero objects). By combinatorial theory we have

$$C(k, m) = \binom{k + m - 1}{m - 1}. \tag{8.21}$$

We adopt the model description (8.2) and seek a simple representation for the model $f(x)$. For simple notation, assume that f is scalar valued with past inputs only, i.e., ($m_y = 0, m_u = m$) with input location x given by (8.4). A straightforward idea is to mimic polynomial Taylor expansion

$$f(x) = \sum_{k=1}^{p} g_k x^k. \tag{8.22}$$

This innocent-looking function expansion is in fact a bit more complex than it looks. The kth power of the m-row vector x is to be interpreted as C-dimensional column vector with each element being a monomial of the m-components $x(i)$ of x with sum of exponents being k:

$$\alpha_r^{(k)} = x(1)^{\beta(k,1)} x(2)^{\beta(k,2)} \cdots x(m)^{\beta(k,m)} \tag{8.23a}$$

$$\beta(k, p) \text{ non negative, such that } \sum_{\ell=1}^{m} \beta(k, \ell) = k \tag{8.23b}$$

$$r = 1, 2, \ldots, C(k, m). \tag{8.23c}$$

In (8.22) g_k is to be interpreted as a row vector with $C(k, m)$ elements

$$g_k = [g_k^{(1)}, \ldots, g_k^{C(k,m)}]. \tag{8.23d}$$

The response $f(x)$ is thus made of $d(p, m) = \sum_{k=1}^p C(k, m)$ contributions ("impulse responses") from each of the nonlinear combinations of past inputs

$$\alpha_r^{(k)} = u_{t-1}^{\beta(k,1)} u_{t-2}^{\beta(k,2)} \cdots u_{t-m}^{\beta(k,m)} \tag{8.23e}$$

$$r = 1, \ldots, C(k, m), \quad k = 1, \ldots, p. \tag{8.23f}$$

This expansion of the model (8.22) is the *Volterra Model* discussed, e.g., by [7, 35]. It has $d(p, m)$ parameters. The reader may recognize this as an explicit treatment of the polynomial kernel (8.13) which does not exploit any basis functions implicit encoding and, hence, does not exploit the kernel trick described in Remark 6.3. This has also some connections with the explicit regularization approaches for linear system identification discussed in Sect. 7.4.4 using, e.g., Laguerre functions.

So, this model has memory length m and polynomial order p. As $p \to \infty$ it follows that $f(x)$ in (8.22), with possibly the addition of a constant function, can approximate any ("reasonable") function arbitrarily well. This universal approximation property is of course very valuable for black box models and created considerable interest in Volterra models. However, it is easy to see that the number $d(p, m)$ of parameters g_k increases very rapidly with m and p and that high-order polynomials in the observed signals may create numerically ill-conditioned calculations. Hence, Volterra models have not been used so much in practical identification problems, unless for small values of m and p.

A remedy for the large number of parameters and ill-conditioned numerics is clearly to use regularization. In [4] it is discussed how to regularize the Volterra model to make it a practical tool. In short, the idea is the following, illustrated for a small example with $p = 2$.

We write the model also adding a scalar g_0 which accounts for a constant component in the output so that one has

$$y(t) = g_0 + g_1^T \varphi(t) + \varphi^T(t) G_2 \varphi(t) \tag{8.24a}$$

$$\varphi^T(t) = [u(t_1), u(t_2) \ldots u(t_m)] \tag{8.24b}$$

$$g_1 = \theta_1 \quad m - \text{dimensional column vector} \tag{8.24c}$$

$$G_2 \ m \times m \text{ symmetric matrix,} \tag{8.24d}$$

where the matrix G_2 is formed from $g_2^{(1)} g_2^{(2)} g_2^{(3)}$ in the expansion (8.22)–(8.23e).

The regularized estimation can now be formed as the criterion

$$\hat{\theta}^R = \underset{\theta}{\arg\min} \|Y - \Phi_N^T \theta\|^2 + \theta^T D\theta \tag{8.25}$$

with

$$\theta = [g_0, \theta_1^T, \theta_2^T]^T \tag{8.26}$$

and θ_2 is an $m(m+1)/2$ dimensional column vector made up from G_2, and Y is the vector of observed outputs $y(t)$ with $t = 1, \ldots, N$. The regression vector Φ_N if formed from the components of $\varphi(t)$ in the obvious way. It is natural to decompose the regularization matrix accordingly:

$$D = \begin{bmatrix} d_0 & 0 & 0 \\ 0 & D_1 & 0 \\ 0 & 0 & D_2 \end{bmatrix} \tag{8.27}$$

and treat the regularization of the constant term, (d_0), the linear term (D_1) and the quadratic term (D_2) in (8.24a) separately. As discussed in Chap. 5, a natural choice of regularization matrices is to let them reflect prior information about the corresponding parameters. That means that d_0 can be taken as any suitable scalar. The θ_1 vector for the first-order term describes a regular linear impulse response, and the prior for that one can be taken as, e.g., the DC kernel reported in (5.40), i.e.,

$$P_1(i, j) = c \cdot e^{-\alpha|i-j|} e^{-\beta \frac{(i+j)}{2}}. \tag{8.28}$$

For the second-order model θ_2 it is natural to treat the second-order nonlinear term in the Volterra expansion as a two-dimensional surface, described by two time-indices τ_1 and τ_2 so that the parameter at τ_1, τ_2 is the contribution to the Volterra sum from $u(t - \tau_1) \cdot u(t - \tau_2)$. This is illustrated in Fig. 8.12. The prior value of this contribution can be formed as the product of two kernels built up from responses in a coordinate system $\mathcal{U}, \mathcal{V}$ after an orthonormal coordinate transformation, corresponding to a rotation

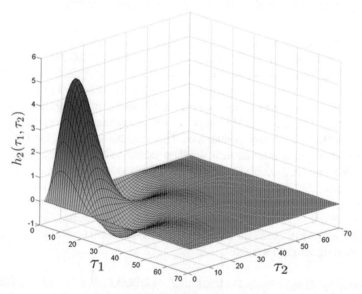

Fig. 8.12 Regularization surface for the second-order term in a Regularized Volterra expansion

of 45° of the original τ_1, τ_2-plane:

$$P_2(i, j) = c_2 P_{\mathscr{V}}(i, j) P_{\mathscr{U}}(i, j) \tag{8.29}$$

$$P_{\mathscr{V}}(i, j) = e^{-\alpha_{\mathscr{V}} \left| |\mathscr{V}_i| - |\mathscr{V}_j| \right|} e^{-\beta_{\mathscr{V}} \frac{|\mathscr{V}_i| + |\mathscr{V}_j|}{2}} \tag{8.30}$$

$$P_{\mathscr{U}}(i, j) = e^{-\alpha_{\mathscr{U}} \left| |\mathscr{U}_i| - |\mathscr{U}_j| \right|} e^{-\beta_{\mathscr{U}} \frac{|\mathscr{U}_i| + |\mathscr{U}_j|}{2}}, \tag{8.31}$$

where $\mathscr{U}_i$ and $\mathscr{V}_i$ refer to the coordinates in the new system. The corresponding prior distribution is depicted in Fig. 8.12. As desired, it is smooth and decays to zero in all directions. The coordinate change is useful to make the surface smooth over critical border lines.

This regularization was deployed in [36], section "Example 5(a) Black-Box Volterra Model of the Brain". Quite useful results were obtained with a regularized model with 594 parameters, thanks to the regularization. An extension for the regularized Volterra models, based on similar idea, is treated in [41], which also provides an EM algorithm to estimate the hyperparameters in the regularization matrices. Another development where the ideas developed in [4] are coupled with kernels implicit encoding can be found in [8].

8.5 Other Examples of Regularization in Nonlinear System Identification

8.5.1 Neural Networks and Deep Learning Models

There are many other universal approximators fon nonlinear systems $f(x)$ than those based on kernels or on the explicit Volterra model (8.22). The most common ones are various neural network models (NNMs), see, e.g., [12, 23]. They use simple nonlinearities connected in more or less complex networks. The parameters are weights in the connections as well as characterizations of the nonlinearities. Like Volterra models they are capable of approximating any reasonable system arbitrarily well given sufficiently many parameters. This means that the NNM typically has many parameters. In simple application there could be hundreds of parameters but some applications, especially in the so-called deep model applications, could have tens of thousands of parameters [18], see also [9, 11, 13, 43] for deep NARX and state-space models. Even if benign overfitting has been sometimes observed also for overparametrized models [3, 19, 30], in general regularization is a very important tool also for estimating such model. Hence, many tricks are typically included in the estimation/minimization schemes.

ℓ_2, ℓ_1 **penalties** They include the traditional weighted ℓ_2 and ℓ_1 norm penalties that we discuss in this book, see, e.g., Sect. 3.6. For example, all estimation algorithms in the *System Identification Toolbox*, [22] are equipped with optional weighted ℓ_2-regularization—also when NNM are estimated.

Early termination It is common to monitor not only the fit to estimation data in the minimization process, but also how well the current model fits a *validation data set*. Then the minimization is terminated when the fit to validation data no longer improves, even when the estimation criterion value keeps improving. This *early termination* technique is in fact equivalent to traditional regularization, as shown in [38].

Dropout or Dilution A special technique common in (deep) learning with NNM is to curb the flexibility of the model by ignoring (dropping) randomly chosen nodes in the network. This is of course a way to control that the model does not become prone to overfitting and provides regularization of the estimation just as the other methods in this book, but by a quite different technique. See, e.g., [17, 28] for more details.

8.5.2 Static Nonlinearities and Gaussian Process (GP)

A basic problem in nonlinear system identification is to handle estimation of a static nonlinear function $h(\eta)$ from known observations

$$\{\zeta(t), \eta(t), t = 1, \ldots, N\}, \qquad \zeta(t) = h(\eta(t)) + noise.$$

A general way to do this is to apply Gaussian Process (GP) estimation, [29], see also Sects. 4.9 and 8.2.1. Then $h(\eta)$ is seen as a Gaussian stochastic process with a prior mean (often zero) and a certain prior covariance function $K(\eta_1, \eta_2)$. The arguments can range both over a discrete and continuous domain. After a number of observations $z = \{\zeta(t), \eta(t), t = 1, \ldots, N\}$, the posterior distribution of the process $h^p(\eta|z)$, can be determined for any η. This is, in short, how the function h can be estimated. As seen in Sect. 8.2.1, it corresponds to a kernel method with the kernel determined by the prior covariance function $K(\eta_1, \eta_2)$.

8.5.3 Block-Oriented Models

A very common family of nonlinear dynamic models is obtained by networks of linear dynamic models $G(q)$ and nonlinear static functions $h(x)$, see Fig. 8.13. The simplest and most common ones are the *Hammerstein Model* $y(t) = G(h(u(t))$ which is obtained by passing the input through a static nonlinearity before it enters the linear system. *Wiener model* $z = G(u), y(t) = g(z(t))$, where the output of a linear system is subsequently passing through the nonlinearity. The important contribution [5] has shown that any nonlinear system with fading memory can be approximated by a Wiener model. See also, e.g., [37] for a survey and [42] for a general approach

Fig. 8.13 Common
block-oriented models.
Green ovals: static
nonlinearity h. Red blocks:
linear dynamic systems.
From top to bottom: Wiener
model, Hammerstein model,
Hammerstein–Wiener model

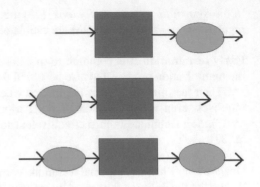

to Hammerstein–Wiener identification allowing coloured noise sources both before
and after the nonlinearities (which may be non-invertible).

Traditionally, the nonlinearities have been parametrized, e.g., as piecewise constant or piecewise linear, as polynomials or as neural nets. Recently it has been more
common to work with nonparametric nonlinearities which are typically modelled
by the GP approach, and whole estimation is then treated in a Bayesian setting. For
example, in [21] the linear part of a Wiener model is parametrized by state-space
matrices A, B in an observer canonical form with suitable priors and the output
nonlinearity $h(z)$ is a Gaussian Process with a prior mean $= z$ ("linear output") and
large and "smooth" prior covariance. To obtain the posterior densities, a particle
Gibbs sampler (PMCMC, Particle Markov Chain Monte Carlo) is employed.

In [32] the same approach is used to model the output nonlinearity, but the linear
part is written as an impulse response, with a prior of the same type as discussed in
Sect. 5.5.1. The whole problem can then be written as

$$y = \varphi(\Phi g), \tag{8.32}$$

where y is the observed output, φ is the output static nonlinearity, g is the impulse
response of the linear system and Φ is the Toeplitz matrix formed from the input. The
problem is then to determine the posterior densities $p(\varphi|y)$ and $p(g|y)$ by Bayesian
calculations. In [31] a similar technique is used for estimating Hammerstein models.

8.5.4 Hybrid Models

A common class of nonlinear models are *Hybrid models* [15, 39]. They change
their properties depending on some *regime variable* $p(t)$ (which may be formed
from the inputs and outputs themselves) [16]. Think of a collection of linear models
that describe the system behaviour in different parts of the operating space and
automatically shift as the operating point changes. To build a hybrid model involves
two steps: (1) find the collection of relevant different models and (2) determine the

areas where each model is operative. This is considered as quite a difficult problem, and approaches from different areas in control theory have been tested. Here we will comment upon a few ideas that relate to regularized identification.

A basic problem is to decide when a change in system behaviour occurs. This relates to *change detection* and *signal segmentation*. A regularization based method to segment ARX models was suggested in [25]. The standard way to estimate ARX models can be described as in Chap. 2:

$$\min_{\theta} \sum_{t=1}^{N} \| y(t) - \varphi^T(t)\theta \|^2. \tag{8.33}$$

This gives the average linear model behaviour over the time record $t \in [1 \ldots, N]$. To follow momentary changes over time, we could estimate N models by

$$\min_{\theta(t), t=1, \ldots, N} \sum_{t=1}^{N} \| y(t) - \varphi^T(t)\theta(t) \|^2. \tag{8.34}$$

This would give a perfect fit with a pretty useless collection of models. To tell that we need to be more selective when accepting new models, we can add a ℓ_1 regularization term, discussed in Sect. 3.6, obtaining:

$$\min_{\theta(t), t=1, \ldots, N} \sum_{t=1}^{N} \| y(t) - \varphi^T(t)\theta(t) \|^2 + \gamma \sum_{t=2}^{N} \| \theta(t) - \theta(t-1) \|_1. \tag{8.35}$$

One could also use the norms in ℓ_p with $p > 1$ as regularizers but it is crucial that the penalty is a sum of norms and not a sum of squared norms. Then, adopting a suitable value for the regularization parameter γ, the penalty favours the terms in the second sum to be exactly zero and not just small. This will force the number of different models from (8.35) to be small and thus just flag when essential changes have taken place.

This idea is taken further in [24] to build hybrid models of PWA (piecewise affine) character. The starting point is again (8.34), but now the number of models is reduced by looking at all the raw models:

$$\min_{\theta(t), t=1, \ldots, N} \sum_{t=1}^{N} \| y(t) - \varphi^T(t)\theta(t) \|^2 + \gamma \sum_{t=1}^{N} \sum_{s=1}^{N} K(p(t)p(s)) \| \theta(t) - \theta(s) \|. \tag{8.36}$$

Here $K(p_1, p_2)$ is a weighting based on the respective regime variables p. This gives a number of, say, d submodels, and they can then be associated with values of the regime variable by a classification step.

These ideas of segmentation, building a collection of d submodels and associating them with particular values of time are taken to a further degree of sophistication in [27]. The idea there is to build *hybrid stable spline* (HSS) algorithm, based on a joint use of the TC (stable spline) kernel, see Sect. 5.5.1, for a family of ARX models like (8.34). The classification of the models is built into the algorithm, by letting the classification parameters be part of the hyperparameters. An MCMC scheme is employed to handle the nonconvex and combinatorial difficulties of the maximum likelihood criterion.

8.5.5 Sparsity and Variable Selection

In all estimation problems it is essential to find the regressors $x_k(t)$, where $k = 1, \ldots, d$, which are best suited for predicting the goal variable $y(t)$. The variables x_k can be formed from the observations from the system in many different ways. It is generally desired to find a small collection of regressors, and statistics offers many tools for this: hypothesis analysis, projection pursuit [14], manifold learning/dimensionality reduction [10, 26, 34], ANOVA, see, e.g., [20] for applications to nonlinear system identification.

The problem of variable (regressor) selection can be formulated as follows. Given a model with n candidate regressors $\tilde{x}_k(t)$

$$y(t) = f(\tilde{x}_1(t), \ldots, \tilde{x}_n(t)) + e(t) \tag{8.37}$$

find a subselection or combination of regressors $x_1(t), \ldots, x_d(t)$ that gives the best model of the system. Note that the NARX model (8.3) is a special case of (8.37) with $x_k(t) = [y(t - k), u(t - k)]$. In principle one could try out different subsets of regressors and see how good models (in cross validation) are produced. That would in most cases mean overwhelmingly many tests.

Instead the ℓ_1-norm regularization discussed in Sect. 3.6.1, leading to LASSO in (3.105), is a very powerful tool for variable selection and sparsity. In what follows each $\tilde{x}_i(t)$ is scalar and is the ith component of the n-dimensional vector $x(t)$. Then, for a linearly parametrized model

$$y(t) = \beta_1 \tilde{x}_1(t) + \cdots + \beta_n \tilde{x}_n(t) + e(t), \tag{8.38}$$

where the best regressors are found by the criterion

$$\min_{B} \sum_{t=1}^{N} \|y(t) - \Phi(t)B\|^2 + \gamma \|B\|_1 \tag{8.39}$$

$$B = [\beta_1, \beta_2, \ldots, \beta_n]^T \tag{8.40}$$

$$\Phi(t) = [\tilde{x}_1(t), \tilde{x}_2(t), \ldots, \tilde{x}_n(t)]. \tag{8.41}$$

This idea to use ℓ_1-norm regularization was extended to the general model (8.37) in [2]. It is based on the idea to estimate the partial derivatives $\beta_k = \frac{\partial f}{\partial \tilde{x}_k}$ in (8.37) analogously to (8.39). In particular, the Taylor expansion of $f(x(t))$ around x^0 is

$$f(x(t)) = f(x^0) + (x(t) - x^0)^T \frac{\partial f}{\partial \tilde{x}} + \mathcal{O}(\|x(t) - x^0\|^2). \qquad (8.42)$$

The partial derivative is evaluated at x^0 and is a column vector of dimension n with row k given by the derivative w.r.t. $\tilde{x}_k$. As anticipated, denote this by β_k. These parameters can be estimated by least squares with

$$\min_{\alpha, B} \sum_{t=1}^{N} \|y(t) - \alpha - (x(t) - x^0)^T B\|^2 \cdot K(x(t) - x^0) + \gamma \|B\|_1, \qquad (8.43)$$

where α corresponds to $f(x^0)$, B is the vector of partial derivatives β_k and K is a kernel that focuses the sum to points $x(t)$ in the vicinity of x^0. The ℓ_1 norm regularization term is added just as in (8.39) to promote zero estimates of the gradients. This will focus on selecting regressors $\tilde{x}_k$ that are important for the model.

With the so-called iterative reweighting, [6], the regularization term can be refined to

$$\gamma \sum_{k=1}^{n} w_k |\beta_k|, \qquad (8.44)$$

where $w_k = 1/|\hat{\beta}_k|$ are based on the estimates from (8.43). This refinement is suggested to be included in the algorithm of [2].

Note that this test depends on the chosen point x^0. It will be a big task to investigate "many" such points. In [1] it is instead suggested to estimate the expected values $E x_i \frac{\partial f}{\partial \tilde{x}_i}$ and $E \frac{\partial f}{\partial \tilde{x}_i}$. This is done using the pdfs for $\tilde{x}_k$ given by $p_i(u)$ and $\frac{dp_i(u)}{dx_i}$ which can be estimated by simple density estimation (involving only a scalar random variable).

A comprehensive study of sparsity and regularization is made in [33]. It works with a more complex model definition, allowing $f : \mathbb{R}^n \to \mathbb{R}$ to be defined over several Hilbert spaces. The bottom line is still based on ℓ_1-norm regularization of partial derivatives and the final learning algorithm is given by minimization of a functional

$$\frac{1}{N} \sum_{t=1}^{N} (y_t - f(x(t)))^2 + \gamma \left(2 \sum_{i=1}^{n} \left\| \frac{\partial f}{\partial \tilde{x}_i} \right\|_N + \nu \|f\|_{\mathcal{H}}^2 \right). \qquad (8.45)$$

Here, $\mathcal{H}$ can be a RKHS, the penalty on each partial derivative is given by

$$\left\|\frac{\partial f}{\partial \tilde{x}_i}\right\|_N = \sqrt{\frac{1}{N}\sum_{t=1}^{N}\left(\frac{\partial f(x(t))}{\partial \tilde{x}_i}\right)},$$

γ is the regularization parameter and ν is a small positive number to ensure stability and strongly convex regularizer.

References

1. Bai EW, Cheng C, Zhao W (2019) Variable selection of high-dimensional non-parametric nonlinear systems by derivative averaging to avoid the curse of dimensionality. Automatica 101:138–149
2. Bai EW, Li K, Zhao W, Xu W (2014) Kernel based approaches to local nonlinear non-parametric variable selection. Automatica 50(1):100–113
3. Bartlett PL, Long PM, Lugosi G, Tsigler A (2020) Benign overfitting in linear regression. PNAS 117:30063–30070
4. Birpoutsoukis G, Marconato A, Lataire J, Schoukens J (2017) Regularized nonparametric Volterra kernel estimation. Automatica 82:72–82
5. Boyd S, Chua L (1985) Fading memory and the problem of approximating nonlinear operators with Volterra series. IEEE Trans Circuits Syst 32(11):1150–1161
6. Candes E, Waking M, Boyd S (2008) Enhancing sparsity by reweighted ℓ_1 minimization. J Fourier Anal Appl 14(5):877–905
7. Cheng CM, Peng ZK, Zhang WM, Meng G (2017) Volterra-series-based nonlinear system modeling and its engineering applications: a state-of-the-art review. Mech Syst Signal Process 87:340–364
8. Dalla Libera A, Carli R, Pillonetto G (2021) Kernel-based methods for Volterra series identification. Automatica 129(1):109686
9. Dalla Libera A, Pillonetto G (2021) Deep prediction networks. Neurocomputing
10. Fukumizu K, Bach FR, Jordan MI (2004) Dimensionality reduction for supervised learning with reproducing kernel Hilbert spaces. J Mach Learn Res 5:73–99
11. Gedon D, Wahlström N, Schön TB, Ljung L (2021) Deep state space models for nonlinear system identification. In: Proceedings of the 19th IFAC symposium on system identification (SYSID), online, July, 2021
12. Haykin S (1999) Neural networks, 2nd edn. Prentice-Hall, Upper Saddle River
13. Hendriks J, Gustafsson FK, Ribeiro AH, Wills A, Schön TB (2021) Deep energy-based NARX models. In: Proceedings of the 19th IFAC symposium on system identification (SYSID), online, July, 2021
14. Huber PJ (1985) Projection pursuit. Ann Stat 13:435–475
15. Juloski A, Heemels WPMH, Ferrari-Trecate G, Vidal R, Paoletti S, Niessen JHG (2005) Comparison of four procedures for the identification of hybrid systems. In: Morari M, Thiele L (eds) Hybrid systems: computation and control. Lecture notes in computer science. Springer, Berlin, pp 354–369
16. Juloski AL, Paoletti S, Roll J (2006) Recent techniques for the identification of piecewise affine and hybrid systems. In: Abdallah CT, Menini L, Zaccarian L (eds) Trends in nonlinear systems and control: in honor of Petar Kokotovic and Turi Nicosia. Birkhäuser

17. Labach A, Salehinejad H, Valaee S (2019) Survey of dropout methods for deep neural networks. arXiv:1904.1336.v2
18. LeCun Y, Bengio Y, Hinton G (2015) Deep learning. Nature 521:436–444
19. Liang T, Rakhlin A (2020) Just interpolate: Kernel ridgeless regression can generalize. Ann Stat 48(3):1329–1347
20. Lind I, Ljung L (2009) Regressor and structure selection in NARX models using a structured ANOVA approach. Automatica 44:305–383
21. Lindsten F, Schön TB, Jordan MI (2013) Bayesian semiparametric Wiener system identification. Automatica 49:2053–2063
22. Ljung L (2013) System identification toolbox V8.3 for MATLAB. Natick, MA: the MathWorks, Inc
23. Nelles O (2001) Nonlinear system identification. Springer
24. Ohlsson H, Ljung L (2013) Identification of switched linear regression models using sum-of-norms regularization. Automatica 49:1045–1050
25. Ohlsson H, Ljung L, Boyd S (2010) Segmentation of ARX-models using sum-of-norms regularization. Automatica 46(6):1107–1111
26. Ohlsson H, Roll J, Glad T, Ljung L (2007) Using manifold learning for nonlinear system identification. In: Proceedings of the IFAC symposium on nonlinear conrtrol systems (NOLCOS, Pretoria, South Africa, August 2007. IFAC
27. Pillonetto G (2016) A new kernel-based approach to hybrid system identification. Automatica 70:21–31
28. Poernomo A, Kang DK (2018) Biased dropout and crossmap dropout: learning towards effective dropout regularization in convolutional neural network. Neural Netw 105:60–67
29. Rasmussen CE, Williams CKI (2006) Gaussian processes for machine learning. The MIT Press
30. Ribeiro AH, Hendriks J, Wills A, Schön TB (2021) Beyond Occam's razor in system identification: double-descent when modeling dynamics. In: Proceedings of the 19th IFAC symposium on system identification (SYSID), Online, July, 2021
31. Risuleo RS, Bottegal G, Hjalmarsson H (2017) A nonparametric kernel-based approach to Hammerstein system identification. Automatica 85:234–247
32. Risuleo RS, Lindsten F, Hjalmarsson H (2019) Bayesian nonparametric identification of Wiener systems. Automatica 108:108480
33. Rosasco L, Villa S, Mosci S, Santoro M, Verri A (2013) Nonparametric sparsity and regularization. J Mach Learn Res 14:1665–1714
34. Roweis ST, Saul LK (2000) Nonlinear dimensionality reduction by local linear embedding. Science 290:2323–2326
35. Schetzen M (1980) The Volterra and Wiener theories of nonlinear systems. Wiley, New York
36. Schoukens J, Ljung L (2019) Nonlinear system identification - a user-oriented roadmap. IEEE Trans Control Syst Technol 39(6):28–99
37. Schoukens M, Tiels K (2017) Identification of block-oriented nonlinear systems starting from linear approximations: a survey. Automatica 85:272–292
38. Sjöberg J, Ljung L (1995) Overtraining, regularization and searching for minimum with application to neural nets. Int J Control 62(6):1391–1407
39. Sontag ED (1996) Interconnected automata and linear systems: a theoretical framework in discrete-time. In: Henzinger TA, Alur R, Sontag ED (eds) Hybrid systems III, volume 1066 of Lecture notes in computer science. Springer, Berlin, pp 436–448
40. Spinelli W, Piroddi L, Lovera M (2005) On the role of prefiltering in nonlinear system identification. IEEE Trans Autom Control 50(10):1597–1602
41. Stoddard JG, Welsh JS, Hjalmarsson H (2017) EM-based hyperparameter optimization for regularized Volterra kernel estimation. IEEE Control Syst Lett 1(2):388–393
42. Wills A, Schon TB, Ljung L, Ninness B (2013) Identification of Hammerstein-Wiener models. Automatica 49:70–81
43. Zancato L, Chiuso A (2021) A novel deep neural network architecture for nonlinear system identification. In: Proceedings of the 19th IFAC symposium on system identification (SYSID), online, July, 2021

Chapter 9
Numerical Experiments and Real World Cases

Abstract This chapter collects some numerical experiments to test the performance of kernel-based approaches for discrete-time linear system identification. Using Monte Carlo simulations, we will compare the performance of kernel-based methods with the classical PEM approaches described in Chap. 2. Simulated and real data are included, concerning a robotic arm, a hairdryer and a problem of temperature prediction. We conclude the chapter by introducing the so-called multi-task learning where several functions (tasks) are simultaneously estimated. This problem is significant if the tasks are related to each other so that measurements taken on a function are informative with respect to the other ones. A problem involving real pharmacokinetics data, related to the so-called population approaches, is then illustrated. Results will be often illustrated by using MATLAB boxplots. As already mentioned in Sect. 7.2, when commenting Fig. 7.8, the median is given by the central mark while the box edges are the 25th and 75th percentiles. The whiskers extend to the most extreme fits not seen as outliers. Then, the outliers are plotted individually.

9.1 Identification of Discrete-Time Output Error Models

In this section, we will consider two numerical experiments with data generated according to the discrete-time output error (OE) model

$$y(t) = G_0(q)u(t) + e(t),$$

where G_0 is a rational transfer function while e is white Gaussian noise independent of the known input u. Using simulated data, we will compare the performance of the classical PEM approach, as described in Chap. 2, with some of the regularized techniques illustrated in this book. In particular, we will adopt regularized high-order FIR, with impulse response coefficients contained in the m-dimensional (column) vector θ and the output data in the (column) vector $Y = [y(1) \ldots y(N)]^T$. So, letting the regression matrix $\Phi \in \mathbb{R}^{N \times m}$ be

© The Author(s) 2022
G. Pillonetto et al., *Regularized System Identification*, Communications and Control Engineering, https://doi.org/10.1007/978-3-030-95860-2_9

$$\Phi = \begin{pmatrix} u(0) & u(-1) & u(-2) & \dots & u(-m+1) \\ u(1) & u(0) & u(-1) & \dots & u(-m) \\ \dots & & & & \\ u(N-1) & u(N-2) & u(N-3) & \dots & u(N-m) \end{pmatrix},$$

our estimator is

$$\hat{\theta} = \arg\min_{\theta} \|Y - \Phi\theta\|^2 + \gamma\theta^T P^{-1}\theta \tag{9.1a}$$

$$= P\Phi^T(\Phi P\Phi^T + \gamma I_N)^{-1}Y; \text{ or} \tag{9.1b}$$

$$= (P\Phi^T\Phi + \gamma I_m)^{-1}P\Phi^T Y. \tag{9.1c}$$

We have already seen in (5.40), (5.41) and (7.30), using MaxEnt arguments and spline theory, that choices for the regularization matrix P can be the first- or second-order stable spline kernel, denoted respectively by TC and SS, respectively, or the DC kernel. They are recalled below specifying also the hyperparameter vector η:

$$\text{TC} \quad P_{kj}(\eta) = \lambda\alpha^{\max(k,j)};$$
$$\lambda \geq 0, \ 0 \leq \alpha < 1, \ \eta = [\lambda, \alpha], \tag{9.2}$$

$$\text{SS} \quad P_{kj}(\eta) = \lambda\left(\frac{\alpha^{k+j+\max(k,j)}}{2} - \frac{\alpha^{3\max(k,j)}}{6}\right)$$
$$\lambda \geq 0, \ 0 \leq \alpha < 1, \ \eta = [\lambda, \alpha], \tag{9.3}$$

$$\text{DC} \quad P_{kj}(\eta) = \lambda\alpha^{(k+j)/2}\rho^{|j-k|};$$
$$\lambda \geq 0, \ 0 \leq \alpha < 1, |\rho| \leq 1, \ \eta = [\lambda, \alpha, \rho]. \tag{9.4}$$

9.1.1   Monte Carlo Studies with a Fixed Output Error Model

In this example the true impulse response is fixed to that reported in Fig. 8.2, obtained by random generation of a rational transfer function of order 10. It has to be estimated from 500 input–output couples (collected with system initially at rest). The input is white noise filtered by the rational transfer function $1/(z - p)$ where p will vary over the unit interval during the experiment. Note that p establishes the difficulty of our system identification problem. Values close to zero make the input similar to white noise and the output data informative over a wide range of frequencies. Instead, values of p close to 1 increase the low-pass nature of the input and, hence, the ill-conditioning. The measurement noise is white and Gaussian with variance equal to that of the noiseless output divided by 50. Two estimators will be adopted:

- *Oe+Or.* Classical PEM approach (2.22) equipped with an oracle. In particular, our candidate models are rational transfer functions where the order of the two polynomials is equal and can vary between 1 and 30. For any model order, estimation is performed through nonlinear least squares by solving (2.22) with ℓ in (2.21) set to the quadratic function. The method is implemented in `oe.m` of the MATLAB System Identification Toolbox. Then, the oracle chooses the estimate which maximizes the fit

$$100 \left(1 - \left[\frac{\sum_{k=1}^{100} |g_k^0 - \hat{g}(k)|^2}{\sum_{k=1}^{100} |g_k^0 - \bar{g}^0|^2} \right]^{\frac{1}{2}} \right), \quad \bar{g}^0 = \frac{1}{100} \sum_{k=1}^{100} g_k^0, \qquad (9.5)$$

where g_k^0 are the true impulse response coefficients while $\hat{g}(k)$ denote their estimates. The estimator is given the information that system initial conditions are null.

- *TC+ML.* This is the regularized estimator (9.1), equipped with the kernel *TC*. The number of estimated impulse response coefficients is $m = 100$ and the regression matrix is built with $u(t) = 0$ if $t < 0$. At every run, the noise variance is estimated by fitting via least squares a low-bias model for the impulse response. Then, the two kernel hyperparameters are obtained via marginal likelihood optimization, see (7.42). The method is implemented in `impulseest.m` of the MATLAB System Identification Toolbox.

We consider 4 Monte Carlo studies of 300 runs defined by different values of p in the set $\{0, 0.9, 0.95, 0.99\}$. As already mentioned, $p = 0$ corresponds to white noise input while $p = 0.99$ leads to a highly ill-conditioned problem (output data provide little information at high frequencies). Figure 9.1 reports the boxplots of the 1000 fits returned by *Oe+Or* and *TC+ML* for the four different values of p. Even if PEM exploits an oracle to tune complexity, the performance is (slightly) better than *TC+ML* only when the input is white noise, see also Table 9.1. When p increases, the ill-conditioning affecting the problem increases and *TC+ML* outperforms *Oe+Or* even if no oracle is used for hyperparameters tuning. This also points out the effectiveness of marginal likelihood optimization in controlling complexity.

This case study shows that continuous tuning of hyperparameters may be a more versatile and powerful approach than classical estimation of discrete model orders. A problem related to PEM here could be also the presence of local minima of the objective. This is much less critical when adopting kernel-based regularization. In fact, *TC+ML* regulates complexity through only two hyperparameters while *Oe+Or* has to optimize many more parameters (function of the postulated model order).

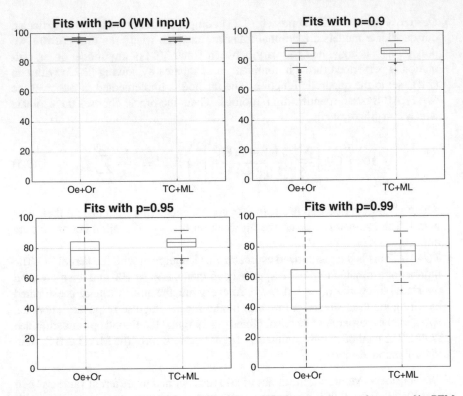

Fig. 9.1 *Experiment with a fixed OE-model* Boxplots of 300 impulse response fits returned by PEM with an oracle to tune discrete model order and by TC with continuous hyperparameters estimated by marginal likelihood optimization. Results are function of the level of ill-conditioning affecting the problem which increases with p (the input is white Gaussian noise for $p = 0$ while the other values define low-pass inputs)

Table 9.1 *Experiment with a fixed OE-model* Average fit, as a function of p, after 300 Monte Carlo runs. The value $p = 0$ corresponds to white noise input and the level of ill-conditioning then increases as p increases

	Oe+Or	TC+ML
$p = 0$	95.8	95.3
$p = 0.9$	85.2	86.3
$p = 0.95$	75.3	83.2
$p = 0.99$	49.9	74.3

9.1.2 Monte Carlo Studies with Different Output Error Models

Now we consider two Monte Carlo studies of 1000 runs regarding identification of several discrete-time output error models. The outputs are still given by

$$y(t) = G_0(q)u(t) + e(t)$$

with e white Gaussian noise independent of u, but the rational transfer function G_0 changes at any run. In fact, a 30th-order single-input single-output continuous-time system is first randomly generated by the MATLAB command `rss.m`. It is then sampled at 3 times of its bandwidth and used if its poles fall within the circle of the complex plane with centre at the origin and radius 0.99.

With the system at rest, 1000 input–output pairs are generated as follows. At any run, the system input is unit variance white Gaussian noise filtered by a second-order rational transfer function generated by the same procedure adopted to obtain G_0. The outputs are corrupted by an additive white Gaussian noise with a SNR (the ratio between the variance of noiseless output and noise) randomly chosen in [1, 20] at any run. In the first experiment, the data set

$$\mathscr{D}_T = \{u(1), y(1), \ldots, u(N), y(N)\}$$

contains the first 200 input–output couples, i.e., $N = 200$, while in the second experiment all the 1000 couples are used, i.e., $N = 1000$.

Starting from null initial conditions, at any run we also generate two different kinds of test sets

$$\mathscr{D}_{test} = \{u^{new}(1), y^{new}(1), \ldots, u^{new}(M), y^{new}(M)\}, \quad M = 1000.$$

The first test set is especially challenging since noiseless outputs are generated by using unit variance white Gaussian noise as input. In the second test set the input has instead the same statistics of that entering the identification data, hence making easier its prediction.

The performance of a model characterized by $\hat{\theta}$, and returning $\hat{y}^{new}(t|\hat{\theta})$ as output prediction at instant t, is

$$\mathscr{F}(\hat{\theta}) = 100 \left(1 - \sqrt{\frac{\sum_{t=1}^{M} \left(y^{new}(t) - \hat{y}^{new}(t|\hat{\theta}) \right)^2}{\sum_{t=1}^{M} (y^{new}(t) - \bar{y}^{new})^2}} \right), \quad M = 1000, \quad (9.6)$$

where $\bar{y}^{new}$ is the average output in $\mathscr{D}_{test}$ and $\hat{y}^{new}(t|\hat{\theta})$ are computed assuming zero initial conditions (otherwise high-order models could have the advantage to calibrate the initial conditions to fit $\mathscr{D}_{test}$). The prediction fit (9.6) can be obtained by the MATLAB command `predict(model,data,k,'ini','z')` where `model` and `data` denote structures containing the estimated model and the test set $\mathcal{D}_{test}$, respectively.

In what follows, we will use also estimators equipped with an oracle which evaluates the fit (9.6) for the test set of interest. Different rational models with orders between 1 and 30 are tried and the oracle selects the orders that give the best fit. We are now in a position to introduce the following 6 estimators:

- *Oe+Or1*. Classical PEM approach (2.22), with quadratic ℓ in (2.21), equipped with an oracle which uses the first test set (white noise input). As said, candidate models are rational transfer functions whose order can vary between 1 and 30. For any order, the model is returned by the function `oe.m` of the MATLAB's System Identification Toolbox [14].
- *Oe+Or2*. The same procedure described above except that the oracle maximizes the prediction fit using the second test set (test input with statistics equal to those of the training input).
- *Oe+CV*. The classical approach now does not use any oracle: model order is estimated by cross validation by splitting the identification data into two sets with the first and the last $N/2$ data contained in $\mathcal{D}_T$. The prediction errors are computed assuming zero initial conditions. The model order minimizing the sum of squared prediction errors (computed assuming zero initial conditions) is chosen. Finally, the system estimate is computed using all the data in $\mathcal{D}_T$ by solving (2.22) with quadratic loss.
- *{TC+ML,SS+ML,DC+ML}*. These are three regularized FIR estimators of the form (9.1) with order 200 and kernels *TC* (9.2), *SS* (9.3) and DC (9.4). Marginal likelihood optimization (7.42) is used to determine the noise variance and the kernel hyperparameters (2 for *SS* and *TC*, 3 for *DC*). The regularized FIR models are estimated using the function `impulseest.m` in the MATLAB's System Identification Toolbox [14].

9.1.2.1 Results

The MATLAB boxplots in Fig. 9.2 contain the 1000 fit measures returned by the estimators during the first experiment with $N = 200$ (left panels) and the second experiment with $N = 1000$ (right panels). Table 9.2 reports the average fit values.

In the top panels of Fig. 9.2 one can see the fits of the first test set. Recall that *Oe+Or1* has access to such data to optimize the prediction capability. Interestingly, despite this advantage, the performance of all the three regularized approaches is close that of the oracle while that *Oe+CV* is not so satisfactory. This is also visible in the first two rows of Table 9.2.

The bottom panels of Fig. 9.2 show results relative to the second test set which is used by *Oe+Or2* to maximize the prediction fit. Since training and test data are

Table 9.2 *Identification of discrete-time OE-models* Average fit, after 1000 Monte Carlo runs, as a function of the test set type and the identification data set size ($N = 200$ or $N = 1000$). Results in the first and last column come from oracle-based estimators which cannot be implemented in practice

	Oe+Or1	TC	SS	DC	Oe+CV	Oe+Or2
1st test set, $N = 200$	52.7	51.9	48.8	51.1	34.8	−11.9
1st test set, $N = 1000$	66.2	63.4	58.5	63.1	−20.9	28.2
2nd test set, $N = 200$	84.8	86.3	85.9	86.8	72.9	87.8
2nd test set, $N = 1000$	93.2	92.9	91.8	93.1	88.6	94.2

more similar, the prediction capability of *Oe+CV* improves significantly but the regularized estimators still outperform the classical approach, see also the last two rows of Table 9.2.

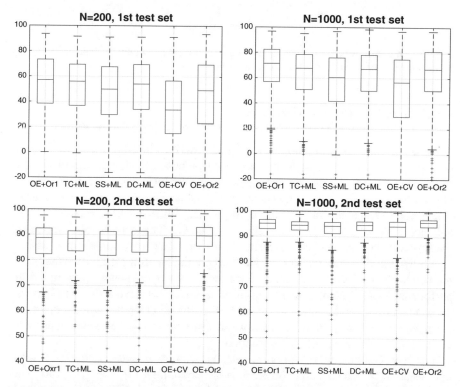

Fig. 9.2 *Identification of discrete-time OE-models Top* Boxplot of the 1000 prediction fits on future outputs with test input given by white noise. The size of the identification data set is 200 (top left) or 1000 (top right). *Bottom* Differently from the results in the top panel, input statistics in the estimation and test data set are the same. The first and last boxplot contained in the four panels contain results from the estimators *Oe+Or1* and *Oe+Or2* which cannot be implemented in practice

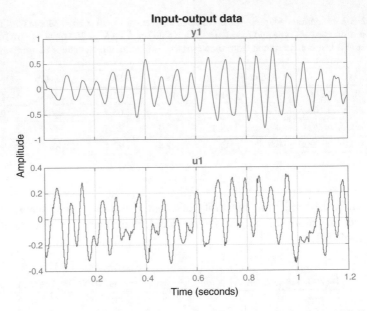

Fig. 9.3 *Robot arm* A portion of the input–output data for the robot arm: the input is the driving couple (bottom) and the output is the tip of the robot arm (top)

9.1.3   Real Data: A Robot Arm

Consider now the vibrating flexible robot arm described in [27], where two feedforward controller design methods were compared on trajectory tracking problems. The input of the robot arm is the driving couple and the output is the acceleration at the tip of the robot arm. The input–output data contain 40960 data points. They are collected at a sampling frequency of 500 Hz for 10 periods with each period containing 4096 data points. A portion of the data is shown in Fig. 9.3. The identification problem of the robot arm was studied in [23, Sect. 11.4.4] with frequency domain methods.

We will build models by both the classical prediction error method and the kernel method with the DC kernel. Since the true system is unknown, to compare the performance of different impulse response estimates we divide the data into two parts: the training and the test set, given by the first 6000 input–output couples and the reaming ones, respectively. Then, we measure how well the models, built with the estimation data, predict the test outputs.

For the prediction error method, we estimate nth-order state-space models without disturbance model and with zero initial conditions for $n = 1, \ldots, 36$. This method is available in MATLAB's System Identification Toolbox [13] as the command `pem(data,n,'dist','no','ini','z')`. The prediction fits computed using (9.6) are shown as a function of n in Fig. 9.4, respectively. An oracle that has access to the test set would select the order $n = 18$, hence obtaining a prediction fit equal to 79.75%. For the kernel method with the DC ker-

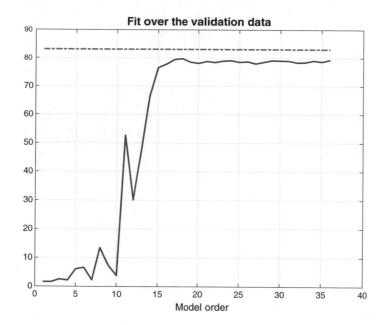

Fig. 9.4 *Robot arm* The solid red line is the fit for the prediction error method for different model order $n = 1, \ldots, 36$. The dash-dot blue line is the prediction fit on the test set for the regularized method with the DC kernel

nel, we estimate a FIR model of high-order 3000 with hyperparameters tuned by optimizing the marginal likelihood. When forming the regression matrix, the unknown input data are set to zero. The prediction fit (9.6) is 83.07% and is shown as a horizontal solid line in Fig. 9.4. The kernel method with the DC kernel is available in MATLAB's System Identification Toolbox [14] as the command `impulseest(data,3000,0,opt)` where, in the option `opt`, we set `opt.RegulKernel='dc'; opt.Advanced.AROrder=0`.

The Bode magnitude plot of the models estimated by PEM and the DC kernel is shown in Fig. 9.5. The empirical frequency function estimate obtained using the command `etfe` in MATLAB's System Identification Toolbox [14] is also displayed.

The measured output and the predicted output over a portion of the test set are shown in Fig. 9.6. If one has concern that a FIR model of order 3000 is quite large, then one could reduce such high-order model by projecting it to a low-order state-space model. Exploiting model order reduction techniques, the fit of a state-space model of order $n = 25$ is 79.8%, still better than the best state-space description that can be obtained by PEM.

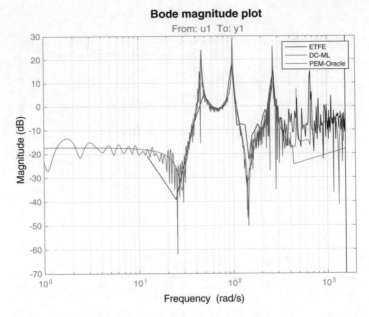

Fig. 9.5 *Robot arm* Bode magnitude plot of the estimated models obtained by the empirical frequency function estimate ETFE (black), the regularized method with DC kernel and hyperparameters estimate by marginal likelihood optimization (blue), and the prediction error method with order $n = 18$ (red)

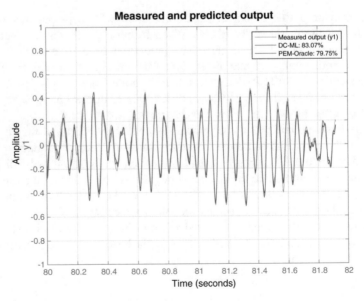

Fig. 9.6 *Robot arm* A portion of the test set (grey) and the predictions returned by the regularized method with DC kernel (blue) and the prediction error method (red)

9.1.4 Real Data: A Hairdryer

The second application is a real laboratory device, whose function is similar to that of a hairdryer: the air is fanned through a tube and then heated at a mesh of resistor wires, as described in [13, Sect. 17.3]. The input to the hairdryer is the voltage over the mesh of resistor wires while the output is the air temperature measured by a thermocouple. The input–output data contain 1000 data points collected at a sampling frequency of 12.5 Hz for 80 s. A portion of the data is shown in the top panel of Fig. 9.7. Since the input–output values move around 5 and 4.9, respectively, we detrend the measurements in such a way that they move around 0. The estimation and test set data are then given by the first and the last 500 input–output couples, respectively.

As in the case of the robot arm, we build models by the classical prediction error method with an oracle, which maximizes the prediction fit, and the regularized approach with the DC kernel, with hyperparameters tuned by marginal likelihood optimization. For the prediction error method, we estimate nth-order state-space models without disturbance model for $n = 1, \ldots, 36$ and with zero initial conditions. The fits, as a function of n, are shown in Fig. 9.8. The best result is obtained for order $n = 5$ and turns out 88.38%. For the kernel method with the DC kernel, we estimate a FIR model with order 70. When forming the regression matrix, we set the unknown input data to zero. The prediction fit (9.6) is somewhat close to that achieved by PEM+Oracle being equal to 88.15%. It is shown as a dash-dot blue line in Fig. 9.8. The test set and the predicted outputs returned by the two methods are shown in Fig. 9.9. One can see that the regularized approach has a prediction capability very close to that of PEM+Oracle.

9.2 Identification of ARMAX Models

In this section we consider the identification of linear systems

$$y(t) = \left\{ \sum_{i=1}^{p} G_{0i}(q)u_i(t) \right\} + H_0(q)e(t). \tag{9.7}$$

Differently from the previous cases, beyond the presence of multiple observable inputs u_i, also the noise model is unknown. In fact, the $e(t)$ are white Gaussian noise of unit variance filtered by a system $H_0(q)$ that has to be estimated from data.

First, it is useful to cast the identification of the general model (9.7) in a regularized context. Without loss of generality, to simplify the exposition, let $p = 1$ with the single observable input denoted by u. Exploiting (2.4), given the general linear model (9.7), we can write any predictor as two infinite impulse responses from y and u, respectively. When using ARX models, we have seen in (2.8) that such infinite responses specialize to finite responses. One has

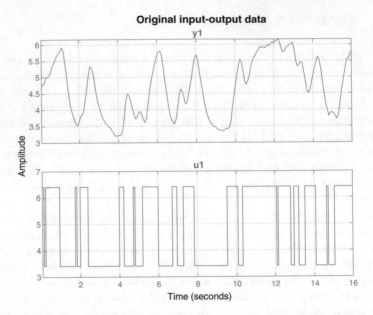

Fig. 9.7 *Hairdryer* A portion of the input–output data for the hairdryer. The input is the voltage over the mesh of resistor wires (bottom panel) and the output is the air temperature measured by a thermocouple (top panel)

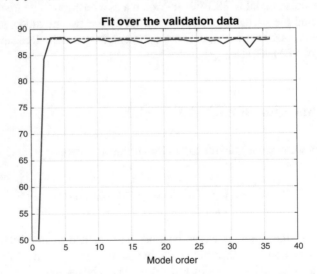

Fig. 9.8 *Hairdryer* The solid line is the fit for the prediction error method for different model order $n = 1, \ldots, 36$. The dash-dot line is the fit for the ReLS method with the DC kernel

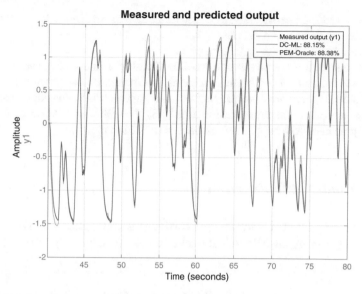

Fig. 9.9 *Hairdryer* The measured output and the predicted output over the test data: the measured output (grey), the ReLS method with DC kernel (blue) and the prediction error method (red)

$$y(t) = -a_1 y(t-1) - \cdots - a_{n_a} y(t - n_a) + b_1 u(t-1) + \cdots$$
$$+ b_{n_b} u(t - n_b) + e(t) = \varphi_y^T(t)\theta_a + \varphi_u^T(t)\theta_b + e(t), \qquad (9.8)$$

where $\theta_a = \begin{bmatrix} a_1 \ \ldots \ a_{n_a} \end{bmatrix}^T$, $\theta_b = \begin{bmatrix} b_1 \ \ldots \ b_{n_b} \end{bmatrix}^T$ and $\varphi_y(t)$, $\varphi_u(t)$ are made up from y and u in an obvious way. Thus, the ARX model is a linear regression model, to which the same ideas of regularization can be applied. This point is important since we have seen in Theorem 2.1 that ARX-expressions become arbitrarily good approximators for general linear systems as the orders n_a, n_b tend to infinity. However, as discussed in Chap. 2, high-order ARX can suffer from large variance. A solution is to set $n_a = n_b = n$ to a large value and then introduce regularization matrices for the two impulse responses from y and from u. The P-matrix in (9.1) can be partitioned along with θ_a, θ_b:

$$P(\eta_1, \eta_2) = \begin{bmatrix} P^a(\eta_1) & 0 \\ 0 & P^b(\eta_2) \end{bmatrix} \qquad (9.9)$$

with $P^a(\eta_1)$, $P^b(\eta_2)$ defined, e.g., by any of (9.2)–(9.4). Letting $\theta = [\theta_a^T \ \theta_b^T]^T$ and building the regression matrix using $[\varphi_y^T(t) \ \varphi_u^T(t)]$ as rows, the estimator (9.1) now becomes a regularized high-order ARX. The MATLAB code for estimating this model using, e.g., the DC kernel would be

```
ao=arxRegulOptions('RegularizationKernel','DC'),
[Lambda,R] = arxRegul(data,na,nb,nk,ao),
```

```
aropt= arxOptions; aropt.Regularization.Lambda = Lambda,
aropt.Regularization.R = R,
m = arx(data,na,nb,nk,aropt).
```

We can also easily extend this construction to multiple inputs. Given any generic p, one needs to estimate $p + 1$ impulse responses with the matrix (9.9) now containing $p + 1$ blocks. If there are multiple outputs, one approach is to consider each output channel as a separate linear regression as in (9.8). The difference is that now also the other outputs need to be appended as done with the inputs.

9.2.1   Monte Carlo Experiment

One challenging Monte Carlo study of 1000 runs is now considered. Data are generated at any run by an ARMAX model of order 30 having p observable inputs, i.e.,

$$ y(t) = \left\{ \sum_{i=1}^{p} \frac{B_i(q)}{A(q)} u_i(t) \right\} + \frac{C(q)}{A(q)} e(t), $$

with p drawn from a random variable uniformly distributed on $\{2, 3, 4, 5\}$. Note that the system contains $p + 1$ rational transfer functions. They depend on the polynomials A, B_i and C which are randomly generated at any run by the MATLAB function drmodel.m. Such function is first called to obtain the common denominator A and the first numerator B_1. The other p calls are used to obtain the numerators of the remaining rational transfer functions. The system so generated is accepted if the modulus of its poles is not larger than 0.95. In addition, letting $G_i(q) = \frac{B_i(q)}{A(q)}$ and $H(q) = \frac{C(q)}{A(q)}$ the signal to noise ratio has to satisfy

$$ 1 \le \frac{\sum_{i=1}^{p} \|G_i\|_2^2}{\|H\|_2^2} \le 20 $$

where $\|G_i\|_2$, $\|H\|_2$ are the ℓ_2 norms of the system impulse responses.

After a transient to mitigate the effect of initial conditions, at any run 300 input–output couples are collected to form the identification data set $\mathscr{D}_T$ and other 1000 to define the test set $\mathscr{D}_{test}$. In any case, the input is white Gaussian noise of unit variance.

Differently from the output error models, in the ARMAX case the performance measure adopted to compare different estimated models depends on the prediction horizon k. More specifically, let $\hat{y}_k^{new}(t|\hat{\theta})$ be the k-step-ahead predictor associated with an estimated model characterized by $\hat{\theta}$. For any t, such function predicts k-step-ahead the test output $y^{new}(t)$ by using the values of the test input u^{new} up to time $t - 1$ and of the test output y^{new} up to $t - k$. The prediction difficulty in general increases as k gets larger. The special case $k = 1$ corresponds to the one-step-ahead predictor given by (2.4), while see, e.g., [13, Sect. 3.2] for the expressions of the generic k-step-ahead impulse responses.

As done in (9.6) we use $\bar{y}^{new}$ denote the mean of the outputs in $\mathcal{D}_{test}$, but now the prediction fit depends on k, being given by

$$\mathcal{F}_k(\hat{\theta}) = 100 \left(1 - \sqrt{\frac{\sum_{t=1}^{M} \left(y^{new}(t) - \hat{y}_k^{new}(t|\hat{\theta}) \right)^2}{\sum_{t=1}^{M} (y^{new}(t) - \bar{y}^{new})^2}} \right), \quad M = 1000. \quad (9.10)$$

In this case, we say that an estimator is equipped with an oracle if it can use the test set to maximize $\sum_{k=1}^{20} \mathcal{F}_k$ by tuning the complexity of the model estimated using the identification data. The following estimators are then introduced:

- *PEM+Oracle*: this is the classical PEM approach (2.22) with quadratic loss equipped with an oracle. The candidate model structures are ARMAX models with polynomials all having the same degree up to 30. For any model order, the MATLAB command pem.m (or armax.m) is used to obtain the system's estimate. of the MATLAB System Identification Toolbox [14].

- *PEM+CV*: in place of the oracle, model complexity is estimated by cross validation splitting $\mathcal{D}_T$ into two sets containing, respectively, the first and the last 150 input–output couples. The model order which minimizes the sum of the squared one-step-ahead prediction errors computed with zero initial conditions for the validation data is selected. The final system's estimate is returned by (2.22) using all the identification data.

- *{PEM+AICc,PEM+BIC}*: this is the classical PEM approach with AIC-type criteria used to tune complexity, as reported in (2.35) and (2.36).

- *{TC+ML,SS+ML,DC+ML}*: these are the three regularized least squares estimators introduced at the beginning of this section which determine the unknown coefficients of the multi-input version of the ARX model. After setting the length of each predictor impulse response to 50, the regularization matrices entering the multi-input version of (9.9) are defined by TC (9.2) or SS (9.3) or DC (9.4) kernels. The first 50 input–output pairs in $\mathcal{D}_T$ are used just as entries of the regression matrix. For every impulse response, a different scale factor λ and a common variance decay rate α (and, in the case of DC, a correlation ρ) is adopted. The hyperparameters are determined via marginal likelihood optimization.

All the system inputs delay are assumed known and their values are provided to all the estimators described above.

The average of the fits $\mathcal{F}_k$ given by (9.10), function of the prediction horizon k, is reported in Fig. 9.10. Since PEM equipped with Akaike-like criteria return very small average fits, results achieved by this kind of procedures are not displayed. The MATLAB boxplots of the 1000 values of $\mathcal{F}_1$ and $\mathcal{F}_{20}$ returned by all the estimators are visible in Fig. 9.11. The average fit of *SS+ML* is quite close to that of *PEM+Oracle* which is in turn outperformed by *TC+ML* and *DC+ML*. This is remarkable also considering that such kernel-based approaches can be used in real applications while *PEM+Oracle* relies on an ideal tuning which exploits the test set. Results returned by PEM equipped with CV are instead unsatisfactory.

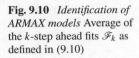

Fig. 9.10 *Identification of ARMAX models* Average of the k-step ahead fits $\mathscr{F}_k$ as defined in (9.10)

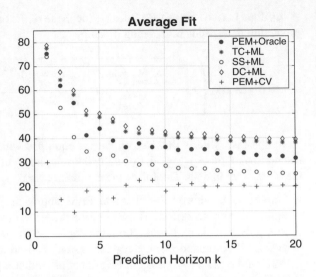

The results outline the importance of regularization, especially in experiments with relatively small data sets. In this case, only 300 input–output measurements are available with quite complex systems of order 30. The classical PEM approach equipped with any model order-selection rule cannot predict the test set better than the oracle. However, this latter can tune complexity by exploring only a finite set of given models. Kernel-based approaches can instead balance bias and variance by continuous tuning of regularization parameters. In this way, better performing trade-offs may be reached.

9.2.2 Real Data: Temperature Prediction

Now we consider thermodynamic modelling of buildings using some real data taken from [22]. Eight sensors are placed in two rooms of a small two-floor residential building of about 80 m² and 200 m³. They are located only on one floor (approximately 40 m²). More specifically, temperatures are collected through a wireless sensor network made of 8 *Tmote-Sky* nodes produced by Moteiv Inc. The building was inhabited during the measurement period consisting of 8 days and samples were taken every 5 min. A thermostat controlled the heating system with the reference temperature manually set every day depending upon occupancy and other needs. This makes available a total of 8 temperature profiles displayed in Fig. 9.12. One can see the high level of collinearity of the signals. This makes the problem ill-conditioned, complicating the identification process.

We just consider multiple-input single-output (MISO) models. The temperature from the first node is seen as the output (y_i) and the other 7 temperatures as inputs

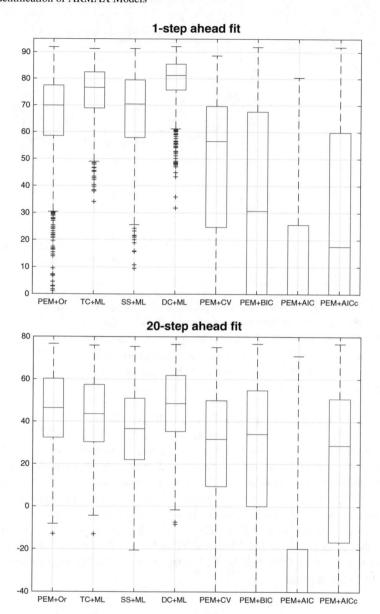

Fig. 9.11 *Identification of ARMAX models* Boxplots of the 1000 values of $\mathscr{F}_1$ (top panel) and $\mathscr{F}_{20}$ (bottom). Recall that *PEM+Oracle* uses additional information, having access to the test set to perform model order selection

Fig. 9.12 *Temperature*
prediction The 8 temperature
readings

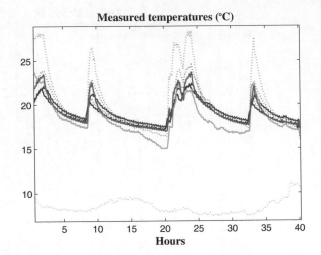

$(u_i^j,\ j = 1, \ldots, 7)$. Data are divided into 2 parts: those collected at time instants $1, \ldots, 1200$ form the identification set while those at instants $1201, \ldots, 2500$ are used for test purposes. With 5 min sampling times, 1200 instants almost correspond to 100 h, a rather small time interval. Hence, we assume a "stationary" environment and normalize the data so as to have zero mean and unit variance before performing identification. Quality of the k-step-ahead prediction on test data is measured by (9.10).

Identification has been performed using ARMAX models with an oracle which has access to the test set. This estimator, called PEM+Or, maximizes $\sum_{k=1}^{48} \mathscr{F}_k$ which accounts for the prediction capability up to 4 h ahead. The other estimator is regularized ARX equipped with the TC kernel with a different scale factor λ assigned to each unknown one-step-ahead predictor impulse response and a common decay rate α. The length of each impulse response is set to 50 and the hyperparameters are estimated via marginal likelihood maximization using only the identification data. This estimator is denoted by TC+ML. Results are reported in Fig. 9.13 (top panel): the performance of PEM+Or and TC+ML is quite similar. Sample trajectories of one-hour-ahead test data prediction returned by TC+ML are also reported in Fig. 9.13 (bottom panel).

9.3   Multi-task Learning and Population Approaches ⋆

In the previous chapters we have studied the problem of reconstructing a real-valued function from discrete and noisy samples. An extension is the so-called multi-task learning problem in which several functions (tasks) are simultaneously estimated. This problem is significant if the tasks are related to each other so that measurements

Fig. 9.13 *Temperature prediction. Top*: the prediction fits for predictions up to 4 h for the estimates obtained by PEM+Or and TC+ML. Recall that PEM+Or is not implementable in practice since it exploits the knowledge of the test set to tune model complexity. *Bottom*: one-hour-ahead test set prediction by TC+ML

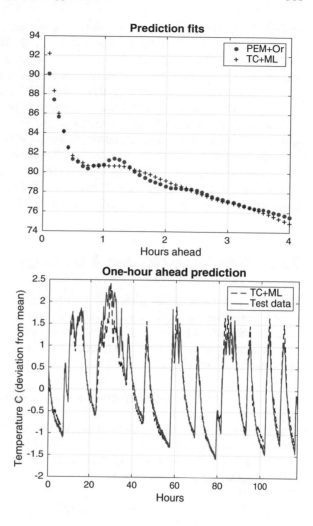

taken on a function are informative with respect to the other ones. An example is given by a network of linear systems whose impulse responses share some common features. Here, a relevant problem is the study of anomaly detection in homogenous populations of dynamic systems [5, 6, 10]. Normally, all of them are supposed to have the same (possibly unknown) nominal dynamics. However, there can be a subset of systems that have anomalies (deviations from the mean) and the goal is to detect them from the data collected in the population. Important applications of multi-task learning arise also in biomedicine when multiple experiments are performed in subjects from a population [9]. Similar patterns are observed in individual responses so that measurements collected in a subject can help reconstructing also the responses of other individuals. In pharmacokinetics (PK) and pharmacodynamics (PD) the joint analysis of several individual curves is often exploited and called population analysis [24]. One class of adopted models is parametric, e.g., compartmental ones [7]. The problem can be solved using, e.g., the NONMEM software, which traces back to the seventies [3, 25], or more sophisticated approaches like Bayesian MCMC algorithms [15, 28]. More recently, machine learning/nonparametric approaches have been proposed for the population analysis of PK/PD data [16, 19, 20].

In the machine learning literature, the term multi-task learning was originally introduced in [4]. The performance improvement achievable by using a multi-task approach instead of a single-task one which learns the functions separately has been then pointed out in [1, 26], see also [2] for a Bayesian treatment. Next, in [8] it has been proposed a regularized kernel method hinging upon on the theory of vector-valued Reproducing kernel Hilbert spaces [18]. Developments and applications of multi-task learning can then be found, e.g., in [11, 12, 17, 21, 29, 30].

9.3.1 Kernel-Based Multi-task Learning

We will now see that multi-task learning can be cast within the RKHS setting developed in the previous chapters by defining a particular kernel. Just to simplify exposition, let us assume that there is a common input space X for all the tasks and consider a set of k functions $\mathbf{f}_i : X \mapsto \mathbb{R}$. Assume also that the following n_i input–output data are available for each task i

$$(x_{1i}, y_{1i}), (x_{2i}, y_{2i}), \ldots, (x_{n_i i}, y_{n_i i}). \tag{9.11}$$

Our goal is to jointly estimate all the unknown functions $\mathbf{f}_i$ starting from these examples. For this aim, first a kernel can be introduced to include our knowledge on the single functions (like smoothness) and also on their relationships. This can be done by defining an enlarged input space

$$\mathscr{X} = X \times \{1, 2, \ldots, k\}.$$

Hence, a generic element of $\mathscr{X}$ is the couple (x, i) where $x \in X$ while $i \in \{1, \ldots, k\}$. The index i thus specifies that the input location belongs to the part of the function domain connected with the ith function. The information regarding all the tasks can now be specified by the kernel $K : \mathscr{X} \times \mathscr{X} \to \mathbb{R}$ which induces a RKHS of functions $\mathbf{f} : \mathscr{X} \to \mathbb{R}$. In fact, we are just exploiting RKHS theory on function domains that include both continuous and discrete components. Note that, in practice, any function $\mathbf{f}$ embeds k functions $\mathbf{f}_i$.

Regularization in RKHS then allows us to reconstruct the tasks from the data (9.11) by computing

$$\hat{\mathbf{f}} = \arg \min_{\mathbf{f} \in \mathscr{H}} \sum_{i=1}^{k} \sum_{l=1}^{n_i} \mathscr{V}_{li}(y_{li}, \mathbf{f}_i(x)) + \gamma \|\mathbf{f}\|_{\mathscr{H}}^2. \tag{9.12}$$

Under general conditions on the losses $\mathcal{V}_{li}$, we can then apply the representer theorem, i.e., Theorem 6.15, to obtain the following expression for the minimizer:

$$\hat{\mathbf{f}}_j(x) = \sum_{i=1}^{k} \sum_{l=1}^{n_i} c_{li} K\left((x, j), (x_{li}, i)\right) \quad x \in X, \, j = 1, \ldots, k \tag{9.13}$$

where $\{c_{li}\}$ are suitable scalars. Adopting quadratic losses which include weights $\{\sigma_{li}^2\}$, i.e.,

$$\mathcal{V}_{li}(a, b) = \frac{(a - b)^2}{\sigma_{li}^2}$$

for any $a, b \in \mathbb{R}$, a regularization network is obtained and the expansion coefficients $\{c_{li}\}$ solve the following linear system of equations

$$\sum_{i=1}^{k} \sum_{l=1}^{n_i} \left[K\left((x_{li}, i), (x_{jq}, q)\right) + \gamma \sigma_{jq}^2 \delta_{lj} \delta_{iq}\right] c_{li} = y_{jq}, \tag{9.14}$$

where $q = 1, \ldots, k$, $j = 1, \ldots, n_q$ and δ_{ij} is the Kronecker delta.

Connection with Bayesian estimation Exploiting the same arguments developed in Sect. 8.2.1, the following relationship between (9.13), (9.14) and Bayesian estimation of Gaussian random fields is obtained. Let the measurements model be

$$y_{ji} = \mathbf{f}_i(x_{ji}) + e_{ji} \tag{9.15}$$

where $\{e_{ji}\}$ are independent Gaussian noises of variances $\{\sigma_{ji}^2\}$. Define

$$y_i = [y_{1i} \cdots y_{n_i i}]^T, \qquad y^k = [y_1^T \cdots y_k^T]^T.$$

Assume also that $\{\mathbf{f}_i\}$ are zero-mean Gaussian random fields, independent of the noises, with covariances

$$\mathrm{Cov}\left(\mathbf{f}_i(x), \mathbf{f}_q(s)\right) = K\left((x, i), (s, q)\right) \qquad x, s \in X,$$

where $i = 1, \ldots, k$ and $q = 1, \ldots, k$. Then, one obtains that for $j = 1, \ldots, k$, the minimum variance estimate of $\mathbf{f}_j$ conditional on y^k is defined by (9.13), (9.14) by setting $\gamma = 1$. Furthermore, the posterior variance of $\mathbf{f}_j(x)$ is

$$\mathrm{Var}\left[\mathbf{f}_j(x)|y^k\right] = \mathrm{Var}\left[\mathbf{f}_j(x)\right] - \mathrm{Cov}\left(\mathbf{f}_j(x), y^k\right)\left(Var\left[y^k\right]\right)^{-1}\mathrm{Cov}\left(\mathbf{f}_j(x), y^k\right)^T. \tag{9.16}$$

In the above formula, in view of the independence assumptions, one has

$$\mathrm{Var}\left[y^k\right] = \begin{pmatrix} V_{11} & V_{12} & \ldots & V_{1k} \\ V_{21} & \ldots & \ldots & V_{2k} \\ \vdots & & \ldots \ldots \ldots & \\ V_{k1} & & \ldots \ldots & V_{kk} \end{pmatrix} + \begin{pmatrix} \Sigma_1 & 0 & \ldots & 0 \\ 0 & \Sigma_2 & \ldots & 0 \\ 0 & 0 & \ldots & 0 \\ 0 & 0 & 0 & \Sigma_k, \end{pmatrix}$$

where each block V_{iq} belongs to $\mathbb{R}^{n_i \times n_q}$ and its (l, j)-entry is given by

$$V_{iq}(l, j) = K((x_{li}, i), (x_{jq}, q)),$$

while $\Sigma_i = \mathrm{diag}\{\sigma_{1i}^2, \ldots, \sigma_{n_i i}^2\}$. In addition

$$\begin{aligned}\mathrm{Cov}\left(\mathbf{f}_j(x), y^k\right) &= \mathrm{Cov}\left(\mathbf{f}_j(x), [\mathbf{f}_1(x_{11}) \ldots \mathbf{f}_1(x_{n_1 1}) \ldots \mathbf{f}_k(x_{1k}) \ldots \mathbf{f}_k(x_{n_k k})]\right) \\ &= [K((x, j), (x_{11}, 1)) \ldots K((x, j), (x_{n_1 1}, 1)) \\ &\qquad \ldots K((x, j), (x_{1k}, k)) \ldots K((x, j), (x_{n_k k}, k))].\end{aligned}$$

Example of multi-task kernel: average plus shift A simple yet useful class of multi-task kernels is obtained by defining K as follows:

$$K((x_1, p), (x_2, q)) = \overline{\lambda}\,\overline{K}(x_1, x_2) + \delta_{pq}\widetilde{\lambda}\,\widetilde{K}_p(x_1, x_2) \qquad (9.17)$$

where $\overline{\lambda}^2$ and $\widetilde{\lambda}^2$ are two-scale factors that typically need to be estimated from data. Such kernel describes each function as the sum of an average function $\overline{\mathbf{f}}$, hereafter named *average task*, and an *individual shift* $\widetilde{\mathbf{f}}_j(x)$ specific for each task. Indeed, if $\overline{\lambda} = 0$ all the functions would be learnt independently of each other. Instead, when $\widetilde{\lambda} = 0$ all the tasks are actually the same. The Bayesian interpretation of multi-task learning discussed above facilitates also the understanding of this model. In fact, once the kernel is seen as a covariance, it is easy to see that, for any i and $x \in X$, each task decomposes into

$$\mathbf{f}_i(x) = \overline{\mathbf{f}}(x) + \widetilde{\mathbf{f}}_i(x)$$

where $\overline{\mathbf{f}}$ and $\{\widetilde{\mathbf{f}}_i\}$ are zero-mean independent Gaussian random fields.

9.3.2 Numerical Example: Real Pharmacokinetic Data

Multi-task learning is now illustrated by considering a data set connected with xeno-biotics administration in 27 human subjects [20]. Such administration can be seen as the input to a continuous-time linear dynamic system whose (measurable) output is the drug profile in plasma. In any subject, 8 measurements were collected at 0.5, 1, 1.5, 2, 4, 8, 12, 24 h after a bolus, an input which can be seen as a Dirac delta. Hence, one has to deal with a particular continuous-time system identification problem where noisy and direct samples of the impulse response are available.

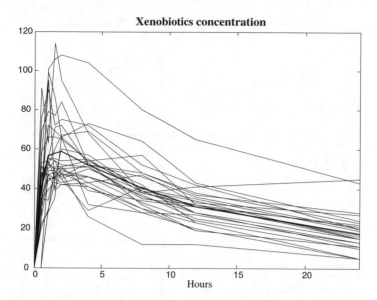

Fig. 9.14 *Multi-task learning* Xenobiotics concentration data after a bolus in 27 human subjects: average curve (thick) and individual curves

In this experiment, noises are known to be Gaussian and heteroscedastic, i.e., their variances are not constant being given by $\sigma_{ij}^2 = (0.1 y_{ij})^2$. The 27 experimental concentration profiles are displayed in Fig. 9.14, together with the average profile. In light of the number of subjects, such average curve is a reasonable estimate of the average task $\bar{\mathbf{f}}$.

The whole data set consists of 216 pairs (x_{ij}, y_{ij}), for $i = 1, \ldots, 8$ and $j = 1, \ldots, 27$, and is split in an identification (training) and a test set. For what regards training, a sparse sampling schedule is considered: only 3 measurements per subject are randomly chosen within the 8 available data. We will adopt the multi-task estimator (9.12) to reconstruct all the continuous-time profiles. In view of the Gaussian and heteroscedastic nature of the noise, the losses are defined by

$$\mathcal{V}_{li}(a, b) = \frac{(a - b)^2}{\sigma_{ij}^2}.$$

For what regards the function model, since humans are expected to give similar responses to the drug, quite close to an average function, the kernel (9.17) is adopted. In addition, it is known that in these experiments there is a greater variability for small values of t, followed by an asymptotic decay to zero. This motivates the use of a stable kernel to model both the average and the shifts. A model suggested in [20] is a cubic spline kernel under the time-transformation

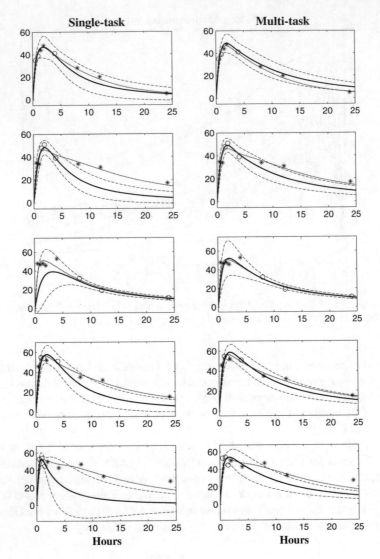

Fig. 9.15 *Multi-task learning* Single task (left) and multi-task (right) estimates of some curves (thick line) with 95% confidence intervals (dashed lines) using only three data (circles) for each of the 27 subjects. The other five "unobserved" data (asterisks) are also plotted. Dotted line indicates the estimates obtained by using the full sampling grid

$$h(t) = \frac{3}{t+3}$$

which defines (9.17) through the correspondences

$$\overline{K}(t, \tau) = \tilde{K}_p(t, \tau) = \frac{h(t)h(\tau)\min\{h(t), h(\tau)\}}{2} - \frac{(\min\{h(t), h(\tau)\})^3}{6}.$$

One can check that this model induces a stable RKHS by using Corollary 7.2. In fact, the kernels are nonnegative-valued and the integral of a generic kernel section is

$$\int_0^{+\infty} \left(\frac{h(t)h(\tau)\min\{h(t), h(\tau)\}}{2} - \frac{(\min\{h(t), h(\tau)\})^3}{6} \right) d\tau$$

$$= \frac{1}{2(t+3)^3} \left((27t + 81)\log(\frac{t+3}{3}) + 13.5t + 67.5 \right)$$

and this result clearly implies

$$\int_0^{+\infty} \int_0^{+\infty} \left(\frac{h(t)h(\tau)\min\{h(t), h(\tau)\}}{2} - \frac{(\min\{h(t), h(\tau)\})^3}{6} \right) d\tau dt < \infty.$$

The initial plasma concentration is known to be zero. Hence, a zero variance virtual measurement in $t = 0$ was added for all tasks. The hyperparameters $\overline{\lambda}^2$ and $\tilde{\lambda}^2$ were then estimated via marginal likelihood maximization by exploiting the Bayesian interpretation of multi-task learning discussed above.

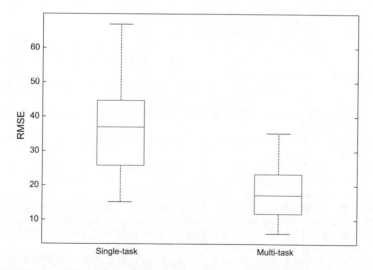

Fig. 9.16 *Multi-task learning* Boxplots of the prediction errors (RMSE) obtained by the single-task approach and by the multi-task approach

The left and right panels of Fig. 9.15 report results obtained by the single- and the multi-task approach, respectively, in 5 subjects. One can see the data and the estimated curves with their 95% confidence intervals obtained using the posterior variance (9.16). Each panel shows also the estimates obtained by employing the full sampling grid. It is apparent that the multi-task estimates are closer to these reference profiles. A good predictive capability with respect to the other five "unobserved" data is also visible. To better quantify this aspect, let I^f and I_j^r denote the full and reduced sampling grid in the jth subject. Let also $I_j = I^f \setminus I_j^r$, whose cardinality is 5. Then, for each subject, we also define the prediction error as

$$RMSE_j^{MT} = \sqrt{\frac{\sum_{i \in I_j} (y_{ij} - \hat{\mathbf{f}}_j(x_{ij}))^2}{5}}$$

with the single-task $RMSE_j^{ST}$ defined in a similar way. Figure 9.16 then reports the boxplots with the 27 $RMSE$ returned by the single- and multi-task estimates. The improvement on the prediction performance due to the kernel-based population approach is evident.

References

1. Bakker B, Heskes T (2003) Task clustering and gating for Bayesian multitask learning. J Mach Learn Res 4:83–99
2. Baxter J (1997) A Bayesian/Information theoretic model of learning to learn via multiple task sampling. Mach Learn 28:7–39
3. Beal S, Sheiner L (1992) NONMEM user's guide. NONMEM Project Group, University of California, San Francisco
4. Caruana R (1997) Multitask learning. Mach Learn 28(1):41–75
5. Chen T, Andersen MS, Chiuso A, Pillonetto G, Ljung L (2014) Anomaly detection in homogenous populations: a sparse multiple kernel-based regularization method. In: 53rd IEEE conference on decision and control, pp 265–270
6. Chu E, Gorinevsky D, Boyd S (2011) Scalable statistical monitoring of fleet data. IFAC Proc Vol 44(1):13227–13232. 18th IFAC world congress
7. Davidian M, Giltinan DM (1995) Nonlinear models for repeated measurement data. Chapman and Hall, New York
8. Evgeniou T, Micchelli CA, Pontil M (2005) Learning multiple tasks with kernel methods. J Mach Learn Res 6:615–637
9. Ferrazzi F, Magni P, Bellazzi R (2003) Bayesian clustering of gene expression time series. In: Proceedings of 3rd international workshop on bioinformatics for the management, analysis and interpretation of microarray data (NETTAB 2003), pp 53–55
10. Gorinevsky D, Matthews B, Martin R (2012) Aircraft anomaly detection using performance models trained on fleet data. In: 2012 conference on intelligent data understanding, pp 17–23
11. Ling H, Wang Z, Li P, Shi Y, Chen J, Zou F (2019) Improving person re-identification by multi-task learning. Neurocomputing 347:109–118
12. Liu Z, Huang B, Cui Y, Xu Y, Zhang B, Zhu L, Wang Y, Jin L, Wu D (2019) Multi-task deep learning with dynamic programming for embryo early development stage classification from time-lapse videos. IEEE Access 7:122153–122163

13. Ljung L (1999) System identification - theory for the user, 2nd edn. Prentice-Hall, Upper Saddle River
14. Ljung L (2013) System identification toolbox V8.3 for MATLAB. The MathWorks, Inc., Natick
15. Lunn DJ, Best N, Thomas A, Wakefield JC, Spiegelhalter D (2002) Bayesian analysis of population PK/PD models: general concepts and software. J Pharmacokinet Pharmacodyn 29(3):271–307
16. Magni P, Bellazzi R, De Nicolao G, Poggesi I, Rocchetti M (2002) Nonparametric AUC estimation in population studies with incomplete sampling: a Bayesian approach. J Pharmacokinet Pharmacodyn 29(5/6):445–471
17. Maurer A, Pontil M, Romera-Paredes B (2016) The benefit of multitask representation learning. J Mach Learn Res 17(1):2853–2884
18. Micchelli CA, Pontil M (2005) On learning vector-valued functions. Neural Comput 17(1):177–204
19. Neve M, De Nicolao G, Marchesi L (2005) Nonparametric identification of pharmacokinetic population models via Gaussian processes. In: Proceedings of 16th IFAC world congress, Praha, Czech Republic
20. Neve M, De Nicolao G, Marchesi L (2007) Nonparametric identification of population models via Gaussian processes. Automatica 97(7):1134–1144
21. Pillonetto G, Dinuzzo F, De Nicolao G (2010) Bayesian on-line multi-task learning of Gaussian processes. IEEE Trans Pattern Anal Mach Intell 32(2):193–205
22. Pillonetto G, Chen T, Chiuso A, De Nicolao G, Ljung L (2016) Regularized linear system identification using atomic, nuclear and kernel-based norms: the role of the stability constraint. Automatica 69:137–149
23. Pintelon R, Schoukens J (2012) System identification: a frequency domain approach, 2nd edn. Wiley, New York
24. Sheiner LB (1994) The population approach to pharmacokinetic data analysis: rationale and standard data analysis methods. Drug Metab Rev 15:153–171
25. Sheiner LB, Rosenberg B, Marathe V (1977) Estimation of population characteristics of pharmacokinetic parameters from routine clinical data. J Pharmacokinet Biopharm 5(5):445–479
26. Thrun S, Pratt L (1997) Learning to learn. Kluwer, Boston
27. Torfs D, Vuerinckx R, Swevers J, Schoukens J (1998) Comparison of two feedforward design methods aiming at accurate trajectory tracking of the end point of a flexible robot arm. IEEE Trans Control Syst Technol 6(1):2–14
28. Wakefield JC, Smith AFM, Racine-Poon A, Gelfand AE (1994) Bayesian analysis of linear and non-linear population models by using the Gibbs sampler. J Appl Stat 41:201–221
29. Xu Y, Li X, Chen D, Li H (2018) Learning rates of regularized regression with multiple Gaussian kernels for multi-task learning. IEEE Trans Neural Netw Learn Syst 29(11):5408–5418
30. Zhou Q, Zhao Q (2016) Flexible clustered multi-task learning by learning representative tasks. IEEE Trans Pattern Anal Mach Intell 38(2):266–278

Index

A

Absolute homogeneity, 224
Absolute Shrinkage And Selection Operator (LASSO), 69
Actuarial science, 168
Admissible estimator, 4, 140
Akaike's information theoretic criterion (AIC), 28, 38
Approximation property of ARX models, 22
ARMAX model, 21, 353
ARX model, 21, 353
Asymptotic covariance matrix of parameter estimates, 26

B

Basis expansion, 155
Bauer's maximum principle, 303, 305
Bayes estimate, 102
Bayes estimate (conditional mean), 99
Bayes estimate with improper prior, 110
Bayes factor, 128
Bayes formula, 96
Bayesian confidence intervals, 124
Bayesian function reconstruction, 120
Bayesian information criterion (BIC), 28, 38
Bayesian interpretation of multi-task learning, 363
Bayesian interpretation of regularization, 97, 267
Bayesian interpretation of ridge regression, 269
Bayesian interpretation of the James–Stein estimator, 105
Bayesian model probability, 270

Bayesian optimal model approximation/reduction, 115
Bayesian VARs, 168
Bias, 26, 37, 137
Bias and variance, 26
Bias and variance decomposition, 2
Bias space, 109, 212
Bias-variance trade-off, 8, 28, 139
BIBO stability, 274
BIC, 28, 38, 270
Black-box model, 21, 314
Block-oriented models, 335
Bochner theorem, 210
Box–Jenkins (BJ) model, 21
Boxplot, 269

C

Cauchy sequence, 185, 225
Cauchy–Schwarz inequality, 225
Change detection, 337
Chi-square random variable, 106
Choice of the loss, 200
Classification problem, 204
Closed-form expression of IIR estimate, 251
Closed graph theorem, 228
Closed set, 228
Closed subspace, 228
Column and row rank, 36
Compactness of the operator induced by a stable kernel, 281
Compact set, 228
Competitive kernels, 270
Complete orthonormal basis, 192
Complete space, 225

Compound loss, 139
Compressive sensing, 70
Conditional mean, 99
Conditional median, 99
Condition number, 8, 43
Confidence ellipsoid, 101
Conjugate priors, 126
Consistency, 219
Continuous functional, 225
Continuous-time model, 24
Convergence rate of the stable spline estima-
      tor, 291
Convex envelope, 78
Convex envelope of the rank function, 79
Convex relaxation, 70
Corrected Akaike's criterion (AICc), 28
Covariance matrix estimation with low-rank
      structure, 80
Covariance of estimated frequency function,
      27
Cramér–Rao inequality, 27
Credible region, 99
Cross-Validation (CV), 30, 38
Cubic spline estimator, 263
Cubic spline kernel, 212, 263
Curse of dimensionality, 319
CVX software, 79

D
Damped Newton iterations, 223
DC kernel, 151, 252, 344
Deep networks, 314, 334
Degrees of freedom, 272, 318
Diagonalized Bayesian estimation problem,
      117
Diagonalized kernel, 191
Differential entropy, 111
Dimensionality reduction, 338
Discrete-time Fourier transform, 19
Distributed lag estimator, 163
Disturbances and noise sources, 19
Dropout or dilution, 335

E
Early termination, 335
Empirical Bayes approach, 107, 108, 124,
      168, 266
Empirical risk, 218
Empirical Risk Minimization (ERM), 219
Entropy maximization, 111
Equivalent degrees of freedom, 54, 118, 152

Equivalent degrees of freedom and maxi-
      mum likelihood estimate, 119
Equivalent degrees of freedom for the DC
      kernel, 153
Ergodic process, 125
Estimation data, 59
Estimation set, 30, 59
Euclidean inner product, 225
Euclidean norm, 36
Excess degrees of freedom, 68
Expected in-sample validation error, 64
Expected Validation Error (EVE), 58, 59
Experimental conditions, 17
Experiment design, 29

F
Fading memory concepts, 325
Fano's inequality, 290
Feature map, 280
Feature map and feature space, 208
Filtered frequency domain smoothness con-
      dition, 153
Finite Impulse Response (FIR) model, 7, 21,
      248
Finite-trace kernels, 279
First-order spline kernel, 188, 212
First-order Stable Spline, 153
Frequency response, 19, 49
Frequentist paradigm, 95
Frobenius norm, 84
Full Bayesian approach, 107, 125
Full conditional distribution, 127
Full rank, 45
Full rank matrix, 36
Function estimation, 33

G
Gaussian kernel, 185, 210, 263, 319
Gaussian process realizations and RKHS,
      261
Gaussian random field, 316, 363
Gaussian regression, 260, 314, 335
Generalization, 219
Generalization and consistency, 218
Generalization and consistency of kernel-
      based approaches, 221
Generalized Cross Validation (GCV), 63,
      272, 273, 318
Generalized ridge regression, 168
Gibbs sampler, 125
Gibbs sampling, 125
Gibbs structure, 175

Green's function, 212
Grey box models, 23, 314

H

Hairdryer experiment, 353
Hammerstein model, 335
Hammerstein–Wiener identification, 336
Hankel matrix, 153, 159
Hankel nuclear norm, 160
Hankel prior, 161
Hat matrix, 53, 118, 273
Heteroscedastic noise, 365
Hilbert and Banach spaces, 225
Hilbert–Schmidt (HS) norm, 305
Hilbert–Schmidt (HS) operator, 304
Hinge loss, 204
Hold out cross-validation, 30, 40, 61
Huber estimation, 76
Huber loss, 76, 202
Hybrid models, 336
Hybrid stable spline, 338
Hyperparameter, 58, 104, 127, 136, 145, 266
Hyperparameter tuning, 58, 345
Hypothesis test, 30

I

Identification data, 17, 24, 34, 64, 135, 320,
    347–349, 356
Identification method, 17
Identification set, 17, 24, 34, 64, 135, 320,
    347–349, 356
Identity kernel, 187
Ill-conditioned matrix, 44, 47
Ill-conditioning, 8, 42
Ill-conditioning and system input, 344
Ill-conditioning in system identification, 47
Ill-posedness, 182
Improper priors, 109
Impulse response, 7
Inclusions between notable kernel classes,
    279
Infinite Impulse Response (IIR) model, 247,
    250
Influence matrix, 53, 273
Inner product, 224
Inner products and norms, 224
Innovations, 25
Input design, 29, 295
Input location, 182
Input space, 182
Instability of spline kernel, 277
Instability of the Gaussian kernel, 277

Instability of translation invariant kernels,
    277, 279
Integral equation, 222
Integral operator approach, 222
Integrated random walk, 125
Interior point methods, 79, 223
Inverse chi-square random variable, 106
Inverse Gamma random variable, 126
Inverse problems, 135
Iteratively reweighted methods, 161
Iterative reweighting, 339

J

James–Stein estimator, 3, 139
James–Stein's MSE, 3
Jointly Gaussian vectors, 100

K

Karhunen-Loève decomposition, 155
Karush–Kuhn–Tucker (KKT) equations,
    223
Kernel as similarity function, 185
Kernel-based multi-task learning, 362
Kernel-based population approach, 368
Kernel defined by a finite number of basis
    functions, 208
Kernel eigenvalues and eigenvectors, 191
Kernel hyperparameters tuning, 266
Kernel implicit encoding, 185, 198, 209
Kernel matrix, 198, 318
Kernel operator L_K, 191, 278
Kernel ridge regression, 200
Kernel section, 183
Kernel selection using marginal likelihood,
    270
Kernels for NARX models, 328
Kernels for nonlinear system identification,
    319
Kernels generated by Fourier basis, 195
Kernel stability: necessary and sufficient
    condition, 275
Kernel stability: sufficient condition, 276
Kernel stability: system theoretic interpreta-
    tion, 276
Kernel trick, 198, 209, 284
Kernel tuning, 266, 316
Kernel width, 263
K-fold cross-validation, 61
Ky-Fan norm, 160

L
L_1 norm, 69, 85, 339
L_2 norm, 85
L_∞ norm, 85
L_1 regularization, 69
Lagrange multiplier, 70
Lagrangian theory, 70, 221
Laguerre and Kautz basis functions, 259, 285
Laplace approximation, 109
LASSO, 69
LASSO for non-orthogonal regression, 71
LASSO for orthogonal regression, 70
Learning from examples, 218
Least squares, 2, 8, 33, 35
Leave-one-out cross-validation, 62
Likelihood function, 98
Linear and bounded functionals, 183, 228
Linear Gaussian model, 97, 101
Linear kernel, 205
Linear plus nonlinear kernel, 327
Linear regression, 22, 33
Linear spline kernel, 212
Linear state-space model, 24
Linear time-invariant (LTI) system, 18
Loss function and noise model, 260
Loss functions, 197
Low rank kernel approximation, 156

M
Manifold learning, 338
Margin, 204
Marginal likelihood, 108, 266, 345
Marginal likelihood estimate, 124, 145, 266, 318, 345
Markov chain, 125
Markov Chain Monte Carlo (MCMC), 125
Markov parameters, 160
Markov's theorem, 2
Matrix inversion lemma, 85
Matrix norm, 83
Maximum A Posteriori (MAP) estimate, 98
Maximum entropy principle, 98, 111
Maximum entropy priors, 111, 148
Maximum entropy stable priors, 150
Maximum Likelihood (ML), 25, 38
Maximum Likelihood Estimate (MLE), 25, 38
McMillan degree, 20
MDL, 28, 38
Mean Squared Error (MSE), 1, 37, 139
Mercer kernel, 183
Mercer representation of a stable RKHS, 283

Mercer theorem, 192
Metric space, 226
Minimal state space realization, 159
Minimax estimators, 286, 288
Minimum Description Length (MDL) criterion, 28, 38
Minimum norm solution, 46
Minnesota prior, 168
Model and predictor, 18, 313
Model approximation/reduction, 114
Model bias, 35
Model error modeling, 167
Model order, 10, 34
Model order selection, 28, 37
Model posterior probability, 270
Model prediction capability, 60
Model quality, 28
Model structure, 17, 18
Model validation, 29
Monomials' encoding, 319
Monte Carlo study with different ARMAX models, 356
Monte Carlo study with different OE-models, 347
Moore–Aronszajn theorem, 184, 209
Moore–Penrose pseudoinverse, 46, 51
MSE decomposition, 26, 52
MSE matrix, 37, 52
Multi-task kernel: average plus shift, 364
Multi-task learning, 222, 362
Multivariate Gaussian, 99

N
NARX, 314
Necessary and sufficient conditions for generalization and consistency, 219
Networks of linear dynamic models, 335
Neural networks, 314, 334
NFIR, 314
Noise spectrum, 27
Nondegenerate Borel measure, 191
Nonlinear model, 24
Nonlinear random surface, 316
Nonlinear stable spline kernel, 326
Nonlinear state-space model, 24
Nonlinear system Identification, 313
Norm, 224
Normal equations, 36
Norm induced by inner product, 225
Nuclear norm, 84, 304
Nuclear norm heuristic, 79
Nuclear norm regularization, 78

Nuclear operator, 304
Null vector condition, 225
Numerical expansion of a stable kernel, 282
Numerical expansion of the first-order stable
spline kernel, 281
Nystrom methöd, 281

O

Occam's factor, 267
Occam's razor principle, 267
One-step-ahead predictor, 20, 313
One-step-ahead predictor for ARX models,
22
Optimal regularization matrix, 57, 102
Optimality in order, 288
Optimism and equivalent degrees of free-
dom, 65
Oracle and test set, 321, 322, 348, 358
Oracle-based estimation procedure, 8, 263
Oracle-based procedure, 345
Orthogonality, 228
Orthogonality property of Bayes estimate,
102
Orthonormal basis expansion, 156
Orthonormal basis in ℓ_2, 284
Outliers, 76
Output Error (OE) model, 148, 166, 247
Output Error (OE) model in continuous-
time, 253
Output kernel matrix, 267, 272, 318
Overfitting, 28, 40, 50, 182

P

Parametrized regularization matrix, 58
Parseval's theorem, 145, 155
Partial realization problem, 160
Particle Gibbs samples, 336
Particle Markov chain Monte Carlo, 336
PEM asymptotic properties, 26
Penalty function, 136
Piecewise affine models, 337
Pointwise evaluator, 183
Polynomial kernel, 210, 319
Polynomial regression, 34, 39, 120
Population approaches, 362
Positive definite kernel, 183
Posterior distribution, 98
Posterior variance, 101
Power spectrum, 49
Predicted Residual Error Sum of Squares
(PRESS), 62, 273
Prediction error, 24

Prediction Error Method (PEM), 23, 25
Prediction fit, 25, 357
Prediction risk, 318
PRESS derivation, 86
Prior distribution, 96, 97, 137
Projection pursuit, 338
Projection theorem, 228
Proportionality principle, 164
Pseudoinverse, 46

Q

Q-boundedness of a kernel, 278
QR factorization, 83
Quadratic loss, 200
Quadratic polynomial kernel, 324

R

Radial basis kernels, 210
Random walk, 120
Rank-deficient matrix, 45
Real pharmacokinetic data example, 364
Regression function/optimal predictor, 215
Regression matrix, 36
Regularization design, 56
Regularization in quadratic form, 50, 51
Regularization in RKHS, 196
Regularization in RKHS and Bayesian esti-
mation, 260
Regularization matrix, 51
Regularization network, 200
Regularization network as projection, 201
Regularization network for linear system
identification, 258
Regularization network representation using
stable kernels, 283
Regularization network using Laguerre or
Kautz basis functions, 259
Regularization parameter, 5, 50, 96, 106,
116, 197, 214, 263
Regularization parameter and condition
number, 52
Regularization term, 50
Regularized ARX, 355
Regularized estimate in RKHS as Bayes esti-
mate, 261, 315
Regularized FIR, 248, 289
Regularized IIR, 251, 263
Regularized impulse response estimation in
RKHS, 252
Regularized Least Squares (ReLS), 5, 33, 50,
106
Regularized NARX, 315

Regularized NFIR, 315
Regularized Volterra models, 331
Regularizer, 5
Relationship between $\mathcal{H}$ and $\mathcal{L}_2^\mu$, 193
Representation of stable kernels, 281
Representer theorem, 197
Representer theorem for continuous-time linear system identification, 254
Representer theorem for discrete-time linear system identification, 251
Representer theorem for nonlinear system identification, 315
Representer theorem with bias space, 213
Representer theorem with linear and bounded functionals, 199
Reproducing kernel, 184
Reproducing Kernel Hilberts Space (RKHS), 183
Reproducing property, 184
Residual analysis, 29
Ridge regression, 10, 51, 136, 208, 269
Ridge regression as Bayesian estimation, 116
Riesz representation theorem, 184, 199, 229
Risk functional, 218
RKHS induced by a diagonalized kernel, 195
RKHS induced by kernel operations, 190
RKHS induced by kernel sampling, 190
RKHS induced by kernel sums or products, 190
RKHS induced by Mercer kernel, 184
RKHS induced by the Stable Spline kernel, 257
RKHS map, 209
RKHSs inclusions in general Lebesque spaces, 278
RKHS stability using Mercer expansions: necessary and sufficient conditions, 284
Robot arm experiment, 350
Robust regression, 76, 202
Robust statistics, 202

S
Sample complexity, 286
Sample complexity and minimax properties of the stable spline estimator, 290
Second-order spline kernel, 212
Semidefinite programming, 79
Separable space, 227
Sherman–Morrison–Woodbury formula, 85
Shift operator, 19, 248

Signal segmentation, 337
Simplest unfalsified model, 29
Singular Value Decomposition (SVD), 42
Singular values, 43
Smoothness in the frequency domain, 146, 165
Sobolev norm, 188, 254
Sobolev space, 188, 194, 222
Space ℓ_1 of absolutely summable sequences, 226
Space ℓ_2 of squared summable real sequences, 225
Space ℓ_∞ of bounded sequences, 227
Space $\mathcal{C}$ of continuous functions, 184, 227
Space $\mathcal{L}_1$ of absolutely integrable functions, 226
Space $\mathcal{L}_2$ of square summable functions, 186, 226
Space $\mathcal{L}_\infty$ of essentially bounded functions, 227
Sparse estimation, 70, 205
Sparsity and variable selection, 338
Sparsity inducing regularization, 73
Spectral decomposition, 155
Spectral decomposition of stable kernels, 281
Spectral decomposition of Stable Spline kernel, 255
Spectral feature map, 209
Spectral map, 192
Spectral norm, 84
Spectral representation of RKHS, 191
Spectral theorem, 281
Spline and Stable Spline kernel, 257
Spline estimator, 214
Spline kernel expansion, 193
Spline kernels, 211
Spline norm, 197
Square summable kernel, 279
Stable diagonal kernel, 145
Stable kernels, 247, 367
Stable prior, 145
Stable RKHS, 274, 279
Stable Spline (SS) kernel, 148, 252, 255, 263, 269
Stable Spline kernel of higher order, 257
Stable Spline norm, 252, 257, 296
Static nonlinearity, 314
Stationary distribution, 125
Statistical consistency of regularization networks, 217
Statistical learning theory, 218
Stein's effect, 3, 139

Stein's estimation in non-orthogonal setting, 6
Stein's lemma, 12, 67
Stein's Unbiased Risk Estimator (SURE), 66, 271, 318
Stochastic embedding, 166
Strictly positive definite kernel, 183
Strong mixing condition, 217
Subderivative, 71
Subdifferential, 71
Subjective/Bayesian estimation paradigm, 95, 96
Subjective probability, 97
Subselection of regressors, 338
Subspace, 227
Summable kernel, 279
Sup-norm, 227
Support vector classification, 204
Support vector regression, 202
Support vector regression for linear system identification, 259

T
Takenaka-Malmquist orthogonal basis functions, 284
Taylor expansion and Volterra models, 331
TC, SS and DC kernel, 344
Temperature prediction experiment, 358
Test set, 138, 320, 330, 347, 350, 353, 356, 365
Tikhonov regularization, 11, 222
Time series, 19
Trace norm, 160
Training data, 17, 24, 33, 34, 64, 135, 218, 320, 347, 349, 356
Training set, 17, 24, 33, 34, 59, 64, 135, 218, 320, 347, 349, 356
Transfer function, 19, 136
Translation invariant kernel, 194, 210
Triangle inequality, 225
Truncated Mercer expansions, 281

Truncated SVD, 46
Tuned Correlated (TC) kernel, 146, 252

U
Unbiased estimate, 26
Unbiased estimation of EVE, 66, 272
Unbiased estimator, 2
Uniform Glivenko Cantelli class, 220
Uniform norm, 227
Universality, 211
Universal kernel, 211, 222

V
V_γ-dimension, 220
Validation data, 38
Validation process, 17
Validation set, 30, 59
Vapnik–Chervonenkis (VC) dimension, 220
Vapnik's ε-insensitive loss, 203
Variable selection, 70
Variance, 37
Vector norm, 83
Vector space, 224
Vector-valued RKHSs, 222
Volterra model, 319
Volterra series, 319

W
Well-conditioned matrix, 44
Wiener model, 335

Y
YALMIP software, 79

Z
0-1 loss, 160, 204
Z-transform, 19

Printed in the United States
by Baker & Taylor Publisher Services